An Introduction to Quantum Fluids

An Introduction to Quantum Fluids

Phuong Mai Dinh
Laboratoire de Physique Théorique de Toulouse, Université Paul Sabatier, CNRS

Jesús Navarro
Instituto de Física Corpuscular, Consejo Superior de Investigaciones Científicas, University of Valencia

Éric Suraud
Laboratoire de Physique Théorique de Toulouse, Université Paul Sabatier, CNRS

CRC Press is an imprint of the
Taylor & Francis Group, an **informa** business

CRC Press
Taylor & Francis Group
6000 Broken Sound Parkway NW, Suite 300
Boca Raton, FL 33487-2742

Printed on acid-free paper
Version Date: 20170821

International Standard Book Number-13: 978-1-1385-5327-9 (Hardback)
International Standard Book Number-13: 978-1-4987-2163-9 (Paperback)

Library of Congress Cataloging-in-Publication Data

Names: Suraud, Eric, author. | Dinh, P. M. (Phuong Mai), author. | Navarro, Jesâus, 1949- author.
Title: An introduction to quantum fluids / Eric Suraud, Phuong Mai Dinh, Jesâus Navarro.
Description: Boca Raton, FL : CRC Press, Taylor & Francis Group, [2017] | Includes bibliographical references and index.
Identifiers: LCCN 2017027616| ISBN 9781498721639 (hardback) | ISBN 149872163X (hardback) | ISBN 9781498721646 (e-book) | ISBN 1498721648 (e-book)
Subjects: LCSH: Quantum liquids.
Classification: LCC QC174.4 .S87 2017 | DDC 530.4/2--dc23
LC record available at https://lccn.loc.gov/2017027616

Visit the Taylor & Francis Web site at
http://www.taylorandfrancis.com

and the CRC Press Web site at
http://www.crcpress.com

To Philippe, Lisa, Mathis, and Evelyne

Contents

Introduction

The term 'Quantum liquids' may sound surprising. It merges two terms, quantum and liquids, commonly accepted in well defined contexts but usually not jointly. Liquids are all around us. Life stems from water, which under usual temperature and pressure conditions on earth, mostly exists as a liquid. About 70 % of the earth surface is thus covered by liquid water, and all of us are dominantly made of water, about 60% when adult and up to almost 80 % in babies. Liquids are thus naturally associated with macroscopic objects and, theoretically speaking, with classical thermodynamics. On the contrary the word 'quantum' immediately evokes microscopic systems, such as atoms or nuclei. And indeed, both systems played a key role in the development of quantum theory in the early 20th century. The difference in scales between oceans, or even our own body, and such microscopic objects as atoms or nuclei, hence seems to provide an obvious barrier to separate both concepts. But this separation is not only a bit arbitrary but also, to a large extent, misleading. Consider for example a classical theory of atoms which is unable to account for the stability of such objects. The electrons in orbit around the nuclear core should classically fall onto it by radiation loss... which would simply mean that matter, microscopic or macroscopic, is unstable, which is obviously wrong. Quantum concepts are thus also at the core of our understanding of macroscopic phases of matter, including liquids of course. The superficial argument on different scales is tus certainly not robust. But this is not enough to claim a direct connection between quantum concepts and liquid phases. This would even be a wrong understanding of the notion of quantum liquids which is the central topic of this book. Then what do quantum liquids correspond to? And why associate these two terms together? Before trying to answer these questions let us ask a few more, which will help understand what we shall actually discuss.

What is common to the Sun, a piece of metal, an atomic nucleus and a few cubic centimeters of helium? At first look these systems are different by all means, starting with their spatial dimensions. How to compare the astronomical size of the Sun with its radius of 700,000 km to the human scale of a piece or metal or helium? And how to fit in the picture the microscopic size of an atomic nucleus with a characteristic length scale of order 10^{-15}m? Energy or temperature scales reflect similar discrepancies with temperatures close to absolute zero for helium droplets up to thousands of degrees at the Sun's surface and even millions of degrees in its interior. These enormous differences simply stem from the fact that these various systems are composed of different constituents involving different energy scales and interactions. Atomic nuclei are made of neutrons and protons, bound together by the strong interaction, while a piece of metal consists of atoms sharing electrons via a complex interplay of Coulomb interactions, and the Sun by atoms of hydrogen and helium interacting via the Coulomb interaction and all bound within the star by gravitation. After all, is it really surprising that such different constituents and interactions lead to such different systems?

But the true point here is not so much to realize that these systems look different but rather to understand to what extent they share common features. And this is indeed the key question where the concept of quantum liquid will show up. Actually, it is always extremely useful to look for similarity among diversity. If such trends can be identified they usually carry a wealth of new information leading to a better understanding of all systems taken separately. Indeed if a common theory, or at least theoretical framework, can be developed for such apparently diverse systems, one can transfer concepts from one system to the other, and, ultimately, build a more unified picture of nature. It is precisely the aim of this book to work in this direction around the concept of quantum liquids.

Finally, what are 'quantum' liquids? These are liquids, in the commonly accepted sense, as experienced in everyday life, for example with water. However, water is liquid when it is not too dense and not too cold or hot. In other words when water molecules are able to move sufficiently to deliver this unique behavior of a liquid. Quantum liquids are liquids not necessarily because of particular temperature or pressure conditions. They are liquids because of the quantum nature of their constituents which make them sufficiently move to deliver a liquid behavior. Of course, this has to be understood, as classical liquids, in relation to the interactions between the constituents. And again as in classical liquids, temperature and pressure are to be accounted for. But the new aspect is that there is no need of temperature or pressure to produce the liquid state. In white dwarfs, which corresponds to the ultimate state of evolution of low mass stars, such as our Sun, the stability of the star is ensured by the pressure exerted by the electrons of the star which form a huge macroscopic phase. And counterintuitively, these electrons, at their scale, can be considered at zero temperature. Similar pictures hold in most of the systems we shall discuss in this book.

One of the goals of this book is thus to speak of quantum physics, but of quantum physics in a somewhat unusual context. Indeed the roots of the development of quantum physics lie in microscopic systems such as atoms or photons. But the field of application of quantum physics is by no means only restricted to these microscopic objects. Quantum theory provides the basis for understating a huge range of systems from the microscopic to the macroscopic world. And quantum liquids precisely constitute some of these systems. We shall encounter them at various scales, from the atomic nucleus to white dwarfs, and in macroscopic objects at human scale such as a piece of metal. We shall thus present these various objects which interestingly enough may appear either as finite objects, such as nuclei, or infinite phases such as in white dwarfs. Both the finite and infinite realizations nevertheless share the same characteristics, even if the finiteness induces some specific trends precisely linked to this finiteness. This never contradict the generic trends common to all these systems.

The book is organized in two parts. The first part entitles 'The Players and the Rules', gathers the necessary basic concepts from classical and quantum physics to understand the forthcoming developments. Key words here are quantum physics of course and many-body properties. By this, we mean in particular the methods developed to understand assemblies of interacting particles (macroscopic or microscopic ones) and we shall address these questions both from the classical/macroscopic viewpoint of thermodynamics (Chapter 1) and from the microscopic (classical or quantum) viewpoint of statistical physics (Chapter 3). Basic notions of quantum theories will also be given in Chapter 2. This first part is complemented by a fourth chapter dedicated to theoretical developments around the many-body problem, including a few technical aspects only superficially addressed in the first three chapters.

The second part of the book is entitled 'The Games' and provides a description of the major quantum liquids, starting from electronic systems in Chapter 5 to liquid helium in Chapter 6 and finally atomic nuclei in Chapter 7. These three chapters present the standard properties of these systems on the basis of the concepts introduced in the first part. The underlying unifying concept is of course that of quantum liquid. The presentation in these chapters focus on rather 'conventional situations involving these systems in their 'natural' state. Two more chapters are included and concern specific situations. They again stem from the quantum liquid nature of these systems, but involving particular situations leading to unexpected behaviors of the quantum liquids. Specifically, we address very high density systems such as encountered in stars and will here consider in Chapter 8 both electronic systems, as observed in white dwarfs, and neutrons systems forming compact stars. Finally, Chapter 9 considers what we call 'superphases' which occur for example at very low temperatures in helium (superfluidity) or solids (superconductivity). A superphase can also be observed in atomic nuclei in an analogous form. These phases are quite involved and exhibit unexpected behaviors which have been, and continue to be, the focus of numerous investigations. The theoretical concepts involved may also be quite elaborate and we have thus kept these more demanding topics for the last chapter. There are finally a few appendices gathered at the end of the book with some basic information, a very short bibliography and some tables on elementary particles and interactions, atomic classification and some widely used units.

To close this short introduction, we would like to remind readers that the ideas presented in this book stem from numerous discussions with colleagues in very many situations. It is clearly impossible to make here an exhaustive list and we shall henceforth avoid this hazardous exercise. We would nevertheless like to mention a few close colleagues who played a specific role during the realization of this book, in particular those who were kind enough to reread parts of this text, often on short notice. We list them here alphabetically: Manuel Barranco, Mohamed Belkacem, Dany Davesne, Nicolas Laflorencie, Pierre Pujol, Paul-Gerhard Reinhard, and Marc Vincendon.

I

The Players and the Rules

CHAPTER 1

Phases, matter and interactions

The notion of fluid in everyday life includes gases and liquids, viewed as substances that have no fixed shape, contrary to solids. It is important for our purpose to characterize fluids in more detail, specify some of their properties, and clarify how they differ from solids. A key issue here is the role of interactions as well as the values of quantities such as pressure or temperature. Clearly, discussing properties of fluids, even in a classical environment as we aim here, thus requires explaining seemingly simple notions such as temperature or pressure. In this chapter we recall some basics of thermodynamics, which precisely addresses such questions, at least from a macroscopic viewpoint. We will also address in some detail the relationship between them through phase diagrams and phase transitions. This is an important issue in order to better understand how a phase of matter does form. But this is also an important concept in this book as we shall face several examples of phase transitions. As a second step and to gain more insight into the nature of these states, we will need to go deeper into matter and discuss its microscopic structure. We shall then briefly recall some basics of the constitution of matter both in terms of constituents and interactions. Interactions in particular turn out to play a key role here to determine the nature of the phase characterizing a system. This will lead us to consider effective interactions which indeed stem from elementary interactions, but which can practically be characterized in simpler, more effective terms, depending on the considered length or energy scales. This will bring us to explore real fluids, in contrast to the (oversimplified) ideal gas in which all interactions are neglected by construction. Finally, we will end this chapter by presenting more 'exotic' or 'extreme' states of matter, and relate some of them with the quantum fluids at the core of this book.

1.1 STATES AND PHASES OF MATTER

Water is observable in everyday life in three states or phases: solid, liquid and gas. It is the only known substance that can exist simultaneously in these three states within the relatively narrow contditions of pressure and temperature curently found in nature, as is illustrated in Figure 1.1.

Figure 1.1: Coexistence of the three different states of water.

Icebergs are in solid state, the ocean in liquid state and the clouds are formed by the gas state or water vapor. Before really discussing what the states of matter exactly are, we need to recall some thermodynamics basics. Thermodynamics provides a macroscopic understanding of matter and its phases in terms of global variables such as temperature or pressure. It also allows one to understand time evolution of macroscopic systems.

1.1.1 Basics of thermodynamics

Thermodynamics was developed in the early 19th century, in particular in relation with industry revolution and thermal machines. These early developments occurred before one gained a proper understanding of the structure of matter. Links between macroscopic quantities and their microscopic origins are established later on by statistical physics, in particular during the 20th century. These latter aspects will be specifically addressed in Chapter 3.

Thermodynamical variables

Thermodynamics is a phenomenological theory of matter, which is based on experimentally defined parameters, such as the pressure P, the volume V, the temperature T, the magnetic field B, etc. This theory does not aim at explaining the microscopic origin of these parameters, but only at providing relationships between thermodynamical parameters which can be measured. The thermodynamical state of a system (also called a 'macrostate') is defined by giving the values of a set of thermodynamical parameters, also commonly called thermodynamical (or state) variables. It has been observed experimentally that three variables usually suffice to completely characterize a system. When the system does not change with time, one says that it is in thermodynamic equilibrium, which is the situation generally assumed here. It has also been observed that these three variables are not independent from each other (even if one does not know the exact relationship between these variables) and are linked by a relation called an equation of state which is a functional relationship among the thermodynamical variables for a system in equilibrium. In the case of a gas characterized by a pressure P, a temperature T and a volume V, the equation of state takes the form $f(P, V, T) = 0$ which indeed reduces the number of independent variables from three to two. There are thus practically only two independent variables. But because the relation between the variables may be involved, it is in practice simpler to keep the three variables as they are, very often because these are intuitive variables, and associate to them a relation linking them. The above case of P, V and T is very often considered in practical applications in gases, liquids and solids.

Thermodynamical transformation, work, heat and energy
All thermodynamical variables are not of the same nature. One distinguishes 'intensive' ones, such as the temperature, which remain identical for each subsystem if one subdivides a given system, from 'extensive' variables, such as the volume, whose values are the sum of the values for each subsystem. Intensive and extensive variables are linked together by pairs, and then called conjugate variables. For example, pressure and volume are conjugate variables to describe the state of a gas. But it is not so surprising. Indeed the first effect of applying an external pressure on a gas is to change its volume. And vice versa if a gas expands in vacuum, its internal pressure will immediately decrease.

Such pairs of variables allow one to define a key thermodynamical quantity, namely the work received by a system under a certain thermodynamical transformation. Defining the work requires following the process in infinitesimal steps, as the conjugate variables a priori themselves evolve in time in the course of the transformation. One then defines the useful concept of infinitesimal transformation, in which some thermodynamical quantities are assumed to change only in very small amounts. For instance, when one expands a gas at external pressure P_{ext} from a volume V to a volume $V + \Delta V$, the received work W is equal to $-P_{\text{ext}}\Delta V$ (the minus sign is conventional, a work *delivered by* the system being intuitively attributed a positive value). More generally speaking, when an extensive variable is infinitesimally varied, the work received by the system is the product of the corresponding intensive parameters by the variation of the extensive variable. The simplest case corresponds to a transformation called 'quasi-static' when the system is (approximately) in equilibrium along the set of infinitesimal transformations leading it from an initial to a final state. This is an important case as it allows one to safely use thermodynamic concepts of equilibrium quantities. To say it in microscopic terms, it corresponds to situations in which microscopic time scales can be overlooked because the microscopic degrees of freedom have had time to relax, thus allowing a purely macroscopic concept to take the lead. The transformation is furthermore said 'reversible' if one can go back to the initial state along the same path. In such a transformation, one has, in particular, mechanical equilibrium at all times. Therefore, the external pressure is equal to the pressure of the system.

Historically, thermodynamics developed in parallel to the industrial revolution and steam machines delivering work. It is then no surprise that the notion of work is complemented by the notion of heat. Roughly speaking, while work appears as a useful form of energy, for example to induce motion, heat is associated to an unorganized form of energy, essentially linked to temperature, itself a measure of the amount of 'useless' microscopic agitation (see Chapter 3). One then defines the heat capacity C of a system as the amount of heat required to increase its temperature by 1 K. More precisely, a variation of temperature ΔT is related to the variation of heat Q by $Q = C\Delta T$. It can be defined at constant volume or constant pressure (represented as C_V or C_P, respectively), per unit mass (specific heat capacity), or per mole (molar heat capacity). Heat and temperature also allow one to define the key concept of entropy S, which provides a measure of the degree of disorder in a system. S has the dimension of a heat divided by a temperature. We will come back to the notion of entropy in Section 3.2.2.

These quantities govern the thermodynamic transformations of a given system, according to three fundamental laws of thermodynamics (see Section 4.1.1). Both work and heat appear as complementary sources of energy. One defines in particular the internal energy U of a system as the sum of both $U = W + Q$. In an isolated system, the total energy is preserved but energy transfers between work and heat are perfectly possible. This is the typical situation encountered in a steam engine. The internal energy U is a particular thermodynamic potential for an isolated system. The relevant thermodynamic potential must be properly selected to characterize a system according to the actual transformation. For example in the case of a system in contact with a thermostat, which ensures a constant temperature in the system, the thermodynamic potential to be considered is the Helmholtz free energy $F = U - TS$.

States of matter

Let us now turn to the states of matter. One often encounters the use of 'phase' in place of 'state'. Some words of caution are in order here. The distinction between phase and state is not just a subtlety of language. When comparing a small piece of ice to a drop of water, one usually refers to the solid or the liquid *state* or *phase* of water. But what occurs when one considers mixtures, for example of oil and water? How can we distinguish a drop containing oil and water, both in the liquid state? Actually, a region of space which exhibits uniform chemical and physical properties (same composition, density, index of refraction, etc.) corresponds to a same phase. The drop mentioned above is thus the mixture of two immiscible phases, oil and water, of the same liquid state of matter. In practice, we very often use 'phase' as a synonym of 'state' but one should be aware that this might constitute a misuse of language.

Now that we have clarified the terms, let us consider the three standard states of matter. Macroscopically, they behave very differently. A solid, without any external constraint, keeps its volume and shape along the time, whatever the container. It then transfers a pressure applied at one of its boundaries to another one. This pressure transfer is all the stronger when the solid is not deformable. A liquid also keeps a constant volume but takes the shape of the container. Note that this is related to the property of a liquid to flow more or less rapidly or, in other words, to its viscosity. We will detail this notion later in this chapter (see Section 1.4.2). A liquid can also transfer a pressure applied at one of its boundaries. The equivalent of a non-deformable solid in this case would be a non-compressible liquid. As for a gas, it has no definite volume nor shape: all its constituents fill thc whole volume offered by the container.

This macroscopic viewpoint has to be complemented by a microscopic perspective where it turns out that interactions play a major role. Indeed, owing to interactions, the constituents (atoms or molecules) of a given substance in the solid state, are only allowed to vibrate about fixed positions of a crystalline lattice (note that this lattice is not necessarily regular). Henceforth, each constituent always keeps exactly the same neighbors. In a liquid, instead, more freedom of motion is possible but the constituents still stay not too far away from each other. This motion nevertheless induces a constant turnover of the neighbors, while keeping in average approximately the same number of neighbors. Therefore, the number density of a substance, namely the number of constituents per unit volume, is practically the same in both solid and liquid states. Usually, the density of the solid is slightly higher than that of the liquid, with the important exception of water. Indeed, ice is less dense than liquid water, allowing pieces of ice to float on the surface of liquid water. The constant change of neighbors in the liquid state also induces the notion of order: a liquid is usually considered as more disordered than the solid state. But this is not necessarily true: a piece of glass can for instance be viewed as a frozen liquid. Finally, in the gas phase, the density usually takes very small values and the forces between the constituents are small enough to allow them to travel almost freely, in any spatial direction. One cannot, in this case, even properly define a number of neighbors. It should finally be noted that a detailed analysis of the neighborhood brings more information than the mere density. We shall come back to that aspect later on (see Section 1.3.3).

1.1.2 Phase transitions and phase diagrams

Phase transitions

By changing one or several variables characterizing a system, one can change its state. For instance, everyone knows that the 'natural' state of water is the liquid. By 'natural', we usually mean at atmospheric pressure and at a temperature of 25°C or so. If ones changes the temperature of water, either by heating it in a saucepan or by putting it into a freezer, one can obtain water vapor or a piece of ice.

By changing external parameters such as the pressure or the temperature, one can go from one state to another, thus inducing a *phase transition*. Let us consider once again the case of water and place a (solid) piece of ice in a heated saucepan: it first melts and becomes liquid, and then boils to produce water vapor. If a lid is put on top of the saucepan, it will be eventually lifted up by the vapor, as long as the liquid is boiling, demonstrating that the total volume increases during the phase transition. Heating the saucepan therefore induces two successive phase transitions. The first one from solid to liquid is called fusion and the reverse process, from liquid to solid, freezing. The second one from liquid to gas is called vaporization and the reverse process, condensation. The last possible transition, namely the direct one between solid and gas does not occur in our saucepan; it requires much higher pressures to occur. It is called sublimation when from solid to gas and deposition when from gas to solid. But let us come back to our saucepan example. If ones measures the water temperature T during these phase transitions, one observes that T remains constant, as long as both states (solid/liquid or liquid/gas) coexist. In that particular case, once the piece of ice reaches 0°C, it starts to melt and stays at that temperature up to full melting. Again, when temperature reaches 100°C, the liquid water obtained from the original piece of ice starts to boil and transform into vapor. This continues at fixed temperature up to a full transformation of liquid into gas/vapor.

Another phase transition observed in everyday life occurs when one opens a bottle of soda: because of the quick drop of pressure in the bottle, the carbon dioxide which was previously dissolved in the soft drink is suddenly changed into gas and many bubbles appear. More generally, for a given substance, one can identify ranges of temperature, pressure and volume for the existence of different states. For example, intuitively, at high pressure, low volume and low temperature, one expects to observe the solid state, while at low pressure, high volume and high temperature, one observes the gas state instead.

Phase diagrams

As discussed above, the three variables P, V and T are linked to each other by the equation of state. In a 3-dimensional (3D) plot, the possible states of the system occupy surfaces. This graphical representation provides a simple way to identify the regions where the solid, liquid or gas states exist, in terms of P, V and T. For the sake of simplicity, we restrict ourselves to pure substances. In case of mixtures, a separation of components appears. The full 3D representation for a pure substance is schematically depicted in the central panel of Figure 1.2. The vertical axis corresponds to P, and the two horizontal axes are V and T. A given black thin line delineates a region of existence of a single state and a region of existence of a mixture of 2 different states. There are three regions corresponding to such a mixture (of solid and liquid, or liquid and gas, or solid and gas) and they are highlighted by shaded areas in the PVT diagram. The black thick line is often called the 'triple line' and corresponds to the line along which the three states may coexist at equilibrium. The line between the liquid region and the liquid-gas one is called the liquid-saturation line or dome because of its bell shape. The maximum of this dome is called the critical point, with critical values P_c, V_c, T_c specific to each substance.

The light lines drawn on the PVT diagram correspond to isotherms, that is paths at constant T. Let us consider for instance the line that goes just below the critical point. If one follows it from low V (or high P), one starts for instance from the liquid state, crosses the saturated liquid line and observes the apparition of the first bubble of gas, goes through the saturated liquid-

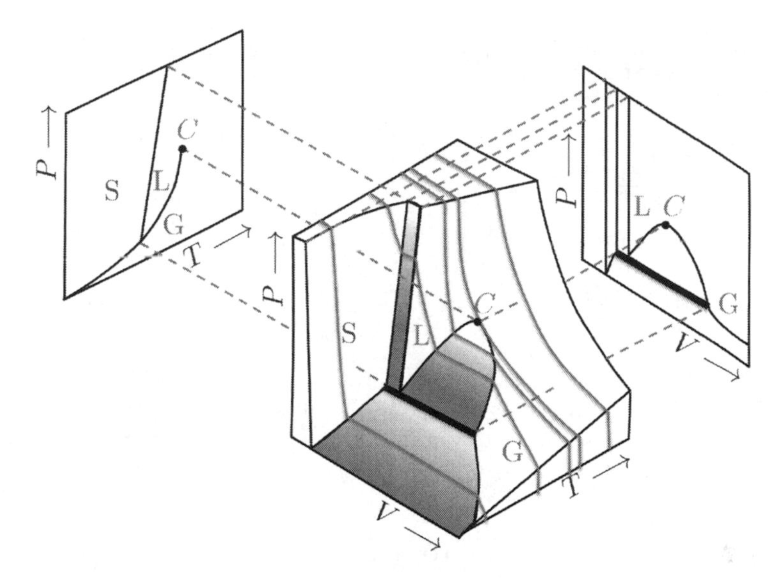

Figure 1.2: Phase diagrams of a pure substance. Center: PVT diagram with regions of existence of the solid (S), liquid (L) and/or gas (G) states. The letter C denotes the critical point, the black thick line is the triple line, and the light full curves are isotherms. Top right: PV diagram. Top left: PT diagram.

gas region (at constant P), crosses the saturated gas line where there is no liquid anymore, and finally explores the gas state only with increasing V (and decreasing P). One can also consider the isotherm going exactly through the critical point as a limiting case. Above this isotherm, there is no more physical distinction between a liquid and a gas: this is the supercritical fluid. On the other side, an isotherm going below the triple line crosses the solid-gas region.

To better visualize phase transitions and areas of existence of a given state, it can be more convenient to consider projections of the PVT diagram, providing 2D phase diagrams. For instance, if one projects the 3D chart along a direction parallel to the T axis, one gets the PV diagram (top right panel), and along a direction parallel to the V axis, one obtains the PT diagram (top left panel). There is a third possibility obtained by projecting along a direction parallel to the P axis and one then gets the TV diagram. Since its features are very similar to those of the PV, we will not discuss the TV diagram in the following. In the PV diagram, the triple line does remain a line, while in the PT diagram, it reduces to a point, called here the 'triple point' at (P_t, T_t), even if in 3D, it corresponds to a line and henceforth with the possibility of different V. Note however that, since the triple line has a finite length, the possible values of V are bound by a minimum and a maximum value.

Let us focus a bit on the PT diagram, often referred to as Clapeyron's diagram. The case of pure CO_2 is shown in the left panel of Figure 1.3 and represents a typical example. In this projection, the three 2-phase regions

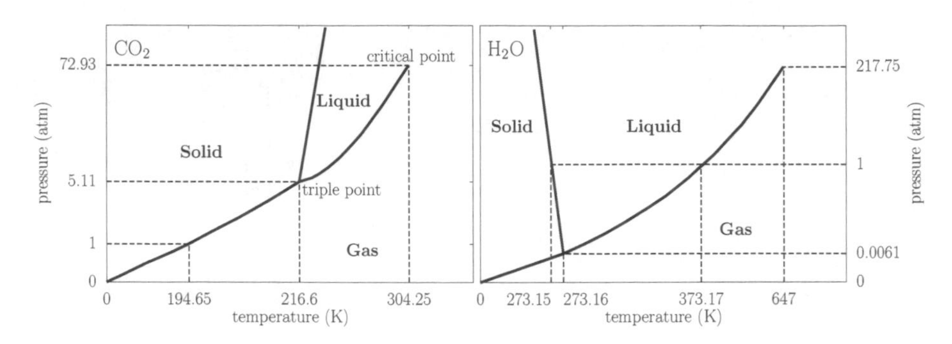

Figure 1.3: Clapeyron's diagram (pressure as a function of temperature) for pure carbon dioxide (left) and pure water (right). The horizontal and vertical scales are neither linear nor logarithmic (data from NIST Chemistry WebBook, http://webbook.nist.gov).

collapse to curves. Indeed, when two states coexist in equilibrium, P and T must remain constant and only V is allowed to vary. For instance, the infinite number of mixtures of liquid and gas that can coexist at a given duplet (P, T) all appear in the Clapeyron's diagram as a single point. We have already mentioned above the same feature for the triple point: the triple line in the 3D chart reduces to a single point, misleadingly called the 'triple point' in this 2D representation. This point is the intersection of three lines: the projection of the solid-gas region as the sublimation (or deposition) line, the projection of the solid-liquid region as the fusion (or freezing) line, and the projection of the liquid-gas region as the vaporization (or condensation) line. The latter line ends at the critical point. Mind that this point is really a point, even in the 3D diagram! There, the pressure P_c, the temperature T_c and the volume V_c are unique and depend only on the considered substance. There is no distinction between a liquid and a gas for $P > P_c$ and $T > T_c$.

The three lines mentioned above delineate the single phase regions. For instance, above the fusion line, that is at high pressures, we have the solid, and below it, the liquid. Note that, for most pure substances, the slope of the fusion line is positive but not for water (see right panel of Figure 1.3), indicating that the solid state of water is less dense than the liquid state. For most pure substances, the triple point lies below standard pressures and temperatures, and the critical point usually much above. This means that in standard ambient conditions (25°C and 1 atm), one can easily observe phase transitions from solid to liquid, and from liquid to gas, if the temperature increases, or from gas to liquid and then from liquid to solid if the pressure increases. However, for CO_2, we have $P_t > 1$ atm. This means that solid CO_2 (also commonly called 'dry ice') can directly sublimate to the gaseous state in standard ambient conditions.

Before moving on, let us quickly comment on the use of the words 'gas' and 'vapor'. Up to now, we have, on purpose, avoided using the word 'vapor' for the description of the low density and high temperature state. The reader can however encounter both words used in different contexts. We can for instance talk about a vapor for a gas below the critical point, meaning that one can observe the coexistence of the liquid and the gas states in standard conditions, as we usually do with water: one would not use the words 'water gas' here. We will not talk about 'oxygen vapor' as well! It therefore seems that the use of the word 'vapor' or 'gas' is more related to our everyday life experience: if the substance is commonly observed as a liquid, that is if the critical point is above standard ambient conditions, one would call the gaseous state a vapor. On the contrary, if a substance is above its critical point at room temperature, as O_2 or N_2, one would rather use the word 'gas'.

1.1.3 Thermodynamics of phase transitions

The thermodynamical description of phase transitions is a rich domain and it is not our goal here to enter such discussions in detail. Still, it is important for further discussions to comment briefly on these notions of phase transitions from the point of view of thermodynamics. In particular, it is interesting to discuss in some detail how state variables evolve during a phase transition. We shall again consider here the case of water (right panel of Figure 1.3) to illustrate the point.

In phase transitions, one usually considers a transformation with one state variable being fixed, and vary one of the two other state variables. The knowledge of the phase diagram of a pure substance allows us to determine at which T, V or P a phase transition occurs. To this end, we choose one of the projections from the PVT diagram in which a state variable maintained constant is depicted as a straight line. For instance, $T = \text{const.}$ would appear as a vertical line in the PT diagram, whereas a horizontal line would correspond to the case $P = \text{const.}$ in the PT and the PV projections. In the right panel of Figure 1.3, which shows the Clapeyron's diagram of pure water, let us consider the line $P = 1$ atm, start from a high temperature (say, $T > 400$ K) and cool down the water at constant P: a first transition from gas to liquid occurs at $T = 373.15$ K, when the line $P = 1$ atm crosses the vaporization line. P and T remain constant as long as the gas and liquid coexist. Not visible on this diagram is the decrease of the total volume of water, since the gaseous state condenses to liquid, more compact than the gas, hence a decrease of V. When water is totally transformed into liquid, one can decrease again T until the line $P = 1$ atm crosses the solidification/freezing line at $T = 273.15$ K for which the liquid starts to freeze into ice. Once again, P and T are constant as long as the solidification is not complete. After this phase transition, the whole water is in the solid phase and one can a priori decrease the temperature until the absolute zero of temperature.

Another peculiarity of water is to consider an isotherm below the triple point at $T_t = 273.16$ K, for instance at $T = 273.15$ K. If we start from a high pressure (say, $P = 10$ atm), water is in the liquid phase. Then, a decrease of pressure causes the freezing of water at $P = 1$ atm, and afterwards, the sublimation of water, that is the transition from solid to gas, at $P < P_c = 0.0061$ atm. In other words, by decreasing P at 0°C, one observes the phase transitions $L \to S \to G$, which is not the order $S \to L \to G$ one usually expects! This is due to the negative slope of the fusion line in pure water.

The energetics of phase transition is also a key issue, widely discussed in thermodynamics. Again, it is not our purpose to enter such discussions in detail here. We will just outline one or two key aspects. We have mentioned above (see Section 1.1.1) that depending on the actual physical situation, there exists a thermodynamic potential best suited to describe the system and its evolution. In the case of phase transitions, one uses the Gibbs free energy defined as $G = U - TS + PV$ including the entropy S and internal energy U. The Gibbs potential is assumed to be minimal in such cases and take equal values in the two phases undergoing the phase transition. The Gibbs potential furthermore allows us to classify phase transitions according to what one calls their order and which provides a measure of the amount of discontinuity faced by variables during a phase transition. To take again the case of the liquid-gas transition, it turns out that volume is discontinuous during the transition, which simply reflects the rearrangement of constituents in a more dilute/compact way when stepping from one phase to the other. Volume does appear in the Gibbs free energy G and it can be shown that it corresponds to a first order derivative of G with respect to pressure. A first derivative of G is thus discontinuous and one then speaks of a first order phase transition. But all phase transitions do not behave the same. Therefore, if all first order derivatives of G are continuous but a second order derivative is discontinuous, one says that the phase transition is of second order. For instance, one can show that a heat capacity can be written as the second derivative of G under certain conditions. A second order transition often corresponds to the discontinuity of a heat capacity at a given temperature. The process can be generalized to any order of derivation/phase transition order. More details are given in Section 4.1.2.

During a first order transition, there is a heat transfer at constant temperature between the system and the environment. The latent heat is the amount of energy exchanged along the phase transition, for example to convert a liquid into gas. In contrast, there is no heat exchange in a second order transition as it happens for example in the ferromagnetic transition of a piece of iron. A ferromagnet is a material which can exhibit spontaneous magnetism if all the atomic magnetic moments point in the same direction. However, if a magnet is warmed up to very high temperatures, it loses its magnetism at the ferromagnetic transition temperature, or Curie point, which is 1043 K (770°C) for iron. This change between magnetic and non-magnetic states at the Curie point is an example of a second order phase transition.

In the 1930s, Landau put forward a general theory of phase transitions in terms of symmetries. For example, a magnet above its Curie point has no magnetic moment. No special direction is singled out, and the system possesses a full rotational symmetry. But below the Curie temperature, a spontaneous magnetization develops. In principle, the system could be magnetized in any direction, but a particular one is spontaneously chosen. The initial rotational symmetry is said to be spontaneously broken at the phase transition. In Landau's theory, such transition is characterized by an 'order parameter' (in that case, the magnetization), which is zero above the critical temperature and non-zero below it. To find the equilibrium state, one has to deduce or guess the form of the Gibbs free energy of the system in terms of the order parameter, and find its minimum.

Let us end this section with the concept of metastable states. The Gibbs free energy can possess, in addition to a global minimum, several local minima. A system can get 'stuck' in such a local minimum for some time because energy barriers temporarily hinder the system to relax to the global minimum. For instance, water liquid at standard pressure can be 'supercooled' below 0°C without turning into ice. This however requires water to be very pure and free of nucleation sites, the latter being needed to initiate crystallization of ice.

1.2 MICROSCOPIC STRUCTURE OF MATTER

We have seen in the previous sections that the existence of a state of matter, and more specifically some properties of liquids, come from the microscopic interactions between constituents, or between the constituents and the external walls of the container in the case of a liquid. To go further, we therefore need to say a few words about how matter is organized and structured, down to the elementary constituents of matter and the fundamental interactions.

1.2.1 Atoms and nuclei

Atomic structure of matter is nowadays a well known and accepted concept, more than two millenniums after its suggestion by Greek philosophers. The experimental notion of atoms, at the level we understand them today, took shape between the late 17th and early 19th centuries with the first quantitative steps of chemistry following Dalton and Lavoisier. The identification of mass conservation in chemical reactions indeed simply reflected conservation of atoms. The atomic theory was further established/accepted in the early 20th century in particular with the measurements of Avogadro's number $\mathcal{N}_A \sim 6.02 \times 10^{23}$ mol^{-1}. In an ideal gas, it would correspond to the number of atoms or molecules in any gas in a container of 22.4 liters and at temperature 20°C and 1 atmosphere pressure ('normal' conditions). At about the same time, the structure of atoms was identified as constituted by a central compact nucleus surrounded by a cloud of electrons. The structure of nuclei was further analyzed during the first decades of the 20th century: its constituents, neutrons and protons, were identified in the early 1930s, as well as the interactions binding them together. Most ensuing developments refined this basically correct picture.

Today's picture of the structure of matter can be summarized as follows. Macroscopic matter (under whatever form, liquid, solid or gas) is constituted of atoms or molecules arranged together according to interactions they exert on each other. Molecules are 'compact' arrangements of atoms in limited number while bulk matter can reach macroscopic sizes. But in both cases the 'elementary' constituent is the atom. The atomic scale corresponds to the angstrom length scale (1 Å$= 10^{-10}$ m) and typical energies of the order of the electron-Volt. The latter energy corresponds to the energy gained by an electron in a potential difference of 1 Volt. Expressed in SI units, 1 eV $= 1.6 \times 10^{-19}$ J. This shows that energies involved in atoms and molecules are really *microscopic* since they are 26 orders of magnitude smaller than the daily human energy requirements, which is about 10^4 kJ.

The atom is constituted of a central compact nucleus surrounded by an electron cloud. While the electron cloud extension basically defines the atom size and lies in the angstrom range, the nucleus size lies 4 to 5 orders of magnitude below, in the fermi range (1 fm $= 10^{-15}$ m). Correlatively, the energy scale lies in the MeV (10^6 eV) range. Electrons orbit around the nucleus by virtue of the electrostatic interaction with the charged nucleus. The similarity between the electrostatic interaction and the gravitation leads to a widely used picture of solar system representation of the atom with electrons orbiting around the nucleus. While the analogy has some pictorial impact, it should remain an analogy and no more, especially because of the strong quantum effects present at that scale. The more correct notion of an electron cloud around the nucleus is a direct consequence of this quantum nature.

Nuclei are themselves constituted of protons and neutrons which are particles of similar masses, and associated energy $m_n c^2 \sim m_p c^2 \sim 940$ MeV, $c = 3 \times 10^8$ m/s being the light velocity, to be compared to the light electron with $m_e c^2 \sim 500$ keV ($m_e/m_n \sim 1/1800$). Atomic masses are thus dominantly concentrated in nuclei. Protons bear a positive electric charge exactly opposite to the electron charge. There are therefore the same numbers of protons and electrons in a neutral atom to ensure electric neutrality. Neutrons, as indicated by their name, are neutral particles. An atom is fully characterized by its electron number Z and its nucleus by Z and the number of neutrons N. Neutrons ensure the stability of the nucleus by more than compensating the electrostatic repulsion between the charged protons. This is achieved by a dominantly attractive interaction called the strong nuclear interaction.

1.2.2 Elementary particles and fundamental interactions

The way down to even more fundamental constituents of matter could be pursued at least one step further with the description of neutrons and protons in terms of quarks. Quarks are one (essential) component of the standard model of elementary particles which also embraces other species such as, notably, electrons. The latter are, in this model, associated to the two quarks (up u, and down d) constituting neutrons ($n = udd$) and protons ($p = uud$)

forming, together with the electron and the electronic neutrino ν_e, the most important family of elementary particles constituting most physical systems in the Universe. A word is in place here, concerning the term 'elementary' used for labeling certain particles. It is used in the case of particles which are not themselves composite particles, namely made of other particles. The most familiar elementary particle is the electron. Neutrons and protons are particles made up from quarks, and therefore they are not elementary. Elementary particles thus provide the most microscopic building blocks of our understanding of matter and its composition. A brief outline of the elementary particles is given in Appendix C, where the two other families of the Standard Model are described.

Binding between quarks, and especially inside neutrons and protons, is ensured by the strong nuclear interaction which is one of the four fundamental interactions. The three other ones are gravitation, the electromagnetic force, and the weak nuclear interaction. Gravitation and electromagnetic interactions exhibit many similarities and they are those which one experiences the most in everyday life. We are all subject to the gravitational interaction that we permanently experience as gravity on the surface of the Earth. In turn, most objects around us, from a metallic tool to a mineral or a vegetable, and our own body as well, are bound by the electromagnetic interaction. The two nuclear interactions (weak and strong) belong to the subatomic world because of the fact that they act on very short distances, of fm order or below.

Gravitation acts on massive particles and the electromagnetic interaction, which combines the electric and magnetic forces, acts on charged particles. This point is less trivial than it looks. Indeed the way one defines massive particles is precisely the fact that they 'feel' (couple to) a gravitational field. Along the same line of reasoning, a charged particle is a particle sensitive to an electric or magnetic field. There is hence a strong duality between these properties (mass and charge) and the characteristic interaction they are associated to. The same holds true for the strong nuclear interaction which defines a class of particles (called hadrons) which are sensitive to it, again establishing a classification of the particles according to their potential coupling to a certain interaction. The strong nuclear force is responsible for the binding of nuclei while the weak nuclear interaction, which bears some relation to the electromagnetic one, is responsible for the radioactive decay of nuclei.

Figure 1.4, in some sense, summarizes the ranges of application for the four fundamental interactions presented here, in terms of length/energy scale. It also shows typical systems at these various scales. Gravitation is negligible below about 1 mm as compared to the electromagnetic force, and even more so at the atomic and subatomic level with respect to the strong interaction. Gravitation mainly acts from the human scale up to astronomical ones (see inner gray arc of circle). The electromagnetic interaction (inner black arc) is important from the nuclear up to the stellar scale. Indeed, the internal structure of stars is a plasma, which is a kind of soup composed of electrons

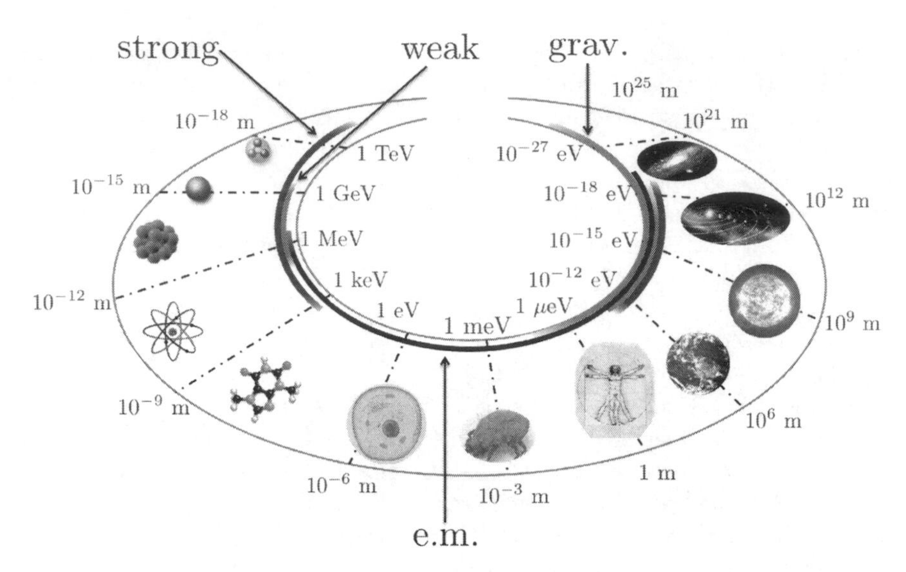

Figure 1.4: Energy and length correspondence and energy ranges of application of the gravitational (inner gray arc of circle), the electromagnetic (inner black arc), the weak nuclear (outer gray arcs) and the strong nuclear (outer black arcs) interactions. Systems at given length scales are given as typical examples (quarks, nucleon, nucleus, atom, molecule, cell, acarid, human being, Earth, Sun, solar system, galaxy).

and ions. The weak and the strong interactions of course act at the nuclear level but also in some stellar objects as white dwarfs and neutron stars.

1.2.3 Effective forces and composite particles

We briefly presented above the most fundamental particles and forces in nature. However they are often useless to describe the physics we encounter every day because, for some of them, their relevance only appears at very high energies. In other words, at lower energy scales, one deals with composite particles which undergo effective forces built from these fundamental interactions.

The first such example is provided by the strong nuclear interaction. Indeed, the interaction between quarks has the peculiar behavior to increase very quickly with distance, so that quarks are always confined in a length scale of 1 fm. The strong interaction is, by nature, the interaction binding quarks together inside composite particles. By extension, one also uses the term of strong interaction to label for example the interaction between two neutrons or neutrons and protons. This is correct to the extent that finally the interaction between two such objects reduces to the sum of the interactions between their constituting quarks, even if the thus constructed interaction exhibits properties which may look different. To explain the stability of a nucleus against the electrostatic repulsion between protons, the strong interaction is rather replaced by the nucleon-nucleon force (also called the residual strong force) which, as a function the the relative distance between nucleons, is repulsive below 0.7 fm, attractive around 1 fm and quickly vanishes above 2.5 fm. The residual strong force therefore appears as an effective interaction of the true strong nuclear force, and between hadrons, composites of quarks. Its range of a few fm imposes the typical size of nuclei.

At a larger length scale, of order 1 Å, lie atoms, which are composed of Z negatively charged electrons and a nucleus of charge $+Ze$, bound together by the electrostatic interaction (also called the Coulomb interaction). Some atoms exhibit similar physical and chemical properties and this is exemplified in the famous periodic table of the elements, where atoms are arranged according to their atomic number Z. Atoms interact with each other via electrostatic interactions between the electrons and the nuclei which compose each atom, resulting in a chemical bond and forming a molecule. This is, at that length scale, the dominant interaction. However, to describe this bonding in certain cases—typically inert atoms such as rare gas atoms—one can also avoid considering the complicated sum of each individual Coulomb interaction by using an effective interaction between the center of mass of the bound atoms. Indeed, although the electromagnetic interaction is in principle of infinite range, in practice long-range effects are suppressed by counteraction of charges of opposite sign. Consider for instance an atom. At large distances, it is electrically neutral, although composed of negative electrons and positive protons. The same holds for the interaction between two atoms. The resulting 'screening' of the Coulomb force leads to an effective interaction which is short-range. The situation very much resembles the mechanism responsible for the residual strong nuclear force discussed above. It furthermore delivers a qualitatively similar effective interaction, repulsive at short distances, attractive at larger distances, and vanishing asymptotically. As there exist different types of bonding (see Section 5.1), there are different types of effective interactions for the description of chemical bonds. The same holds for bonding between molecules and between constituents in a liquid or a solid state. We shall therefore use such effective interactions when necessary. It is however interesting to remember that this is the electromagnetic interaction which is behind such effective interactions and which thus plays a crucial role in physical and chemical properties in most of the everyday life objects.

1.3 FROM IDEAL GAS TO REAL FLUIDS

The discussion on the microscopic structure of matter brought us a bit far away from our starting point on the description of the solid, liquid and gaseous states, and in particular the properties of fluids which are at the core of this book. But it was an important issue to clarify key terms to be used in the following and to emphasize the importance of scales and degree of description. We have seen in particular how the level of description is related to the constituents of a phase and to the interactions between these constituents. It is now time to come back to this central question on the nature and properties of fluids. We shall proceed stepwise, starting from the simplest ideal gas in which interactions are neglected. This model is nevertheless very useful for qualitative reasoning, especially in dilute gases. Including interactions will allow us to step properly into the realm of liquids. Finally, we shall discuss how to pass from gas to liquid, using the widely used van der Waals equation of state.

1.3.1 The ideal gas

The simplest description of any assembly of particles consists in neglecting interactions between them. This is a reasonable approximation in the case of a dilute gas. It is here interesting to recall the widely used model of the ideal gas in which interactions are indeed overlooked. At this stage of the discussion, it has to be viewed as the first step of the more elaborate modeling to come. The richness of this model and its overall use well beyond its limitations is a further strong argument in favor of discussing it.

The ideal gas model relies on the following assumptions/reasoning. Let us consider a large number of N particles in thermodynamical equilibrium in a container, chosen for simplicity as a cubic box of volume $V = L^3$, where L is the side length. The particles are in constant motion and, in the absence of external forces, they are uniformly distributed throughout the container. The number of particles per unit of volume (or number density ϱ) is then the same in any small volume within the box. In addition, all directions of the molecular velocities are equally probable, and the modulus of the velocities can take all possible values. In a sufficiently dilute gas, the particles are most of the time separated by large distances, as compared to their own dimensions. One can henceforth assume that the range of the forces between particles is very short as compared to their typical separation (see Section 1.2.3), and in addition comparable to the size of the particles. This practically means that interactions are reduced to contact collisions. Collisions are considered perfectly elastic, that is the kinetic energy and the momentum are conserved. Because the duration of the collision is very short as compared to the typical time interval between collisions, forces between particles can be safely ignored, at least as a first step. The same assumption can be made to the collisions with the walls of the container, which have an enormous mass as compared to the molecular mass.

To summarize, we can say that we consider point-like particles (occupying no volume) with no internal motion such as vibrations or rotations, nor other internal degrees of freedom (quantum ones in particular). The key assumption underlying the ideal gas model is then to consider that its constituents are non interacting. They are therefore only subject to agitation resulting from the thermal energy stored in the system. Experimentally, all gases behave in a universal way when they are sufficiently dilute. Actually, the ideal gas model corresponds to the case where the interactions become negligible at thermal equilibrium, which is justified in the case of a very dilute gas. The ideal gas thus reflects this limiting behavior, and the remarkable predictive capabilities of the model justify the underlying hypothesis, in spite of some apparent contradictions. In particular, it should be noted here that the notion of temperature itself (see Section 3.1) actually requires interactions between the constituents of a system. It is precisely these interactions which allow the constituents to randomly exchange velocities and progressively smooth out differences. Henceforth, a gas at finite temperature by construction requires interactions. In the ideal gas case, one simply neglects these interactions and directly discusses a gas at thermal equilibrium where the remaining interactions do not play a major role in describing the system.

The equation of state of the ideal gas, linking temperature, pressure and volume/density (see Section 1.1.1), contains the empirical laws of Boyle and Gay-Lussac for dilute gases. The usual form is

$$PV = nRT\,, \tag{1.1}$$

where n is the number of moles in the gas, and $R = 8.31$ J mol^{-1} K^{-1} is the molar gas constant. The number of moles n is related to the number of particles N (either atoms or molecules) by $n = N/\mathcal{N}_A$ with $\mathcal{N}_A$ the Avogadro's number. One can deduce n from the relation $n = m/M$ where m is the total mass m of the system and M its molar mass (usually expressed in g/mol). Alternatively, the equation of state can be written as $PV = Nk_\mathrm{B}T$, where $k_\mathrm{B} = 1.38 \times 10^{-23}$ J K^{-1} is the Boltzmann's constant (since $R = \mathcal{N}_A k_\mathrm{B}$). There is yet another interesting form

$$P = \varrho k_\mathrm{B} T\,, \tag{1.2}$$

where $\varrho = N/V$ is the number of particles per volume unit, or number density. The latter form is interesting as it eliminates the extensive variable V to the benefit of number density, which provides a somewhat more universal version of the equation of state, without explicit reference to a container. Incidentally, the large and extremely small values of the constants R and k_B in SI units show their importance for the macroscopic and microscopic descriptions, respectively. It should be kept in mind that sometimes the number density is written as n, which is the symbol we have adopted for the number of moles, and the symbol ϱ is used for the mass density, which we shall write ϱ_m throughout this book to avoid any confusion with the number density ϱ.

The equation of state of an ideal gas is extremely simple because of the neglect of interactions. Nevertheless, it is not only valid for dilute gases but, as experiment shows, it also provides a reasonable starting point to model dense real gases. Although its use in the case of much denser systems, such as liquids, is not justified, it remains in some cases useful to get rough estimates of orders of magnitude. The proportionality between pressure and temperature for example reflects our everyday experience. To come back to one of our earlier examples, steam in a pan lifts the lid at high enough value of temperature, precisely because the pressure exerted by steam under the cover can compensate its weight. Similarly, the pressure increases with the density of a system, which reflects the many collisions of the particles with the walls. This is for example the case of tires of a car in which the air compressed to about two atmospheres can sustain a vehicle up to several tons.

1.3.2 Interactions and real fluids

We have so far neglected interactions between particles in the ideal gas. The role of interactions is however essential to explain not only the properties of real gases, but also for example the phase transition between a gas and a liquid. Throughout the 19th century, there were several attempts to modify the ideal gas equation of state including qualitative interaction effects. It is J.D. van der Waals who succeeded in writing a simple equation of state which turned out to provide a powerful tool of investigation, yet rather simple, of real gases/fluids. We shall come to it below, after discussing interactions a bit more in detail.

The intermolecular interaction
As discussed in Section 1.2.3, interactions between atoms or molecules can be described as an effective force stemming from all 2-body Coulomb interactions between their constituents. This effective interaction can in general be discussed in terms of potential energy E_P. The term potential means that this energy can potentially turn into another form of energy, typically kinetic energy. It is characterized by a strong repulsion at very short distances ($E_P \gg 0$), when the two particles are very close to each other. In the other extreme, for particles infinitely far apart, the interaction has no effect and E_P vanishes. Finally at intermediate distances, the repulsion and the attraction counteract, forming a well in the potential energy and thus exhibiting a minimum. The net result is an attractive interaction, allowing bonding, with a negative E_P.

There exist many models employed to describe the intermolecular interactions. The simplest case consists in assuming that molecules are impenetrable; one completely ignores the attraction and considers that each molecule behaves as a rigid sphere of diameter σ. The corresponding potential energy is infinite for $r < \sigma$ (hard core repulsion), and null for $r > \sigma$. A more realistic model is to add a weak attraction between molecules in the region $r > \sigma$. A very popular model is to write that part as proportional to $-1/r^6$, the negative sign indicating the attraction. Although phenomenological, this choice is well founded. The interaction is basically an electrostatic effect, which results from the charge distribution within the molecules. Each molecule induces a dipole moment in the other one, so that the interaction between two molecules can be described classically as a dynamical interaction between dipole moments. It turns out that the attractive part of the potential energy at long distances is proportional to $-1/r^6$. This model keeps the dipole-dipole attractive part and modifies the hard core repulsion. This is the popular Lennard-Jones potential, which is written as

$$V(r) = 4\epsilon \left[\left(\frac{\sigma}{r}\right)^{12} - \left(\frac{\sigma}{r}\right)^{6} \right] \quad . \tag{1.3}$$

It contains two parameters, ϵ and σ, whose values for a variety of species are determined by various techniques. One should point out that there is no reason to expect that the short-distance repulsion will follow a power law form or, more generally, will have an analytical form. The resulting energy has the shape displayed in Figure 1.5. There are other choices which include more inverse powers and an exponential repulsion as well.

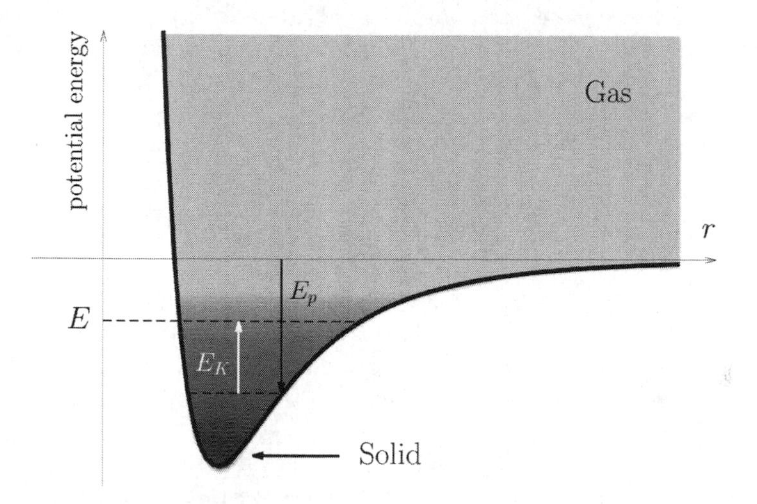

Figure 1.5: Potential energy as a function of intermolecular distance r (thick curve) and qualitative ranges of r corresponding to phases of matter, dark gray for solid and light gray for gas (see text for details).

From intermolecular potential to phases

The above discussed intermolecular potential provides a simple tool to understand the appearance of the various possible phases (solid, liquid, gas) in relation to the importance of the interaction. This is strongly linked to the existence of a minimal potential energy at a given distance, the latter corresponding to the equilibrium distance. Our reasoning relies on the energy potential shown in Figure 1.5. If we consider an isolated system, the total energy E of the system is conserved. This energy reads as $E = E_K + E_P$ where E_K and E_P are the kinetic and the potential energies respectively. Now consider a given value of E. We draw a horizontal line of constant energy E and look at the intersects with the thick curve. These intersects define the range of r accessible by the system. For instance, at a distance r taken in this range, the potential energy E_P corresponds to the downward arrow and the kinetic counterpart to the upward arrow. It is fortunate that the latter arrow is upward, meaning that $E_K > 0$, as it should be. Incidentally, we note that if E is smaller than the minimal potential energy, there exists no intersect and correlatively no physical solution. Depending on their total energy, the interacting molecules then explore different areas of the potential energy interaction.

We now relate these accessible areas with states of matter. We will see in Chapter 3 how the macroscopic temperature of a system is related to the microscopic kinetic energy of its constituents. Here let us admit this proportionality relationship, which implies that the larger E_K, the higher the temperature. Therefore, at low E, the accessible kinetic energies are small and this corresponds to low temperatures. At the same time, the range of accessible distances remains small and located around the equilibrium distance. Excursions far from this minimum are not possible and the particles in interaction can only vibrate about their equilibrium position: this results in a high density of matter and the system freezes into a solid phase (dark area). This is observed for most substances, with some noticeable exceptions discussed later on in this book. On the contrary, the case $E > 0$ (in fact, E larger than the asymptotic value of the potential energy) appears at high temperatures. There, the intermolecular distance still remains bound from below, due to the short-range repulsion between molecules, but there is no maximal r. The intermolecular distance can take arbitrarily large values, much larger than the equilibrium distance, because the effect of interactions tends to fade away. This results in a low density state and particles can move almost freely with respect to each other: the phase is a gas, as depicted by the light gray area above the horizontal axis in Figure 1.5.

Between these two extreme cases (solid and gas) lies the liquid state which, to some extent, interpolates from the solid to the gas (see faded gray region). In liquid, interactions are still strong enough to restrict the range of accessible r but molecules can explore relative distances significantly larger than the equilibrium distance. In a crystal, neighbors remain the same forever and the solid is highly ordered. In a liquid instead, there is a constant turnover of neighbors, bringing some disorder in the system. The density is slightly smaller than in a solid (with the important exception of water, see the floating iceberg in Figure 1.1) but remains of the same order of magnitude. On the other side, liquids and gases also bear some similarities. We remind readers that no distinction between the two exists above the critical point. We will discuss more quantitatively the typical intermolecular distances in a liquid and in a gas via the notion of mean free path in Section 3.1.4.

1.3.3 The effects of interactions and the pair correlation function

Let us go even further in the comparison between the three states of matter by considering pairs of molecules in a given state. We look at the probability $P(\mathbf{r}_1, \mathbf{r}_2)$ to find one molecule at point $\mathbf{r}_1$, knowing that there is another one at another point $\mathbf{r}_2$. If the molecules are independent, we just have the product of the two individual probabilities, $P(\mathbf{r}_1, \mathbf{r}_2) = P(\mathbf{r}_1)P(\mathbf{r}_2)$, and they are thus said to be *uncorrelated.* Therefore, the probability of joint detection at two points separated by a certain distance $r = |\mathbf{r}_1 - \mathbf{r}_2|$, divided by the product of the single detection probabilities at each given point, is strictly equal to 1. This ratio is known as the pair correlation (or distribution) function $g(r)$:

$$g\left(|\mathbf{r}_1 - \mathbf{r}_2|\right) = \frac{P(\mathbf{r}_1, \mathbf{r}_2)}{P(\mathbf{r}_1)P(\mathbf{r}_2)} \quad , \tag{1.4}$$

and is often considered in many-body physics. A value of $g(r)$ different from 1 reflects a correlation between pairs of particles separated by a distance r. In classical physics, two particles are correlated only if they interact with each other. In other words, saying that they are free is equivalent to saying that they are uncorrelated. Correlation and interaction are thus intimately linked. We will see in Section 2.5.5 that this is not the case for quantal particles. For the time being, let us discard any quantum effect from the following discussion.

Conventionally, the pair correlation function is denoted by $g^{(2)}(\mathbf{r})$ and not simply $g(r)$. To get the latter function, one has to average over all possible angles and then evaluate $g^{(2)}(\mathbf{r})$ by counting the particles in a spherical shell of radius r and an infinitesimal width $\mathrm{d}r$, and centered at $\mathbf{r} = \mathbf{0}$. By doing so, one obtains the angle-averaged pair correlation function which coincides with $g(r)$.

Experimentally, $g(r)$ is measured by neutron scattering or X-ray diffraction. The incident beam is described by some wavevector and after collisions with the particles of the studied system, this wavevector is deflected. One detects this deflection angle by placing a detector at a given angle around the scattering system. Let us denote by $\mathbf{q}$ the vector difference between the incident and the scattered wavevectors. The detector measures a so-called differential cross-section. It is differential in the sense that it depends on $\mathbf{q}$. Assuming the scattering to be elastic, this cross-section is proportional to the *structure factor* $S(\mathbf{q})$. Details on the mathematical relation of S with g are given in Section 4.3.2. We simply give here its expression in the case of a homogeneous and isotropic system. In that case, S depends on $q = |\mathbf{q}|$ and g on r only. We then have the following relation:

$$S(q) = 1 + \frac{4\pi\varrho}{q} \int_0^{+\infty} r\, g(r)\, \sin(qr) \mathrm{d}r \quad . \tag{1.5}$$

Theoretically, $g(r)$ is evaluated by means of molecular dynamics. One takes a snapshot of the system, chooses one reference particle and builds a histogram of the particles at distance r of the reference particle. One repeats the same procedure for all particles of the system for the same snapshot and finally averages all histograms. The right part of the figure compares simulated $g(r)$ for a solid, a liquid and a gas. These states are schematically shown in the left panel, with the nearest neighbors (gray circles) of a reference particle (black circle) in the case of a solid, a liquid or a gas. Note that, due to the intrinsic

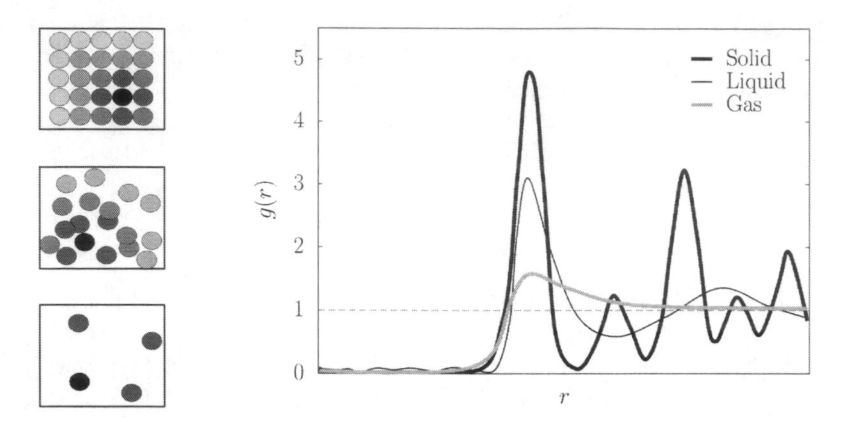

Figure 1.6: Left : schematic view of the distribution of particles in the solid, liquid and gaseous states, with first and second nearest neighbors indicated by dark and light gray balls, respectively. Right : pair correlation function $g(r)$ of solid (thick black curve), liquid (thin black curve) and gaseous (thick gray curve) argon, obtained from molecular dynamics.

volume of the particles, seen here as hard spheres, there are obviously values of r which are excluded. Anyway, if any r were accessible, which would mean that the number density ϱ is a continous function of r, the pair correlation function would read as $g(r) = \varrho(r)/N$ where N is the total number of particles.

Since $g(r)$ is normalized to the total number of particles, it asymptotically goes to 1. This reflects the fact that there is no correlation between very distant atoms, as in an ideal gas. We have $g(0) = 0$, due to the short-range repulsion. At short distances, clear peaks are visible, and only one in the gas phase. A peak indicates a particularly favored separation distance for the neighbors to a given particle. $g(r)$ therefore brings information on the atomic or molecular structure of the system. More precisely, as the name of this function indicates, $g(r)$ provides the distances for which two particles are correlated because they interact with each other. In other words, peaks in $g(r)$ indicate the range of these interactions, rather strong in a solid or a liquid, and almost vanishing in a gas.

There are noticeable differences between a solid and a liquid though. As already stated above, in a crystalline solid, classically, particles possess fixed positions according to the lattice structure. The first peak in $g(r)$ precisely corresponds to the relative position of the first nearest neighbors (dark gray balls in left panels of Figure 1.6), the second peak to the second nearest neighbors (light gray balls in the same figure), and so on, and the probability of finding a particle between successive nearest neighbors is strictly zero. Note however that in the case of a crystal, reasoning on $g(r)$ is too simplistic. Indeed, the shape of the correlation function in the case of a crystal does depend on the direction of $\mathbf{r}$. Therefore, one rather deals with $g^{(2)}(\mathbf{r})$ instead of $g(r)$. Taking the vector $\mathbf{r}$ along one of the edges of a unit cell of the crystal, the pattern of the pair correlation function will be as a series of narrow Gaussian-like peaks located at each lattice site.

In a liquid instead, there is no favored direction and we thus restrict the following discussion to the angle-averaged $g(r)$. One observes the same kind of peaks as in a solid but the pattern is smoother. The first peak in $g(r)$ is located at the same distance as in the solid. However, the height is smaller in the liquid. This reflects the constant turnover of neighbors mentioned in Section 1.3.2, while neighbors remain the same in time in a crystal. This turnover also produces a $g(r)$ which never vanishes between the peaks. Note finally that the distance between peaks is greater than in the solid, revealing a smaller interaction strength, even if it is of the same order of magnitude. On the contrary, the gas phase exhibits only one peak, showing the absence of any long-range order.

The pair correlation function provides useful structural information, particularly in disordered systems, as liquids but also glasses or amorphous solids. One can say that $g(r)$ gives access to the average structure of the disorder. It can also bring information about dynamical change of the structure. Note however that transport properties, that is how fast particles move, are not accessible in $g(r)$. We finally mention that there also exist higher order correlation functions, relating 3, 4, ... particles in a given system. Here we have only considered the correlation between pairs of particles which already provides valuable structural information. Moreover, we have only considered the case of homogeneous systems, that is with identical particles. The case of multicomponent systems, which actually constitute most of glasses and materials, is also possible and one can define a pair correlation function in that case.

1.3.4 The van der Waals equation of state

Van der Waals attempted to find a simple qualitative way to incorporate interaction effects into the equation of state of the ideal gas. As said above, the simplicity of the ideal gas model stems from neglecting interactions. This assumption is justified as long as the system remains sufficiently dilute, namely at small density ϱ. In that case, the intermolecular distances are so large that the interaction becomes negligible. In the case of real gases, one has to include an attractive part at long and medium distances, plus a repulsive part at short distances.

Deriving the equation of state

We aim here at explaining in simple terms how to incorporate dominant interaction effects to reach the equation proposed by van der Waals. We first consider the attractive part of the potential energy which originates from the mutual electric polarization of two molecules. The attraction implies a tendency to form a bound state. Actually, if the attraction is strong enough, it leads to a self-bound system (either in liquid or solid states). Van der Waals made the simplest calculation of the effect of this attraction on the pressure. Molecules at the surface of a fluid are pulled inwards by the attractive component of the interaction and the attraction will globally result in a decrease in the pressure exerted by the molecules on the walls of the container. This (differential) attraction is also the microscopic origin of the surface tension discussed in Section 1.4.1. The amount of the decrease is proportional to the number of pairs of molecules, within the interaction range, in a layer near the walls. This number is roughly proportional to the squared number density $\varrho^2 = N^2/V^2$. Since the number of particles N and the range of the interaction are constant, the pressure P of the system is decreased by a term proportional to $1/V^2$. Van der Waals then made the replacement $P \to P + a/V^2$ into the ideal gas equation of state to account for this decrease of pressure near the walls, with a some new parameter (related to the interaction).

The short-range repulsion is, in turn, related to the finite size of molecules which are not considered as point-like any longer. The extreme case is to assume impenetrable molecules: around the center of each molecule, there is a certain volume where no other molecule can penetrate. In practice, this means that in the equation of state, we should replace the volume V with some effective value $V - b$, where b is a constant characteristic of the system under consideration. Assuming the molecules to be rigid spheres of diameter σ, the minimum possible distance between the centers of two molecules will be precisely σ. Within such an oversimplified picture, one simply has $b = 2N\pi\sigma^3/3$, which is four times the total volume of all molecules in the gas. The equation of state proposed by van der Waals therefore finally takes the following form:

$$\left(P + \frac{a}{V^2}\right)(V - b) = nRT \ . \tag{1.6}$$

Alternatively, it can be written in terms of the number density as

$$P = \frac{\varrho k_{\mathrm{B}} T}{1 - \tilde{b}\varrho} - \tilde{a}\varrho^2 \quad , \tag{1.7}$$

with $\tilde{a} = a/N^2$ and $\tilde{b} = b/N$.

The equation of state and the Liquid-Gas transition

The van der Waals equation of state leads to a more complex relationship than the ideal gas law. But it remains simple enough to be used in many common situations. The complication brought by the equation of state turns out to be extremely rich and fundamental for physics. Note first that the domain of applicability of this equation is, of course, much larger than the one of the ideal gas law, especially in terms of density. We shall thus choose, for the following discussion, the density as a natural variable. By contrast, regarding the temperature, the limitations of the ideal gas persist: at low temperature, quantum physics dominates and van der Waals equation cannot take into account these quantum effects. But temperature remains a good indicator when it is high enough, and we therefore choose it as a parameter.

It is now interesting to discuss the form of the van der Waals equation of state and in particular how it allows us to analyze the liquid-gas phase transition. Indeed, with the choice of density as a variable and temperature as a parameter, the van der Waals equation corresponds to a set of curves relating pressure and density at a given temperature: this is the network of isotherms represented in Figure 1.7. In the case of ideal gas, the same set of

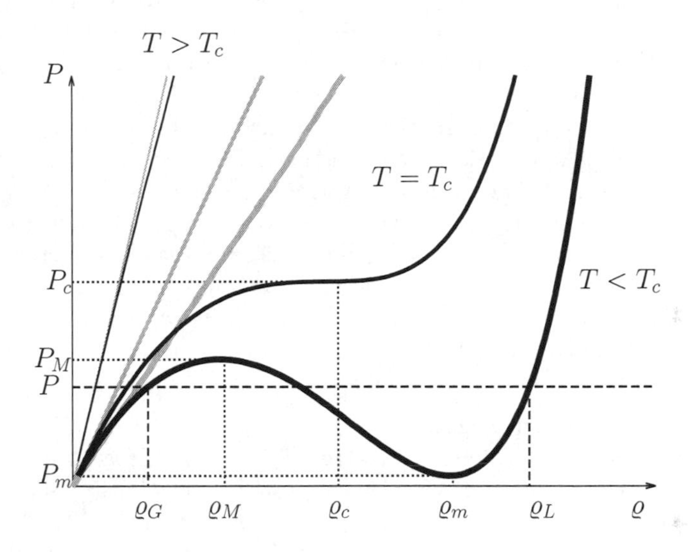

Figure 1.7: Pressure P as a function of number density ϱ at different temperatures (above, equal and below a critical temperature T_c) for the ideal gas (gray lines) and for the van der Waals equation of state (black curves).

isotherms would simply reduce to straight lines with slopes proportional to the temperature (gray lines). Notice also that the agreement between the ideal gas and real gases described by the van der Waals equation of state is naturally recovered within the limits of low densities and/or high temperatures. The van der Waals isotherms however are richer than the ideal gas ones. As we can see from the figure, there exists in particular a critical value T_c of temperature, which separates different behaviors of these isotherms. Above T_c, the isotherms are monotonic curves, approaching the ideal gas isotherms at high enough temperatures. But for $T < T_c$ the isotherms have a typical 'N' shaped form, characterized by a maximum at (P_M, ϱ_M) and a minimum at (P_m, ϱ_m), as indicated in the figure. When T approaches T_c from below, $\varrho_M \to \varrho_m$ and at critical temperature T_c, we have $\varrho_M = \varrho_m$.

Let us now explore Figure 1.7 to understand the gas-liquid transition. Indeed, when the pressure of the system P lies between P_M and P_m, there are two densities that match the same pressure value, as indicated in Figure 1.7 by the intersects of an isotherm and a horizontal line representing $P = \text{const.}$

The corresponding densities are labeled ϱ_G and ϱ_L in the figure. There is actually a third solution with an intermediate value of density but it does not represent a physical state, as it is associated to a situation with negative compressibility (pressure decreasing with increasing density, negative slope of the isotherm). How to interpret these two states points of densities ϱ_G and ϱ_L? The fact that the same value of pressure is associated to two different densities reflects the coexistence of two states of matter at two densities. The equality of pressure reflects the mechanical equilibrium of the two states with respect to each other. This coexistence corresponds for instance to the mixture of vapor bubbles (gas) and liquid water boiling in a saucepan. More generally speaking, the van der Waals equation of state, which first deals with real gases, gives access to the liquid-gas coexistence in real gases. As a consequence, it provides a satisfactory description of a liquid. And this phase coexistence is related to having considered interactions between components (the straight isotherm of the ideal gas cannot lead to the phenomenon of coexistence of such phases). In addition, the disappearance of the N-shape for temperatures above T_c allows us to interpret this threshold value as the temperature of the critical point introduced in the phase diagrams (see Section 1.1.2) beyond which there is no distinction between liquid and gas phases.

The law of corresponding states

One important aspect in physics is to identify universal laws that can encompass properties from different systems. We will here identify such universal laws with the help of the van der Waals equation of state.

Let us first come back on the coexistence of gas and liquid at a given $P < P_c$ and for $T < T_c$, with the two densities ϱ_L and ϱ_G, as discussed above. We can equivalently make the same reasoning with the volume instead of the number density. By measuring V_L and V_G, we can plot in a 3D diagram the triplets (P, T, V_L) and (P, T, V_G). Actually, by varying P and T, the various obtained triplets define the points of the liquid-saturation dome we have mentioned in the 3D phase diagram, see Section 1.1.2 and Figure 1.2. And the maximum of this dome is precisely the critical point at (P_c, T_c, V_c). Figure 1.8 displays the temperature T of a mixture of liquid and gas at a given pressure, as a function of V_L and V_G, that is the volume of the liquid and the gas respectively at equilibrium, for different species. Different experimental data (symbols) are compared. The remarkable point of this figure is that, if one uses reduced variables defined as $T_r = T/T_c$ and $V_r = V/V_c$ instead of using the absolute values of temperatures and volumes, all data superimpose on the curve. This curve is coined 'universal' because it describes the same behavior in very different systems.

One can deduce this theoretical curve from the van der Waals equation of state and by exploiting the critical point. Indeed, this point is associated to the isotherm $P = P(V)|_{T=T_c}$, going through the critical point. This isotherm is the curve in Figure 1.7 that does not exhibit a minimum/maximum shape, but possesses an inflection point instead. In mathematical terms, an inflection point corresponds to a point at which curvature changes sign and is character-

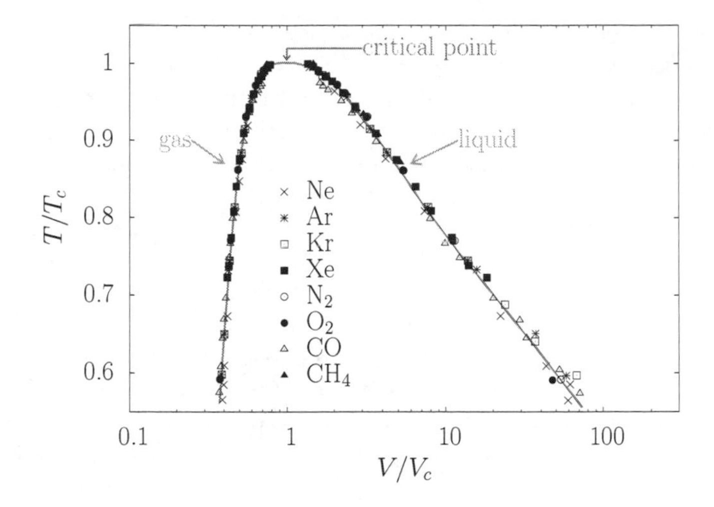

Figure 1.8: Reduced temperature T/T_c as a function of reduced volumes V/V_c in the case of the equilibrium coexistence of liquid and gas at a given pressure below P_c, for various species as indicated. The theoretical curve is compared with experimental measurements (symbols), reproduced from E. A. Guggenheim, J. Chem. Phys. 13 (1945) 253, with the permission of AIP Publishing.

ized by a vanishing second derivative along the isotherm. As the critical point furthermore corresponds to merging minimum and maximum (both characterized by a vanishing first derivative at fixed temperature), the critical point is finally characterized by vanishing first two derivatives at fixed temperature. The resulting two conditions $(\partial P/\partial V)_T = 0$ and $(\partial^2 P/\partial V^2)_T = 0$ together with the van der Waals equation itself then provide a system of three equations to determine the three unknowns P_c, V_c, T_c in terms of parameters a and b. Simple algebra gives $P_c = a/(27b^2)$, $V_c = 3Nb$ and $k_B T_c = 8a/(27b)$.

To go one step further, it is interesting to rewrite the van der Waals equation by scaling each variable to its own critical value, that is in terms of reduced variables $P_r = P/P_c$, $V_r = V_m/V_c$, $T_r = T/T_c$, where V_m is the volume of one mole. The equation of state thus reads:

$$\left(P_r + \frac{3}{V_r^2}\right)(3V_r - 1) = 8T_r \quad . \tag{1.8}$$

In such a way, the parameters a and b, which are specific to a substance, do not appear any longer. This reduced form is known as the law of corresponding states. This principle states that fluids at the same reduced pressures and temperatures have the same reduced volume. Figure 1.8 is an application of this law to coexistent phases and nicely illustrates its universality. This also indicates that the van der Waals equation, despite its apparent simplicity, can describe qualitatively and quantitatively many properties of real fluids.

When dealing with real gases one defines the so-called compression factor as the ratio $Z = PV/RT$. For one mole, one can deduce from the van der Waals equation that at the critical point the value of the compression factor is $Z_c = 3/8 = 0.375$. This value is independent on the parameters a and b which are specific of the gas under study, and obviously provides a universal quantity in that sense. It is then interesting to compute experimental compression factors for various gases and see how they behave with respect to van der Waals's estimate. For example, ^{4}He, Ne or H_2 have values of $Z_c \simeq 0.30$. Common gases such as N_2 or CO_2 possess a Z_c a bit below 0.3 (respectively 0.29 and 0.27), still not so far away from the van der Waals value. There are less favorable cases, though, such as water with $Z_c \simeq 0.23$. However, in spite of obvious discrepancies with the theoretical values, there is a clear qualitative, and even quantitative agreement between experimental Z_c and the van der Waals value. The van der Waals equation is therefore a clear improvement on the ideal gas law, for which there exists no critical point by construction.

1.4 SURFACE TENSION AND VISCOSITY, SOME KEY PROPERTIES OF LIQUIDS

After having discussed in detail the link between the gas and the liquid phases through the van der Waals equation, we now present three important physical properties specific to the liquid state, namely surface tension, viscosity and vorticity.

1.4.1 Surface tension

The notion of surface tension

A liquid occupies a fixed volume but, due to gravity, this volume takes the shape of the container. This may look contradictory to the formation of dewdrops on a leaf. One may also wonder why a liquid released in space would take a spherical shape. We all know that water slowly leaking from a tap falls drop by drop. All these observations are related to the surface tension. To discuss this physical quantity, one implicitly considers a liquid phase in contact with a gas phase and defines the surface of the liquid as the liquid-gaseous interface. In the following, we will only consider the case of a liquid in contact with air but the gas can also be vacuum with a tiny number of molecules inside.

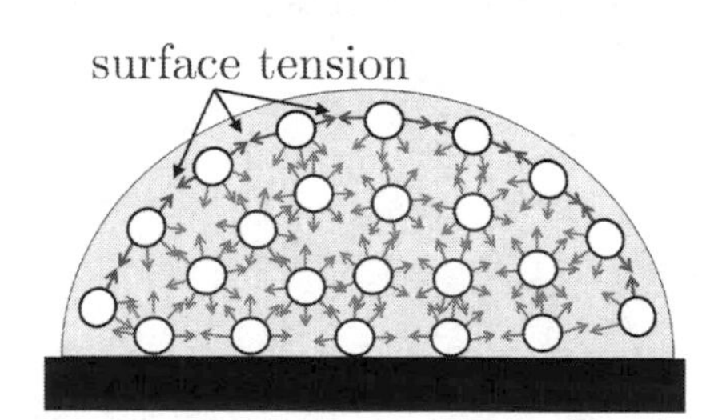

Figure 1.9: Schematic view in a liquid drop deposited on a surface (black) of the cohesive forces (arrows) between the constituents of a liquid (spheres).

We have explained above the importance of interactions between constituents in the liquid state. In the bulk of the liquid, a given constituent undergoes interactions in every direction, as shown in Figure 1.9, and the net force is zero (except for an externally driven motion, the liquid thus remains at rest). On the contrary, at the surface, a constituent does not have the same number of neighbors in all spatial directions. Some other particles *inside* the liquid are exerting attraction on it but there is no attraction from outside the liquid: the net force is therefore non-zero and points inwards. This constitutes the microscopic origin of the surface tension and is accounted for in the van der Waals equation of state by a reduction of the pressure by the term a/V^2. As a consequence, a liquid tends to minimize its surface area, as the surface practically corresponds to a lack of binding as compared to bulk. In the absence of other forces (such as gravity), a liquid thus acquires a spherical shape. This explains the shape of liquid droplets: when a liquid is stretched, as for instance when one opens a tap, the surface tension of water first tends to keep the water drop bound to the aperture. But when the drop mass becomes too large, gravity wins, water streams and eventually breaks up into nearly spherical drops.

The dimension of a surface tension γ is a force per unit length (expressed in N/m) or, equivalently, an energy per unit area. One usually uses a derived unit, that is dyn/cm (1 dyn= 10^{-5} N). The surface tension of water in contact with air is 73 dyn/cm at 20°C, which is a relatively high value, as compared with that of ethyl alcohol (23), gasoline (22), or glycerin (63). Mercury in contact with air possesses on the contrary a very large value of 486 dyn/cm at 20°C. Surface tension explains why some insects, such as the water strider, are able to walk on water: the surface tension is large enough to resist their small weight so that they can literally walk on the water surface without sinking. The existence of dewdrops on a leaf or on a rainproof tent results on the one hand from the high surface tension of water, and on the other hand from a low wettability of the surface the water droplet is deposited on. As already explained, surface tension comes from cohesive forces in the liquid. Wetting results from adhesive forces between the atoms of the liquid and those in the solid: the stronger these forces, the more wettable the surface by this liquid and the more the liquid spreads on the surface. Conversely, a fabric of low wettability will 'expel' water which will remain under the form of droplets rather than spread on the surface and possibly penetrate it. Another interesting example of wettability is provided by the leaves of plants or trees. Leaves are covered by a film, called the plant cuticle, which is composed of lipidic and waxy polymers. One function of the plant cuticle is to provide a low wettability and to form a protection to external stresses. This is why water beads easily form on leaves. Rainproof treatment of materials aims at decreasing their wettability. Note however that one should not touch a tent covered in rain beads because the surface tension can be easily broken up and water can then drip through the mesh of the fabric.

The surface tension γ strongly depends on temperature, as an increase of the latter enhances microscopic motion which can more easily counterweight the attractive forces between constituents. For most liquids, γ decreases linearly with T. One also expects to have a vanishing surface tension at the critical point, as there is no apparent distinction between a liquid and a gas at this point. More quantitatively, there exists a phenomenological law, the Eötvös rule, relating γ and T for $T > T_c$, which reads:

$$\gamma = \frac{k_E}{V_m^{3/2}}(T_c - T) \quad , \tag{1.9}$$

where V_m is the molar volume, T_c the critical temperature, and k_E is the Eötvös constant. The latter one is usually taken as 2.1×10^{-7} J K^{-1} mol$^{-2/3}$ for most substances. Note that γ vanishes at the critical temperature, inline with the fact that above this temperature, there is no distinction between liquid and gas and it is senseless to talk about surface tension any longer. For water in contact with air, $\gamma = 68$ and 59 dyn/cm at 50 and 100 °C respectively. This explains why it is easier to wash clothes in hot rather than in cold water: as the surface tension of water decreases, water becomes a better wetting agent. One also usually uses laundry detergents which contain surfactants, namely agents which decrease the surface tension of water even more and which allow water to better penetrate clothes fibers.

Capillarity

Surface tension in a liquid also occurs in a different context which nevertheless relies on the same physics: capillary action or capillarity is the ability of a liquid to flow in narrow spaces (fibers, narrow tubes, narrow openings) without the need of external forces, and often opposite to external forces such as gravity. A daily example is the ability of a dry paper towel to absorb a liquid, which is drawn into the narrow holes between the fibers of the towel. Once again, wettability, that is competition between cohesive and adhesive forces in a liquid, is at the origin of capillarity. In the example of a dewdrop on a leaf, surface tension (=cohesive forces) dominates the adhesive forces of water molecules with the waxy surface of the leaf. In a narrow tube instead, adhesive forces between water molecules and glass can take the lead.

Take a cylindrical glass tube and place it vertical in liquid water. At the interface between water and air, inside the tube, instead of having a horizontal interface at the same height as in the water container, one observes a concave meniscus (cap like shape with central part above edges) positioned at a higher height h and forming an angle θ with the wall of the tube, called the contact angle. This effect results from a competition between cohesive and adhesive forces on the one hand (leading to a net upwards force) and the gravity pulling the liquid down. Both upwards and downwards forces balance each other at equilibrium. The actual height h above an open surface of a liquid (which can take negative values) and the shape of meniscus (concave or convex) evaluated through θ depend on the nature of liquid, precisely on this balance between cohesive and adhesive forces. For a given liquid, the section of the capillary also plays a key role by controlling in particular the height h: the narrower the tube, the larger the effect.

There are various applications of capillarity in everyday life. For instance, when one uses a fountain or a ball pen to write, ink is extracted from a reservoir by capillary action. The fact that one is able to write on a sheet of paper with such a pen is also due to capillarity since ink flowing from the pen is drawn into the fibers of the paper. As already mentioned, it is exactly the same phenomenon which occurs when one uses a paper towel to draw a liquid into its fibers. Similar mechanisms are observed in tree trunks which, to a large extent, are composed of the same material as paper, namely cellulose. There again exist (very thin) capillaries inside the trunk, which transport water from the roots to the leaves by capillary action. If water were in contact with air, capillarity would lead to a height of 1.2 m, which is obviously not sufficient to explain typical tree heights, or even those of redwoods in California with a 100 m height. Capillarity is not sufficient here: among other effects, evaporation of water from leaves creates a decrease of pressure, which allows water to rise even more against gravity. At some point, gravity nevertheless wins and water cannot reach any arbitrary height.

Even more surface tensions
We end this section by mentioning that, as there exists a surface tension at the interface of a liquid with a gas, similar physics occurs at the interface between two immiscible liquids. For instance, when one shakes a mixture of oil and water, small spherical droplets of oil form, suspended in water. If one waits a bit, one observes that small oil droplets merge to form larger and larger drops. After a while, the two phases finally separate completely and one observes water phase at the bottom of the container and oil phase floating on water, since water density is higher than that of oil. The latter is known to be hydrophobic (namely trying to 'avoid' water, while on the contrary a hydrophilic liquid will tend to mix with water). Water molecules at the interface with oil molecules therefore cannot 'wet' the oil surface and possess, as in contact with air, fewer neighbors than in the bulk of water liquid. As a result, these molecules form a highly ordered water shell around the hydrophobic substance and tend to minimize the number of water molecules involved in this shell. This explains the formation of spherical oil droplets which tend to merge, or of a horizontal interface between water and oil when the two phases are completely separated. In the case of two immiscible phases, one talks about interfacial tension rather than of surface tension, but the physics behind is very similar.

1.4.2 Viscosity of liquids

At variance with the equilibrium concept of surface tension, viscosity addresses dynamical aspects when one considers liquid flow. We are more precisely interested here in what happens at the interface of the liquid and a solid surface in relative motion. Everyone has already experienced that water flows into a vertical straw more easily than honey, so that one usually describes honey as 'thicker' than water. This behavior is related to viscosity, which is a measure of the resistance of a liquid to flow. Microscopically, it comes from collisions between particles moving at different velocities.

More quantitatively, let us consider a liquid at rest between two horizontal plates with a surface $\mathcal{A}$ large enough so that effects at the borders of the plates can be ignored. Now let us move only the upper plate horizontally at constant speed v along the x-axis which defines the unit vector $\mathbf{e}_x$. By doing so, one applies a shearing force to the fluid. In response to this shear stress, the liquid resists the imposed motion and the force needed to counterweight this resistance is a measure of its viscosity. Ideally, if v is not too large, the liquid shears, which means that adjacent horizontal layers move parallel to each other. The friction between them produces forces against their relative motion. Henceforth, each layer of fluid moves slower than the one just above it, and the speed of the fluid linearly decreases with the distance to the moving (upper) plate. Correlatively, the fluid applies to the upper plate a force in the direction opposite to its motion, that is towards $-\mathbf{e}_x$ (an opposite force is also applied to the lower plate). Therefore, to keep the upper plate moving at constant velocity $v\,\mathbf{e}_x$, one has to apply to it an external force $F\mathbf{e}_x = \eta\mathcal{A}\partial v/\partial z\,\mathbf{e}_x$, where z is the vertical distance to the lower plate and the constant of proportionality η the dynamic (shear) viscosity. In other words, η is the ratio between the shear stress $F/\mathcal{A}$ over the velocity gradient of the fluid $\partial v/\partial z$, also called the shear rate:

$$\eta = \frac{F/\mathcal{A}}{\partial v/\partial z} \quad . \tag{1.10}$$

Its unit is the Pa·s but the centiPoise (cP), where the Poise (P) is defined as 1 P = 0.1 Pa.s is more commonly used. Viscosity of liquids can vary between 1 and more than 10^4 cP. For instance, at room temperature, viscosity can take the following values (in cP): 1.002 for liquid water, 3 for blood or milk, 84 for olive oil, 1420 for glycerin, 10^4 for honey, 10^5 for sour cream... The world record is probably held by bitumen which apparently appears solid. In the famous pitch drop experiment started in 1927, tar pitch takes about ten years to form a single drop. This experiment is still running nowadays... A liquid is commonly considered as viscous when its viscosity is greater than that of liquid water. As for gases, they usually exhibit a viscosity 1000 times smaller than that of water. Typical values are thus $\eta \simeq 15 - 20 \times 10^{-3}$ cP for air, dioxygen, helium, etc., at room temperature.

The viscosity of a fluid also strongly depends on temperature: a syrup can flow more easily when heated, and motor oil thickens during winter (and this can significantly affect the performance of cars). In general, the viscosity of a 'simple' liquid (opposite to 'complex' fluids as toothpaste, mixture of water and cornstarch, etc.) increases when temperature decreases. Indeed, decreasing the temperature of a liquid tends to reduce the kinetic energy of the particles and to increase their 'closeness'. Therefore, this increases the cohesive forces in the liquid state and causes more friction. For instance, η_{water} is 1.79, 1, 0.65, and 0.28 at 0, 20, 40, and 100°C respectively. On the contrary, the viscosity of a gas increases with temperature. Indeed, one should keep in mind that viscosity quantifies the resistance of a fluid to flow *as a whole*. Particles in a gas are most of the time flying freely in vacuum. But increasing the temperature T allows the particles to undergo more collisions which tends to blur collective motion of the gas and which then induces a higher viscosity. Indeed, according to kinetic theory of gases (see Section 3.1), $\eta \propto \sqrt{T}$ and this leads for instance to values of $\eta_{\rm air}$ (in cP) at 0.0168, 0.0189, 0.0215, 0.025 at 20, 100, 200, 400°C respectively.

A liquid always possesses a positive viscosity η which increases as temperature decreases. Liquid ^{4}He also follows this rule but only down to a temperature of about 2.8 K. Below this temperature, η suddenly drops. A liquid which exhibits no resistance to shear stress is coined *superfluid*. At very low temperatures, ^{4}He indeed becomes superfluid, as will be discussed in Chapter 9.

1.4.3 Vorticity in liquids

Note however that η alone is not able to fully characterize a fluid flow in the sense that dynamical effects are not all accounted for in η. The dimensionless parameter which is usually used to quantify flow properties is the Reynolds number defined as:

$$\mathrm{Re} = \frac{m\varrho v L}{\eta} \quad , \tag{1.11}$$

where v and L are respectively a characteristic velocity and a characteristic length scale, and m is the mass of the particles constituting the fluid. This empirical parameter is the ratio of the inertial force over the shearing force. In other words, it quantifies how fast a fluid can move relative to how viscous it is. The Reynolds number is usually used as a practical guide to the transition between a laminar flow and a turbulence flow. The first case, for Re lower than 2100 to 2300, corresponds to that described above in the definition of the shear viscosity (mind that, there, the velocity inside the fluid should be low). Low Re are also achieved with high η. On the contrary, large Re (> 4000) are obtained in fluids which are not very viscous. Actually, the equations commonly used in hydrodynamics (the famous Navier-Stokes equations) are the first order of a systematic expansion in 1/Re, the zeroth order corresponding to the 'ideal' fluid. Fluids with a large Reynolds number are thus coined 'good fluids'. On the other hand, they are often the seat of turbulence flow.

Turbulence flow, as opposed to laminar flow in which particles in a fluid move in parallel layers, often manifests itself by the formation of vortices. A vortex corresponds to a spatial region in which the fluid flow rotates about an axis line. The presence of vortices however does not imply that we are in a turbulent regime. Note also that one can have locally some rotational motion even if the velocity field lines are straight... To better quantify local rotational motion, one usually introduces the concept of vorticity which describes a local rotational motion of a point in a fluid, if an observer were to move along with this point. We will avoid the mathematical formulation of vorticity and only give in the following an intuitive approach of it.

Let us consider a float as depicted by the cross in Figure 1.10, and label one of its ends by a black dot. If the fluid does not shear (top row), the float

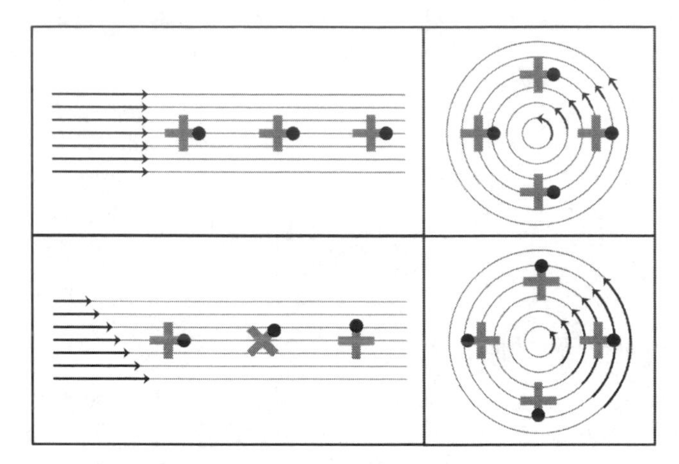

Figure 1.10: Laminar flow with shear vorticity (bottom row) or without (top row), exemplified by the rotation or non-rotation of a cross-shaped float along the flow. The velocity field is depicted in each case by thick arrows.

keeps the same orientation along the flow and there is no vorticity. In the case of shearing instead (bottom row), even if the velocity field lines are (locally) parallel to each other, the float does rotate and we are in presence of vorticity.

The two right panels of Fig 1.10 show two illustrative examples of vortex. The velocity field is highlighted by the thick arrows along the thin velocity lines. There is no local rotational motion if each point of the float undergoes the same velocity v, as in the top right panel. To avoid any rotation of the float, the parts of the float which are closer to the rotation axis must move faster than those which are farther away. One can show that the velocity v should be inversely proportional to the distance r from the center of the vortex. This case corresponds to a vortex without vorticity (also called irrotational vortex). The other extreme case is when $v \propto r$, as depicted in the bottom right panel. This kind of vortex is also called a 'rigid-body' rotation because elements of fluid are forced to move as if they all belong to the same solid body. The fluid shears and there is a non-vanishing vorticity: we then have a rotational vortex. Therefore, one should not mistake vortex for vorticity. This is particularly clear in the bottom left panel in which the velocity field lines are all horizontal. Still, the float does rotate because of the non-vanishing vorticity here.

Vortices can be created in a liquid by rotating the fluid container. Indeed, the tank can, if rotated at the right speed, transfer part of its rotational energy to the liquid. In the examples above, we have seen that shearing is needed to establish vorticity. This is particularly clear in the case of the rotational vortex in the bottom right panel of Figure 1.10, since it can be seen as a solid body rotation. Such a rotation would correspond to the case of an extremely viscous fluid so that particles cannot change their position relative to each other. In other words, while vorticity is a property of the flow and viscosity a property of the fluid, to establish vorticity in a liquid, one needs a non-vanishing viscosity. Later on, we will come back to this relationship in connection with liquid helium and Bose-Einstein condensates in Chapter 6.

1.5 'EXTREME' STATES OF MATTER

1.5.1 More travels in the phase diagram

In Section 1.1, we presented the so-called 'standard' states of matter, namely solid, liquid and gas. We also considered phase transitions while maintaining one of the three state variables constant, either P, T or V. And we have discussed, at least qualitatively, various possible paths in the phase diagram, allowing to step from one phase to the other. This discussion, however, has not addressed several specific aspects that are worth being at least mentioned. The aim of this section is to consider some of these overlooked questions. We shall take as a guideline limits or extreme variations of state variables, namely pressure, temperature and volume. This will allow us to identify new, interesting aspects on the properties of phases of matter.

The limit $P \to 0$ is conceptually conceivable: it corresponds to a perfect vacuum. In high vacuum experiments, P ranges from 10^{-7} to 0.1 Pa (we remind readers that 1 atm $= 1.013 \times 10^5$ Pa). If $P < 10^{-7}$ Pa or $P < 10^{-10}$ Pa, one rather talks about ultrahigh or extreme ultrahigh vacuum respectively. The best vacuum produced on Earth is about 10^{-10} Pa. In interstellar space, the pressure can be several orders of magnitude smaller, corresponding to a density less than 1 atom of hydrogen per m^3.

We have also seen that above the critical temperature T_c, no distinction between the liquid and the gas states exists. But there remains a question on how high temperatures can be effectively reached. In cosmology, which models the origin and the evolution of the Universe, the highest possible temperature should be given by the Planck temperature defined as $T_P = \sqrt{\hbar c^5 / G k_B^2} = 1.4 \times 10^{32}$ K, where $\hbar$ is the reduced Planck's constant, c the light velocity, G the gravitational constant, and k_B the Boltzmann's constant. Above this limit, theories unifying quantum mechanics and gravitation are mandatory (see also Appendix C). Still, at much lower temperatures, interesting physics can happen. Let us consider a gas and increase its temperature T. We have up to now considered the impact of temperature *globally*, namely within assuming that the constituents of the gas (atoms, molecules) are unaffected by the temperature. This is clearly a disputable assumption above sufficiently high temperatures. Indeed, above a certain value of T, the thermal energy of electrons becomes higher than their binding energy to their parent atom/molecule (typically, a few eV, equivalent to some 10^4 K). Then electrons can have enough energy to leave the nuclear potential, and the atom/molecule is *ionized*. We therefore obtain a mixture of a gas of electrons and a gas of positive ions, the ensemble remaining electrically neutral: this is called a *plasma*. Plasmas are such important objects that we devote a section to them.

1.5.2 Plasmas

Plasmas are often considered as the fourth state of matter. They by definition consist of a neutral mixture of electrons and ions. Note that particles in these gases are not free since they are charged: as they move, because of Maxwell equations, electric currents and magnetic fields are created and induce strong interactions between all charge carriers. Even if one usually describes a plasma as a fluid, that is as an interacting gas with no definite shape or volume, a plasma has peculiar characteristics which makes it definitely different in particular from the usual gas: due to the presence of a large amount of charge carriers (either electrons or ions), a plasma is very responsive to electromagnetic fields.

Plasmas are extremely abundant in the Universe: it is commonly accepted that more than 99 % of the observable matter is in the plasma state, either in stars, interplanetary space, interstellar space or intergalactic space. On Earth, one encounters plasmas every night in neon signs, or more occasionally on television screens. We should also mention the ITER project (International Thermonuclear Experimental Reactor) located in Cadarache (France), which aims at controlling energy production from the fusion of a deuterium (^{2}H or D) nucleus and a tritium (^{3}H or T) nucleus. To produce these nuclei, one starts from a gas of atomic D and T at very high temperature (1.5×10^7 K), so that atoms completely ionize and form a very hot plasma. Lasers, thanks in particular to their remarkable properties of collimation, can also deposit a high amount of energy in a small and well defined volume of matter, energy high enough to ionize the atoms in the target volume.

One key characteristic of a plasma is its degree of ionization which is a measure of the percentage of ionized species. It can be written as $\alpha = \varrho_I/(\varrho_I + \varrho_N)$, where ϱ_I is the density of ions and ϱ_N the one of neutral atoms. The electron density reads $\varrho = \langle Z \rangle \varrho_I$ where $\langle Z \rangle$ is the average charge state of the ions. The degree of ionization α depends on the type of atoms, the temperature and the density. For artificial plasmas created by the application of an external electric field, α also depends on the strength of the electric field. There exist many different types of plasmas and we do not want to discuss all of them. We present instead in Figure 1.11 some regions of electron density and temperature/energy, in analogy with standard phase diagrams, either on Earth (metals, laser, fusion) or in the Universe containing plasmas. Note that we have on purpose distinguished the case of the solar core from other astrophysical objects, such as the interplanetary space, the chromosphere, the solar corona, etc. Indeed, matter, even in a plasma state, exhibit in such

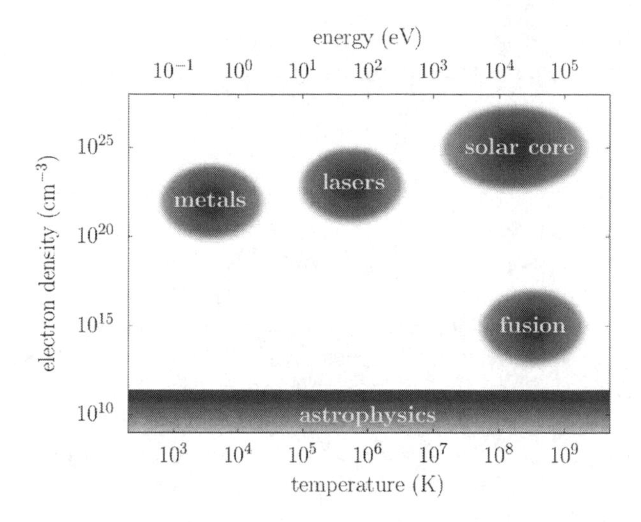

Figure 1.11: Regions in a density-temperature (lower horizontal axis) or density-energy (upper horizontal axis) plane in which matter is in the plasma state (see text for details).

situations very low densities (less than 10^{10} cm^{-3}) and therefore, the role of interactions is not essential and brings us too far away from the liquid state. On the contrary, in the center of the Sun, or of a star in general, the gigantic mass involved there produces tremendous gravitational forces, allowing to reach very high densities. At such high temperatures, the binding energy of electrons in hydrogen atoms can be overcome and atoms are ionized.

Among the cases displayed in Figure 1.11, there is an interesting one belonging to everyday life that we want to discuss a bit more. Indeed, without going to extreme temperatures, sort of ionized matter exists at room temperature in metals. Indeed, we will see that most of the properties of metals, as for example electric conductivity, are due to a partial delocalization of electrons over the crystalline structure of the metal. One sometimes talks about the electron Fermi sea for the description of such a delocalization. This sharing of electrons between different atoms is actually at the root of the chemical bonding, especially in metals. The description of electrons either in a chemical bond or in a metal however strongly relies on quantum mechanics and on the fact that they are *fermions*, a notion which will be discussed at length in Chapter 2. Typical number densities in metals on Earth are about $\varrho \sim 10^{22} - 10^{23}$ electrons per cm^3 (that is a mass per unit volume of about 1 g/cm^3). Chapter 5 will more specifically address the characteristics of electrons in standard matter, which constitute, to some extent, a *quantum electron fluid*. In the case of metals, the origin of the plasma state is not temperature but the fact that electrons are only slightly bound to their parent atom. The presence nearby of other atoms and the associated interactions, is then sufficient to 'ionize' atoms which altogether share the same ensemble of electrons. In turn, these electrons interact among themselves and with the ionic background.

1.5.3 Variations on the solid state

Plasmas are typically associated to the liquid state, mostly because of the usually high temperatures involved. We have just seen that, to some extent, metals are also displaying a plasma behavior. This has brought us back to the realm of moderate to low temperatures where the solid state is a major player. Let us then discuss a bit more the solid state. In most substances, this state is attained by either increasing P or decreasing V since this tends to localize the constituents in a crystalline order. But one should not consider the solid state as being unique. The case of carbon is a typical example for which one uses in everyday life graphite in a pencil, or wears a diamond jewel. Graphite and diamond are both solid states of the same chemical element, carbon. We present in Figure 1.12 the Clapeyron's diagram of carbon. This phase diagram

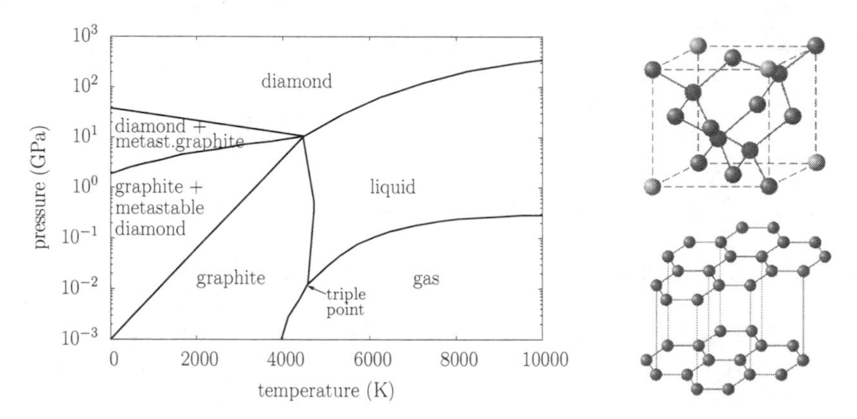

Figure 1.12: Left: Clapeyron's diagram of pure carbon with different solid states, diamond and graphite. Right: atomic structure of diamond (top) and graphite (bottom).

is much more complex than those encountered for example in Figure 1.3. In particular, one notices that there exist two stable solid states, graphite at low P and diamond at high P. They constitute two allotropes (property of some chemical elements to exist under two or more different forms) of carbon, which exhibit different crystalline structures, as displayed in the right part of the figure. They therefore possess very different physical and chemical properties, for example graphite is black, possesses a low density and can be easily eroded, while diamond is transparent, is very dense and has the highest hardness. There are however ranges of P and T where both can coexist, one state being in a metastable equilibrium (see Section 1.1.3). For instance, in standard ambient conditions, graphite is the most thermodynamically stable state, and diamond is metastable. This means that diamond, although one commonly admits that 'diamonds last forever', should actually transform into graphite. However, the kinetics of such a transformation is extremely low (almost 'forever' on a human time scale), because of a very high activation energy barrier (it requires a very high energy to pass from one state to the other). Several substances possess different allotropic (solid) forms, such as oxygen, silicon or iron, to cite a few.

Even more complex situations can occur when one does not observe a single liquid-solid transition when the density of a pure substance sample increases, but rather a cascade of transitions implying the formation of states of increasing order. Let us for example consider rod-like molecules condensed in a (solid) crystal. The molecules have very little freedom to move and the sample is invariant under translations in space. Intuitively, the order of such a crystal is high: the molecules are positioned in a regular lattice and are all aligned along the same direction, called the director. In the liquid state instead, the order is low since the molecules can have any orientation in space (this also holds in the gas state). Because of the elongated shape of the molecules, one can imagine intermediate cases, where the molecules are only slightly twisted with respect to the director. To quantify these situations and to relate them to the notion of order, an *order parameter* S is introduced and can be defined as $S = \langle \frac{1}{2}\left(3\cos^2\theta - 1\right)\rangle$, where θ is the angle of the molecule with respect to the director, and the average is performed over all molecules in the sample. In a perfect crystal, we have $S = 1$, while in a liquid, all angles are possible, and one has $S = 0$ (the sample is isotropic). For $0 < S < 1$, the sample is slightly anisotropic and possesses at the same time some properties of the solid and other ones of the liquid: it henceforth constitutes a mesophase, more known as a liquid crystal. There exist many different types of liquid crystals and a specific classification based on symmetry considerations. But it is not the aim of this book to present here an exhaustive list. Briefly, the existence of mesophases relies on molecules which must have a rigid elongated geometry (with usually several aromatic cycles), and which are easily polarizable or exhibit a permanent electric dipole. This allows the molecules to be oriented when an external electric or magnetic field is applied. The most common applications of liquid crystal are LCD (Liquid Crystal Displays) that have invaded everyday life in any electronic device.

1.5.4 Towards the absolute zero of temperature

Another way to access the solid state is to decrease the temperature T. Indeed, since temperature is the macroscopic expression of the microscopic 'thermal' motion of particles, by annihilating this motion, one can imagine that they will freeze in a crystal. The ideal gas law, presented in Section 1.3.1, states that, at constant pressure, the volume of a gas decreases with temperature, and should vanish around −240°C. But, as T becomes lower and lower, one expects to observe condensation and then solidification of the initially gaseous sample. Actually, as long as T decreases, the constitutive particles keep losing energy and vibrational motion. Conceptually, one can imagine an ultimate lowest temperature at which the particles are completely at rest. This defines the absolute zero of (thermodynamical) temperature which is, by international agreement, taken as -273.15 °C $= -459.67$ °F $= 0$ K, and which is thermodynamically not accessible. For the time being, the coldest temperature produced on Earth in a gas is of the order of the nanokelvin, while in a liquid it is of order mK. These values are to be compared to the coldest temperatures observed in the Universe of the order a few K.

Saying that particles are perfectly at rest at 0 K has been experimentally demonstrated to be wrong, as physicists tried to reach this ultimate limit. Indeed, decreasing T amounts to decrease volume V and hence to fully localize them in space. However, quantum mechanics states that one cannot perfectly localize a particle in space and perfectly cancel its momentum at the same time. The remaining energy at 0 K is the so-called *zero-point energy* (see the discussion in Section 2.5.2). Therefore, it is not correct to say that at 0 K, particles are completely at rest because they possess an intrinsically minimal energy. In most substances, interactions are usually strong enough to fix particles in a crystalline structure and to allow only small vibrations about the crystalline positions. There exists a famous exception, ^{4}He, which is liquid at 0 K or, more precisely, *superfluid*, because the zero-point energy turns out to be greater than the forces between He atoms. Actually, physics at very low temperatures is very rich: some substances which do not exhibit any magnetic properties at standard ambient conditions can possess a magnetization at low temperatures. Even non-conventional states of matter can appear at very low temperatures, such as superconductivity or superfluidity. Some of these aspects will be addressed in Chapter 9. Quantum mechanics (see Chapter 2) is here mandatory to study such states, and one aim of this book is precisely to discuss the quantum origin of such phases.

1.5.5 High densities

Let us finally end this section by exploring the limit $V \to 0$, which is more or less equivalent to the limit $P \to \infty$ or $\varrho \to \infty$. This limit would mean that one can compress matter to a vanishing volume of infinite density. This is of course not possible because the constituents of matter (see Section 1.2) do have an intrinsic volume. Indeed, matter is composed of atoms, the latter ones consisting in a dense nucleus and an electronic cloud. If we approach one atom to another one closer and closer, the electrons will repel each other, first because of the electrostatic repulsion, and second because of their quantal nature (see quantum pressure in Chapter 2). This quantum repelling is macroscopically observed in particular during the late stages of evolution of not too massive stars (such as our Sun). These stars become what one calls white dwarfs, also called degenerate dwarfs. In these stellar remnants, densities can reach $\varrho \sim 10^{28} - 10^{29}$ cm^{-3}, that is about 1 million more than metals on Earth. To explain the stability of such dense objects, one has to invoke quantum properties of an infinite electron gas (see Chapters 2 and 5 for more details).

Finally, the highest densities in the Universe are observed in nuclei ($\varrho \sim 10^{38}$ cm^{-3}) or in neutron stars ($\varrho \sim 10^{39}$ cm^{-3}). One can consider a nucleus as a droplet of a highly quantal fluid of nucleons, while matter in a neutron star is often described as an ideally infinite nuclear fluid. At variance with the electron gas which is usually considered as infinite in a metal (but also as a droplet of electron fluid in what is called a cluster, see Chapter 5), nuclear matter naturally appears in a finite form as nuclei. The case of nuclear matter will be discussed at length in Chapter 7.

1.6 CONCLUDING REMARKS

Liquids constitute what one could call a 'central' state of matter. They lie in between gases and solids. To a rather good approximation, gases, especially when sufficiently dilute, can be understood in rather simple terms with a basic overlook of interactions and correlatively appear as rather 'unstructured'. Solids, on the contrary, exhibit well defined structures, potentially regular and stemming from a dominant role of interactions. Liquids somewhat share properties of both gases and solids. The constant motion of constituents, typical of gases, remains a major characteristic of liquid state. But interactions play a major role to 'organize' components of liquids. This competition between 'free' motion and organization is at the core of the notion of liquids and we shall see all over this book that this also remains true in quantum liquids.

As discussed in this chapter, one can pass from one state to the next by changing physical conditions. We have seen that global properties of a system such as volume/density, temperature and pressure provide useful parameters to tune the nature of the phase (gas, liquid or solid) a system of particles adopts. This then allows one to explore the phase diagram of a substance and access the various phases. Stepping from one phase to the next proceeds via phase transitions, which involve the discontinuity of some physical properties such as for example the specific heat.

It is also important to realize that liquids are observed, at least at our human scale, both as infinite phases, like an ocean, or as finite droplets, like in rain. Water droplets contain still an enormous number of constituents and thus might look like an infinite phase. But the finiteness of the droplet is reflected in properties such as the surface tension which are generic of any droplet, whatever the number of its constituents. Finally it is also important to keep in mind some key dynamical properties of liquids, such as for example viscosity, which characterizes the characteristics of liquid flow. This by the way brings us back to our starting point on interactions which definitively play a central role in liquids and which shall constitute a key issue all throughout this book.

CHAPTER 2

Quantum basics

Quantum fluids are fluids, indeed. We have seen in the previous chapter what it means from a classical physics point of view. But quantum fluids are by definition ruled by quantum physics, as well. The goal of this chapter is to introduce some key concepts of quantum physics necessary to understand what makes a fluid a quantum one or rather what quantum property makes an ensemble of particles a quantum fluid. We are here precisely at the edge of our basic intuition of quantum physics. Indeed, one usually associates the notion of quantum object to a microscopic nature and it may thus a priori seem paradoxical to link the word quantum to a macroscopic object, such as the notion of fluid. Thinking it over a bit deeper, quantum physics is certainly present at all scales and we are confronted with quantum mechanics through the many electronic devices which have invaded our everyday life. And it actually turns out that, as we will see later on, most of these quantum devices are directly connected to the notion of quantum fluids, the case of electron fluids being certainly the most common one in such devices. Quantum physics has for sure deeply modified our everyday life. But it has also changed our perception of the world, and as we shall see in this chapter, it is precisely this new vision of the world which allows us to understand the concept of quantum fluids underlying so many physical systems.

At the core of quantum mechanics lies the apparent paradox of the wave-particle duality. While the classical picture clearly separates space localized particles from space delocalized waves, quantum mechanics brings here a different view. Indeed, any physical object can be seen either as a particle or a wave or both. Stated that way the wave-particle duality may look a bit confusing. A maybe simpler way to formulate it is to state that a physical object may behave as a particle or as a wave, depending on the physical situation. Of course there is here no possible 'choice' of nature. Rather, the physical object will behave as what is well identified in classical physics as a particle in certain circumstances and will behave as a wave otherwise. This mostly sets the limits of the underlying classical picture of a particle or a wave which mostly provides some 'simple' ansatz of an otherwise complex object. Of course both pictures are directly related to each other, the relationship between the two representations also giving a clue to what picture is most relevant in a given physical situation. This particle-wave duality turns out to be at the core of the notion of quantum fluids, as we will see in this chapter. We shall therefore devote some time to detail its meaning and consequences, focusing in particular the discussion on the key concept of zero point motion. This is a crucial concept to understand the possible fluid nature of ensembles of quantum particles.

Quantum particles possess a particular property which make them behave in a specific way when put together. They are indistinguishable, which, stated in simple terms, means that one cannot decide which is which when looking at two particles. In other words quantum mechanics precludes to identify them with flags, as classical particles can be. This property is intrinsically linked to the fully quantal characteristics of particles provided by their spin. Depending on the value of the spin, ensembles of identical particles will behave in one way or another which naturally creates two classes of particles, bosons and fermions. We will then have to distinguish in the following two classes of quantum fluids, which, as we shall see, will indeed exhibit different types of properties.

The present chapter will provide an introduction to the key concepts necessary to understand and characterize fluids of quantum particles. The chapter is by no means meant as an introduction to quantum mechanics, a topic widely covered in many dedicated books. It mostly aims at discussing some key concepts necessary to understand what a quantum fluid is and where it stems from. For this we shall rely on qualitative discussions and a few simple technicalities, in the spirit of the book.

2.1 CLASSICAL CONCEPTS FOR QUANTUM PHYSICS

We shall first recall some basic concepts in classical physics, namely particles and waves, and comment on the puzzling questions they raise in quantum physics. Then the particle-wave duality will be discussed in some depth.

2.1.1 Particles and waves as separate objects

The first, and central, key notion to be discussed here is the particle-wave duality, which lies at the core of quantum mechanics and which has often been seen as paradoxical. In order to better analyze it we first start by recalling what is behind the two *classical* notions of particle and wave. To analyze them in relation to each other, we shall take an illustrative example which puts them in direct relation and thus allows us to directly compare their properties. Let us take the simple example of a pebble thrown into a pond. Everyday life tells us that a pebble is a good example of particle. The pebble, once thrown, follows a parabolic trajectory, typical of point like particles, once the pebble is small enough and launched without dedicated rotational effect to simplify the analysis. The motion is well accounted for by the laws of classical mechanics, and this up to the instant the pebble hits the surface of the pond before disappearing inside water. Mathematically speaking the motion of the pebble is fully characterized by its (center-of-mass) position $\mathbf{r}$ and momentum $\mathbf{p}$, both functions of time $\mathbf{r} = \mathbf{r}(t)$, $\mathbf{p} = \mathbf{p}(t)$. Important for our discussion is to note here that the pebble, while in the air, is well localized in space, at any instant of its trajectory.

The hit with water provokes a perturbation which propagates at the surface of the water. This manifests itself as wavelets with regular (in space and time) vertical oscillations on the surface. This is seen as concentric circles on the water surface, centered at the pebble's impact point. The distance between two successive circles exhibiting similar perturbations, defines the wavelength, usually denoted by λ, and provides a spatial characteristic of the phenomenon. The temporal characteristic of the surface oscillations is provided by the frequency, often denoted ν, and obtained as the ratio between the velocity of propagation c of the circles and the wavelength ($\nu = c/\lambda$). Most important for our purpose is the fact that, after some time, the perturbation in terms of concentric circles, reaches any point of the pond surface. The thus formed wave is defined over the whole surface of the pond, hence fully delocalized. It can mathematically be characterized by a wave amplitude $A(\mathbf{r}, t)$ which is a function of both time and space, and which measures the vertical position of a point of the pond as compared to the surface at rest. For the sake of simplicity, we will restrict ourselves to the one-dimensional (1D) case and instead of the 3D vector $\mathbf{r} \equiv (x, y, z)$ we shall consider only one of its components, for instance x, as a spatial coordinate. We can occasionally consider situations in 2D or 3D, and in these cases specify the dimensionality of the problem. The simplest and most common wave amplitude has one of the forms $\sin(kx - \omega t)$, $\cos(kx - \omega t)$, $\mathrm{e}^{i(kx-\omega t)}$, $\mathrm{e}^{-i(kx-\omega t)}$ or some linear combinations of them, which are generically referred to as 'plane waves'. Here, $k = 2\pi/\lambda$ is the wave number and $\omega = 2\pi\nu$ is the angular velocity, all of them being constants characterizing the undulatory motion, together with a multiplicative constant to characterize the intensity of the wave. Note finally that both pebble motion and surface perturbation are perfectly understood in terms of classical physics, the former as a *localized* particle, the latter as a *delocalized* wave.

2.1.2 The peculiar stationary waves

The example of the surface wave on the pond is interesting in many respects and will allow us to discuss several aspects of waves. As a first simple remark, let us note that, although the pond is large, especially as compared to the size of the pebble, it nevertheless possesses a finite extension. At some stage of the evolution of the wave, it hits the boundary of the lake. Boundary conditions are thus, not surprisingly after all, a key aspect of the evolution of a wave. And we shall see that these boundary conditions are responsible for many important behaviors. In the case of the waves that we shall use in quantum mechanics, boundary conditions will show up again and play a major role to understand many physical behaviors.

As a starter let us consider a case from classical physics and see how boundary conditions may crucially affect even the form of the waves themselves. We keep aside the case of the pond for a while and consider another everyday life example. Let us take a rope, fixed at one extremity on a wall and held by our hand at the other extremity. In order to generate an undulatory motion of the rope it is enough to have our hand oscillate with a very small amplitude, which leads to oscillations propagating along the rope. But experience shows that these oscillations in general quickly die out, except in some very specific cases when we impart oscillations with well defined frequencies. The situation is exactly the one observed in any stringed music instrument. In order to maintain oscillations the length of the rope needs to be an integer number of times half the wavelength λ, as illustrated in Figure 2.1. The situation is

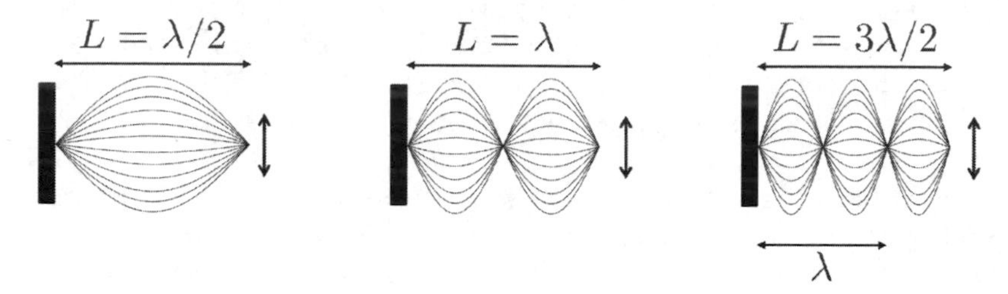

Figure 2.1: Visualization of the wavelength λ of standing waves in a rope of length L having one end fixed and the other one manually stirred at precise frequencies.

similar to sound waves inside music instruments such as a flute in order to hear a proper sound. Such particular waves are known as *stationary* waves.

The end points of the rope then correspond to points with zero amplitude (nodes) and remain as such forever, while points of maximum amplitude again remain as such forever (the amplitude oscillates but remains maximal at these points). The term stationary used to characterize such waves precisely refers to this 'immobility' (in space) of nodes and points of maximum amplitude which differs from the usual view of a propagating wave in time and space. In stationary waves the propagation thus appears as purely temporal with no spatial motion. This mathematically means that the space and time components of the waves factorize, namely that at any time and for space point we can write its amplitude as $A(x,t) = a(t)b(x)$, i.e. the product of a function depending solely on time and another which solely depends on position. As we shall see below, these stationary waves will play a crucial role in our analysis of quantum waves.

2.1.3 Interference, diffraction... the signature of waves

A key characteristic of waves is the way they superpose. The phenomenon is known as interference and again it is totally different from what occurs with particles. Let us come back to our example with pebbles and ponds. Now we throw two pebbles into the pond. The double hit on water generates two series of concentric circles reflecting the propagation of the two surface waves created by the two impacts. In this case the total perturbation becomes a bit more involved as both waves superimpose in space and time. A schematic view of the propagating waves in the case of one or two peebles is shown in Figure 2.2.

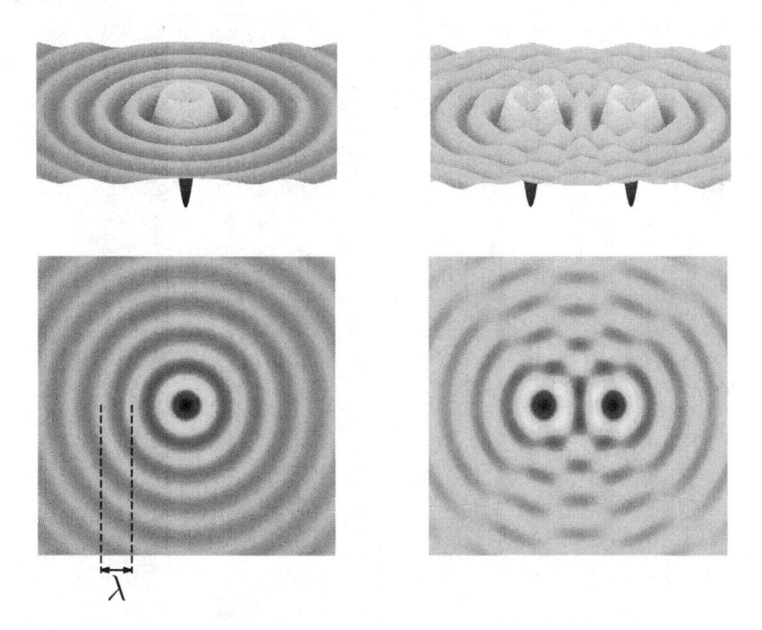

Figure 2.2: Concentric waves generated by the impact of a pebble (left column) and two pebbles (right column) on the surface of a pond. Top row: 3D view. Bottom row: top view. In the case of the impact of a single pebble, the wavelength λ is indicated and corresponds to the distance between two consecutive maxima (or minima).

To be a bit more specific, let us consider the energy of the set of waves. In the case of particles, the total energy of the two particles would simply be (within neglecting their interaction) the sum of the energies of each particle. In the case of waves, the situation is different. The energy of a wave at a given point $\mathbf{r}$ at a given time t is proportional to the square of its amplitude $|A(\mathbf{r},t)|^2$. The amplitude of a wave resulting from the addition of two waves with amplitudes A_1 and A_2, is the sum of their amplitudes $A = A_1 + A_2$. Therefore, the energy of the combination of the two waves produced by two pebbles is not the sum of the energies of each separate wave because $A^2 = |A_1 + A_2|^2 = |A_1|^2 + |A_2|^2 + A_1^* A_2 + A_1 A_2^* \neq |A_1|^2 + |A_2|^2$. The sum of the two last terms (recall that the amplitudes are in general complex quantities) produces the well-known phenomenon of *interferences*, which is a clear signature of waves. Depending on the particular points and times, the interference of the two waves leads to very intense (constructive interference) or very weak (destructive interference) patterns which are known as interference fringes.

Two points are worth being noted here. First, the effect is clearly a consequence of the delocalized nature of waves. Second, it is mathematically associated to the composition of amplitudes and not energies. The important aspect for our discussion is nevertheless the distinction, very clear in this example, between the behavior of particles and that of waves. In the classical picture, particles are well localized objects defined by a position and a velocity. Two non-interacting particles have two well identified positions and velocities; the total energy of the set of two non-interacting particles is just the sum of the energies of each particle.

Diffraction is another characteristic feature of waves, as interferences, again at full variance with behaviors of particles. Diffraction typically occurs when a wave hits an obstacle, as for example light hitting a screen pierced by a hole. Two different situations, depicted in Figure 2.3, can occur. If the hole size is

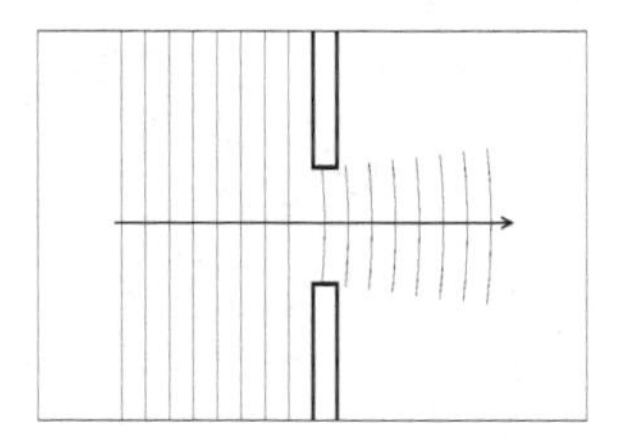

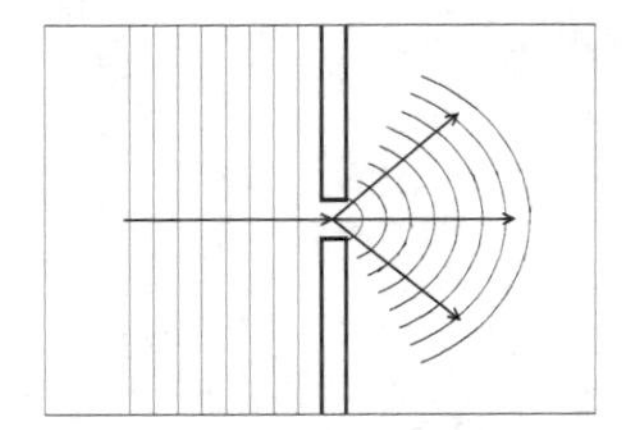

Figure 2.3: Diffraction of a wave, symbolized by vertical lines, arriving to a screen pierced with a hole whose size is larger than (left panel) or similar to (right panel) its wavelength.

much larger than the wavelength of the incoming wave (left panel), nothing special occurs. The light goes through the hole in the screen without visible perturbation. Putting a second screen beyond the pierced one, we will see a net spot of the size of the hole. If we were hitting the screen with particles and not light we would see exactly the same pattern. Particles facing the hole would travel without deviation, the other ones would be stopped by the screen.

Now let us consider the case in which the size of the hole is comparable to the wave's wavelength, as shown in the right panel. The wave is now directly affected by the presence of the hole, precisely because its wavelength matches the size of the hole. In the figure, this size is small enough to see a new spherical wave centered precisely at the hole. The pattern created by the diffracted wave on a second screen is an alternate series of more or less intense circular fringes, as in interferences. Indeed because the hole affects the wave at the scale of its spatial periodicity it does have direct impact on its spatial modulation, whence the fringes. For a large hole differences are washed out because one averages out the differences over a large space region as compared to spatial variations of the wave. Of course diffraction does not occur for classical particles. Either the hole is large enough to let them pass the obstacle or too small and they are then stopped by the screen.

2.1.4 Puzzles: black body radiation and photoelectric effects

Diffraction and interference are typical behaviors of waves, reflecting the fact that they simultaneously explore the whole space, at variance with particles. In classical terms, particles and waves therefore basically appear as irreconcilable or at least concerning different types of objects. And surprisingly enough, in the microscopic world both concepts become deeply united by quantum mechanics. This apparent paradox is known as the particle-wave duality. The particle or undulatory nature of light goes back at least to the confronting views of Huygens and Newton. The diffraction experiments of Young in 1803 settled the discussion in favor of the wave interpretation. The controversy was renewed by Einstein in one of the several seminal works published in 1905. It was based on the hypothesis made by Planck five years before, concerning the black body radiation. Planck was able to explain the black body radiation measurements assuming that absorption and emission of light is produced in discrete 'quanta'. For light of frequency ν, the energy is transferred in multiples of $h\nu$, where h is the celebrated Planck's constant ($h = 6.62 \times 10^{-34}$J/s). Einstein went a step forward by postulating that light is indeed constituted of elementary 'energy packets', which nowadays are called photons. This hypothesis provided a simple explanation to the highly debated puzzle of photoelectric effect. In order to better appreciate the step forward, let us briefly review what were the pending issues at that time.

We first start with Planck's analysis of the black body radiation. Note that the term 'black' used here does not refer to a real color but to a physical situation. A typical prototype of black body is a box pierced with a small hole. The latter suffices to capture outside light but because it is small enough retains most of the absorbed light. Indeed, because of the small size of the hole, the entering light is multiply reflected inside the box and has almost no chance to exit the box through the tiny hole because of the quasi-impossibility of reemission of the absorbed light as such, whence the 'invisibility' of the box's interior and the term 'black body'. But the multiple reflections in the interior will heat up the box to a certain temperature T and the box will in turn radiate back the thus accumulated energy. The emitted light can therefore have any wavelength, visible or not, the spectrum of the emitted radiation being fully determined by the box temperature T: this is the so-called black body radiation. We have all experienced such a phenomenon. The everyday life example of black body radiation is the case of the heated metal (such as in a standard electric cooktop). By increasing temperature, the (often grayish or black when cool) heated metal first turns red, then successively exhibits an orange and yellow color, and finally turns white (white warmed metal). The color of the heated metal is here fully determined by its temperature.

To explain the observations on the black body radiation, specifically the dependence of the radiated energy with frequency and temperature, Planck assumed that the absorption and emission of radiation by the walls of the box only occur in a discontinuous manner, as integer multiples of a minimal energy $h\nu$, for each wave frequency ν. The situation is a bit similar to what would occur if considering that water getting out from a tap does it in terms of well identified droplets with a well defined size (which is actually not exactly the case for tap water!).

At about the same time as the work of Planck on black body radiation the scientific community was also deeply interested in the photoelectric effect, which again concerned a puzzling behavior of electromagnetic radiations. Under some specific conditions a metallic plate hit by light may emit electrons. This by itself is not surprising as metals contain loosely bound mobile electrons which may likely be extracted from the metal provided they absorb enough energy from an electromagnetic radiation. The puzzle rather lies in the characteristics of the emission. Emission is instantaneous and increasing light intensity leads to more emission but does not increase the maximum electron kinetic energy. Even more surprising, emission is suppressed at low frequency (red light) and whatever the light intensity, and it occurs at high frequency (blue light) even for small light intensity. The yield is then limited but the process leads to possibly higher energy electrons than what is attained with higher intensity and smaller frequency light. These characteristics do not match what is expected from an undulatory description of light, following Maxwell's theory. Following Einstein's hypothesis about the corpuscular nature of light, the photoelectric effect—and by construction the black body radiation—becomes easily understandable. When an electromagnetic wave hits a metal what counts is not the wave intensity but the energy $h\nu$ that the photon transfers to an electron in the metal. Indeed, electrons are bound to the metal, and there is a minimum energy to extract them, known as the work function, which is the analogous in bulk metals to the ionization potential in atoms and molecules. Only when the photon energy $h\nu$ is greater than the work function, the electron extraction from the metal is possible. This simply explains the observed threshold effect. Even more so, extraction may become instantaneous by direct energy transfer from the photon to the electron. When increasing light intensity one increases the photon flux but not its energetic properties so that one expects indeed more electrons emitted but not more energetic ones.

2.2 PARTICLE AND WAVE

The photon interpretation of light is the first, key example of particle-wave duality. Actually the duality of light was somehow considered when optics was classified in 'physical' (or 'wave') optics, and 'geometrical' (or 'ray') optics. The former describes properties as interferences or diffraction, directly related to its undulatory nature. Geometrical optics describes light propagation in terms of rays, an abstraction which looks very much like the trajectories of a particle. This is an excellent approximation, known to be valid when the wavelength is small compared to the size of structures with which the light interacts. But nobody related this approximation with real particles. In the first quarter of the 20th century, the existence of unexpected discrete quantities showed up also in various physical systems. These include the Ritz classification of atomic spectral lines, the specific heats of solids, the energy losses of electrons on collisions with atoms... The old controversy about the particle/wave nature of light was renewed but started on new grounds.

We are facing here a situation in which two representations of light are advocated: the usual undulatory one for manipulating light with standard optical devices and the corpuscular one needed to understand the coupling to electrons and the ensuing emission from the metal. We are here at the core of the particle-wave duality. Indeed light 'behaves' as a wave in some circumstances (interference, diffraction) but may as well 'behave' as a particle when coupling to the electrons of a metal. The duality reflects the fact that light is a complex object which can be modeled as a set of particles in some circumstances and as a wave in other ones. The actual behavior, or rather the simple picture we put on it (particle or wave) does depend on the physical situation. In spite of its richness the photon hypothesis nevertheless remains somewhat problematic as it does not provide a unified picture merging both particle and wave representations. As such, both representations still look hardly conciliable with the above discussed dilemma of localized particles against delocalized waves.

2.2.1 The duality resolved: de Broglie relation

In 1924, de Broglie proposed a beautiful scheme, rooted in classical mechanics justification of geometrical optics, to interpret Einstein's hypothesis about the photon, and the special theory of relativity as well. For the purposes of this book, we ignore relativistic effects. Both particle and wave pictures are related at a deeper level, and the well-known de Broglie relation connects the wavelength of the wave representation of a system to the momentum p characteristic of its particle description

$$\lambda_B = \frac{h}{p} \tag{2.1}$$

and it is usually referred to as the de Broglie wavelength.

The particle wave duality turns out to be a general property of any physical object, not an exception specific to the case of light/photons. Any particle possesses an associated wave and vice versa. This for example meant that well known particles such as electrons might exhibit a wave behavior, as was indeed confirmed a few years later in two independent experiments by Davisson and Germer and by Thomson. Actually both experiments were similar to those showing diffraction of X-rays. In both cases the characteristic diffraction pattern perfectly matched the one obtained for a wavelength h/p for electrons of momentum p, which gave a direct quantitative support to de Broglie hypothesis. The case of electrons, again, is not specific and it turns out that any particle can be represented by a particle or a wave, both being related to each other by de Broglie relation.

The de Broglie wavelength λ_B provides a qualitative estimate of whether a physical object behaves as a wave or as a particle. When λ_B is comparable to typical distances of the system the latter is likely to require wave representation (we are in the typical diffraction condition), while when λ_B is much smaller than typical distances the particle representation is likely to apply. In any case the important issue, in particular for the purpose of this book, is the fact that any physical object is represented by a wave whose characteristics are directly related to a particle representation. This, as such, provides the key to the particle-wave duality.

2.2.2 The unavoidable wave: Heisenberg relation

Now that the particle and wave representations are properly linked, it is interesting to come back to the pending question of localization/delocalization discussed above. Indeed, de Broglie relation links the wavelength λ_B characteristic of a (delocalized) wave to a momentum p typical of a (localized) particle. And it states that the product of the two quantities is Planck's constant h. Let us see whether this may clarify the question of localization/delocalization.

We take a simple example to illustrate the point. Let us assume that we measure in a detector the velocity of electrons, at some space point. This might be for example electrons emitted from a metal by photoelectric effect. Because the detector is located at a well defined space point and because we count the number of bypassing electrons, we may assume that the wave amplitude, or wave function $\psi(x)$, of a bypassing electron has a finite space extension, say Δx, namely that it possesses a certain degree of locality. A simple way to represent such a 'localized' wave is to construct a wavepacket made out of monochromatic plane waves. This construction is illustrated in Figure 2.4. A monochromatic plane wave in 1D is defined by a wave vector k (rather called a wave number in 1D), k being a real number, or equivalently by its wavelength $\lambda = 2\pi/k$. Such a plane wave in general reads as $\exp(ikx)$ and a typical example is shown in the top row of the figure. It is fully delocalized in space, reflecting its very nature of wave. In the middle row of the figure are plotted four plane waves with close wave numbers (light gray lines). Their

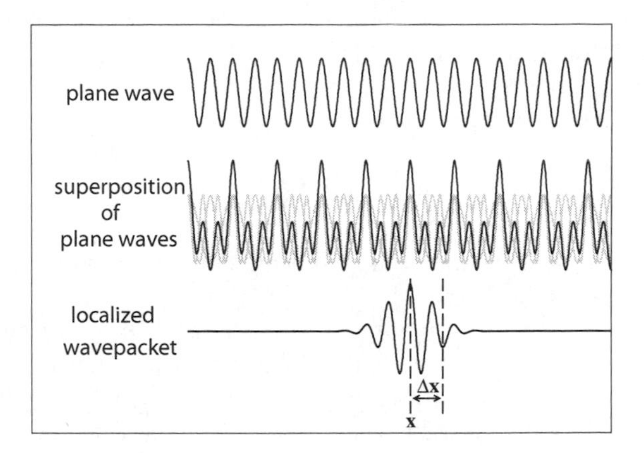

Figure 2.4: In the top panel is plotted a single plane wave (actually its real part), along the horizontal x-axis. In the middle panel are plotted several plane waves (light gray), and their superposition (black). A localized wavepacket is plotted in the bottom panel.

superposition (black line) exhibits interferences, with smaller and greater values in some regions of space. Finally, in the lower figure we show an infinite superposition of plane waves, with a weight function $c(k)$. Mathematically, it is expressed as

$$\psi(x) = \int \mathrm{d}k \, c(k) e^{\mathrm{i}kx} \quad , \tag{2.2}$$

which is nothing else than a Fourier transform. With a proper choice of the weight $c(k)$, one can pile up plane waves to produce a wave with a well defined center of mass x and width Δx.

We now have at our disposal two representations of the wavepacket: a real space representation $\psi(x)$ in terms of position x and a momentum representation, the function $c(k)$, in terms of the set of k values used to build the linear combination delivering $\psi(x)$. Simple rules of Fourier transform allow one to relate both k and x representations. If the spatial extension of the wavepacket is Δx in real space, the associated extension in k space is Δk such that the product $\Delta x \Delta k$ is lower bounded. It turns out that the minimum bound is $1/2$, occurring for a Gaussian weight function $c(k)$, which is the usual choice. The localized wavepacket shown in Figure 2.4 corresponds precisely to that choice.

The next interesting point is that the de Broglie relation $p = h\lambda$ allows us to link wavelength and momentum. Indeed, the wavelength associated to a given k being $\lambda = 2\pi/k$, one can associate a momentum p to each wave such that $p = h\,k/2\pi$. The ratio $h/2\pi$ appears so often in quantum mechanics that it has been assigned a specific symbol $\hbar$ and name (h bar), the reduced Planck's constant. The momentum then reads $p = \hbar k$. Following the de Broglie's interpretation, the simple wave property relating space and wave momenta leads to the famous Heisenberg relation

$$\Delta x \Delta p \geq \frac{\hbar}{2} \tag{2.3}$$

which sets a well defined limit to how much localization can be expected from a quantum system: the better the space resolution ($\Delta x \to 0$), the lesser the momentum resolution ($\Delta p \to \infty$), and vice versa. The implications of these results are deep. First, it tells us that, at variance with a truly classical object, one cannot measure simultaneously with an arbitrary accuracy both position and momentum of a particle. This is formally true in all situations, but practically meaningless in some situations, in which the particle *does indeed* behave classically (see Section 2.2.3). Second, Heisenberg relation clearly supports the undulatory character of the representation of a particle. Indeed while it is conceivably easy to associate a position and a momentum to a particle (in the classical sense), it is obviously less clear for a wave, whose various spatial components possess various velocities. The somewhat surprising result provided by the Heisenberg relation does not lie in the relation itself but rather in the interpretation thereof. It is the interpretation of the de Broglie's relation in conjunction with mathematics of the Fourier transform which makes the Heisenberg relation so important (and a priori surprising).

It should finally be noted that the space-momentum Heisenberg relation introduced above is just an example of the general Heisenberg relation which applies to several sets of such conjugate variables, conjugate in the sense of the Fourier transform. Indeed, from Eq. (2.2), we deduce that x and k, or equivalently x and $p/\hbar$, are conjugate. The reader might be surprised to encounter such conjugate variables and is probably more familiar with the Fourier transform that relates the time t and the frequency ν domains, widely used in electronics for instance. The Heisenberg relation applied to the couple (ν, t) reads as $2\pi\Delta\nu\,\Delta t \geq 1/2$. Using for example that the energy of a photon is $E = h\nu$, we immediately get:

$$\Delta E \Delta t \geq \frac{\hbar}{2} \quad . \tag{2.4}$$

It means in particular that the much celebrated energy conservation principle (in a closed system) can accordingly be 'temporarily violated': ΔE is 'uncertain' within Δt such that $\Delta E \Delta t \sim \hbar$ (incidentally, notice that the factor $1/2$ entering Heisenberg's inequalities is omitted to estimate orders of magnitude).

As a final word, if the couples of variables (x, p) and (E, t) are usually not considered as conjugate variables in classical physics, this is precisely because of the very small value of $\hbar$. Indeed, the limit of classical physics is reached by setting $\hbar \to 0$. Plugged into (2.3) or (2.4), that would mean that there exists no constraint in the simultaneous measurement of the position x and the momentum p, of that of the energy E at a given time t, consistent with our everyday life experience.

2.2.3 Quantum... or classical particles?

Heisenberg relation, as de Broglie's one, provides a rule of thumb to estimate how much a system is 'dominated' by quantum effects. We have seen that estimating de Broglie wavelength λ_B gives a first estimate of 'how much' a system is quantal. If λ_B is much smaller than any characteristic spatial scale of the system, the undulatory character of the system tends to disappear to the profit of its corpuscular character. In turn, the undulatory character dominates when λ_B is of the order of magnitude of the characteristic lengths of the system, such as for example in the case of diffraction pattern. Heisenberg relation provides a similar diagnostics. The rule of thumb is then simply to estimate typical spatial and momentum scales and compare their product to the reduced Planck's constant $\hbar$.

Let us take examples to fix ideas. We start with an example from everyday macroscopic life and consider a pebble of 100 g mass, thrown with a velocity of order 10 m/s. The associated momentum is $p = mv \sim 1$ kg.m.s^{-1} and the de Broglie wavelength $\lambda_B = h/p \sim 6 \times 10^{-34}$m, virtually *zero* at the scale of the pebble. No experiment could detect the associated wave, and the pebble behaves only as a classical particle. Let us now take an example from the microscopic world, and consider an electron (mass $\sim 9 \times 10^{-31}$kg). Remember that its discovery in the experiment by Thomson showed that the electron bevahed as a classical particle. Suppose that it moves in an electric potential of a few volts, with a velocity of about 10^7 m/s (so that relativistic effects can be ignored). Its de Broglie wavelength is about 1 Å. It is of order the size of atoms, so that its wave behavior has to be definitely accounted for. Electrons in atoms are fully quantum objects which cannot be described classically, at variance with the previous case of the macroscopic pebble.

Now that we have introduced this key concept of the particle wave duality and seen the potential importance of undulatory effects, let us see how we can develop a formalism allowing us to integrate this wave concept in the description of nature.

2.3 WAVE FUNCTIONS AND THE SCHRÖDINGER EQUATION

We have seen in the previous section that we needed to account for the undulatory character of particles (to make the 'paradox' short). At the core of quantum mechanics thus lies this key concept that the state of any object has to be described by means of a wave amplitude or *wave function*, providing the wave properties of the physical system when necessary. This holds true for *any* system, in *any* situation. Still, depending on the physical situation the undulatory character may be clearly visible or not (see discussion in Section 2.2.3). The basic ingredient of the description of a system is therefore its wave function that we shall denote $\psi(x,t)$ and which is a function of position x and time t. We here choose to start with a 1D presentation with only one spatial coordinate x. The generalization to full 3D is straightforward and will be used later on when necessary. The description of the system in terms of a wave function opens the door to the undulatory description, provided we possess an associated wave equation.

2.3.1 The Schrödinger equation

The crucial ingredient of the undulatory description is the wave equation followed by the wave function. It is the celebrated Schrödinger equation. It reads, in the non-relativistic case, in one dimension and for a single particle

$$\mathrm{i}\hbar\frac{\partial}{\partial t}\psi(x,t) = H\psi(x,t), \tag{2.5}$$

where H is the Hamiltonian operator, which is written as

$$H = -\frac{\hbar^2}{2m}\frac{\partial^2}{\partial x^2} + V(x,t) \tag{2.6}$$

As in classical physics, the Hamiltonian is the sum of kinetic and potential energy terms. We identify the latter in the form of a general local and time-dependent potential $V(x,t)$. The operatorial character of the Hamiltonian appears in the kinetic term. Classically it is given by the expression $p^2/2m$ for a particle of mass m and momentum p. In the quantal version, the momentum is replaced by the differential operator $p = -\mathrm{i}\hbar\partial/\partial x$. Alternatively, we could have decided to use the momentum p instead of position x as the free variable. In this so-called p-representation, the wave function becomes $\widetilde{\phi}(p,t)$, and the position becomes a differential operator $x = -\mathrm{i}\hbar\partial/\partial p$. The kinetic term has the classical $p^2/2m$ form, while the potential term $V(x,t)$ becomes harder to manipulate and interpret. In the following we shall deal with the x-representation.

The Schrödinger equation then provides an evolution equation for the wave function $\psi(x,t)$. We have seen that the wave function naturally integrates both spatial and momentum information, whose evolution has thus to be integrated in the wave equation. This is achieved by stepping into the realm of complex numbers which allows us to handle simultaneously two real quantities.

2.3.2 A free particle in 1D

The simplest conceivable case is the case of a free particle, i.e. no potential acting on it. For then, the Schrödinger equation simply reads:

$$\mathrm{i}\hbar\frac{\partial\psi(x,t)}{\partial t} = -\frac{\hbar^2}{2m}\frac{\partial^2\psi(x,t)}{\partial x^2}. \tag{2.7}$$

To solve it, we factorize the spatial and temporal components of the wave function as $\psi(x,t) = \phi(x)\varphi(t)$. This reminds us of the well-known stationary solutions of standard wave equations in which spatial and temporal evolutions are separated and in which some points of space correspond to zero wave amplitude (nodes) and some other ones to maximum amplitude (see Section 2.1.2 and Figure 2.1 for a classical standing wave). With this factorization, the equation is written as

$$\mathrm{i}\hbar\frac{1}{\varphi(t)}\frac{\mathrm{d}\varphi(t)}{\mathrm{d}t} = -\frac{\hbar^2}{2m}\frac{1}{\phi(x)}\frac{\mathrm{d}^2\phi(x)}{\mathrm{d}x^2}$$

The left hand side being a function only of t, and the right one only function of x, the solutions for $\varphi(t)$ and $\phi(x)$ are obtained by equating the previous equations to a same constant. The solution is a plane wave of the form:

$$\psi(x,t) = \mathrm{e}^{-\mathrm{i}(\omega t \pm kx)}, \tag{2.8}$$

where the proper $\pm$ sign should be chosen to describe a wave traveling to the left or to the right. The common constant for the equations for $\varphi(t)$ and $\phi(x)$ imposes the relation $\hbar\omega = \hbar^2 k^2/(2m)$. This is the energy of the stationary waves, with k taking *any* value.

The fact that the properties of a particle are described by means of a complex wave function is not a problem. A proper interpretation of its physical meaning should lead to predicting results of physical observables in terms of real numbers. In this respect, we shall see later on that the squared modulus $|\psi(x,t)|^2$ of a general wave function is interpreted as a density of probability of finding the particle at point x at any time.

The integral over all the space of $|\psi(x,t)|^2$ should then be normalized to unity:

$$\int_{-\infty}^{+\infty} \mathrm{d}x |\psi(x,t)|^2 = 1\,. \tag{2.9}$$

In the case of a plane wave, one gets infinity instead, delivering an infinite probability when integrated over space. The remedy is to consider a finite interval of length L and normalize the wave function as

$$\int_L \mathrm{d}x |\psi(x,t)|^2 = 1\,. \tag{2.10}$$

In other words, the plane wave is taken as:

$$\psi(x,t) = \frac{1}{\sqrt{L}}\mathrm{e}^{-\mathrm{i}(\omega t \pm kx)}\,. \tag{2.11}$$

Analogously, in a 3D problem, one would consider a finite volume $V = L^3$ instead of the full space, and write the plane wave as $\psi(x,t) = \frac{1}{\sqrt{V}}\mathrm{e}^{-\mathrm{i}(\omega t \pm \mathbf{k}\mathbf{x})}$. The length L (or the volume V) is assumed to be very large, taking the infinite limit when necessary. In any case, plane waves are widely used as a basis to express wave functions as linear combinations of plane waves. More details can be found in Section 4.2.1.

Interestingly, one finds that the associated density is constant, which means that the particle is *anywhere* (and conversely 'nowhere'). This is physically sound as the absence of interaction/potential means that there is the same probability to find the particle at any place. But mind that the case is a bit touchy as the associated total probability of presence is infinite. In practice the case never occurs, which circumvents the difficulty.

The more interesting case is the realistic case with a non-vanishing potential V. But solving the Schrödinger equation is in general difficult. The standard strategy consists in searching for stationary solutions because these stationary solutions have a trivial time evolution (recall the stationary solutions of the rope in Section 2.1.2). Because the Schrödinger equation is a linear one for ψ, it allows for wave superposition, which is the key to interference and diffraction pattern. Once one has obtained the stationary solutions (see Section 4.2.2 for more details), the latters provide a natural basis on which to expand the actual solution, which is then simply computed as a linear combination of stationary solutions. Even with this simplification, in general the solution of the Schrödinger equation in that case is complicated and can only be attained numerically. A few simple cases can be treated analytically and are nevertheless quite illustrative. When needed, we shall therefore only consider the simplest case of an infinite potential well.

2.3.3 The physical interpretation of wave functions

The solution $\psi(x,t)$ of the Schrödinger equation, satisfying the appropriate boundary conditions, provides the *complete* quantal description of a particle of mass m subject to a potential $V(x,t)$. In that respect, it is the quantal analogue of the classical trajectory $x(t)$, solution of Newton's or equivalent equation. The knowledge of $x(t)$ also induces the perfect knowledge of the classical momentum $p(t) = m\,\mathrm{d}x/\mathrm{d}t$. However, we have seen that this fully localized description is not possible at a quantal level due to Heisenberg uncertainty; see Eq. (2.3). And indeed, the wave function $\psi(x,t)$ provides a delocalized picture, consistent with the particle-wave duality. Keeping in mind the localized wavepackets displayed in Figure 2.4, one can intuitively expect the wave function to be 'large' where the particle is *likely* to be, and small elsewhere. The words 'complete' and 'likely' in the previous sentences contain two basic postulates of quantum mechanics. The predictions about the behavior of a quantum system are probabilistic, and this information is complete. The reliability of the postulates is established by logical consistency and comparison with experiments.

The wave function represents an amplitude of probability of presence of the particle system at one place and one instant. The probability per unit volume of finding the particle at point x at time t is:

$$P(x,t) = |\psi(x,t)|^2. \tag{2.12}$$

Integrating this probability over all the space, $\int dx P(x,t)$, we should obtain 1, so that the wave function has to be normalized consequently. It can happen that the wave function $\psi(x,t)$, even if it is a solution of the Schrödinger equation, cannot be normalized over the whole space (take for instance the case of a plane wave $\psi(x,t) = \cos(kx - \omega t)$). In this case, one considers a finite, but large, volume V for the normalization of the wave function. By large, we mean sufficiently large so that the influence on the wave function of the boundaries of this volume can be neglected. One can eventually extend V to an infinite volume if necessary.

Let us come back to the probability $P(x,t)$. It also directly provides the density of matter (in the classical sense of fluid dynamics) as

$$\varrho(x,t) = |\psi(x,t)|^2. \tag{2.13}$$

In the present case, the normalization to 1 then reflects the fact that the wave function represents exactly *one* particle.

The probability density $P(x,t)$ makes it possible to calculate weighted average values of relevant physical quantities. For instance, the mean position is $\langle x \rangle = \int dx P(x,t) x$ and would correspond to the position of a classical point-like particle (note however that $\langle x \rangle = 0$ if the system is centered at the coordinate origin). This establishes a clear link between the particle and wave pictures. Second, it should be noted that, as in the case of waves, the key ingredient of the description is an amplitude of probability (a wave amplitude), not a probability (an intensity). This is in particular essential as it opens the door to interference and/or diffraction pattern, which will possibly build up at the side of $\psi(x,t)$. It is interesting to note on the above expression that the average squared position, which is computed from the wave function, possesses a clear probabilistic character. It is computed as an average over a probability density $P(x,t)$ just as in any standard statistical average. The key point is that the probability density $P(x,t)$ is directly expressed from the wave function $\psi(x,t)$ giving a specific role to the latter quantity which appears as the key ingredient to compute any measurable quantity characterizing the system (called an observable in the quantum language). We have seen above the case of the probability of presence and the average squared position. But the rule holds for any quantity (1 for the presence, x for the position...) following similar expressions. As a result any measurement on a quantum system appears as probabilistic which has deep formal consequences.

Before proceeding, two remarks are in place. First, the pure undulatory picture is itself limited and does not allow a full description of a quantum system. A more general formulation indeed exists but requires some more elaborate mathematics, not needed for our purpose. To make a long story short let us say that the undulatory picture basically misses spins which cannot be represented by spatial coordinates. We will thus adopt a practical viewpoint adding spins by hand to our wave functions when needed. And for the sake of simplicity we shall furthermore basically remain within the so-called x representation $\psi(x,t)$ of the wave function in which the variables entering the wave function are position and time. The second remark is even more practical and concerns terminology. Now that we have seen in detail how particle and wave representations are interlaced we shall avoid separating both notions. We shall use the generic term of *particle* to label the typical constituents of the systems we shall consider. And the term will implicitly assume that these particles are to be represented by wave functions, and possess an undulatory character. One may alternatively consider the ensemble of particles as a whole, which can be labelled by the generic term of *system*, again to be characterized by a wave function, itself depending on the coordinates of all its constituents. As a starting point we shall consider the simpler case of a one particle system.

In general, the expectation value of a quantity $f(x)$ which depends on position x is calculated as the mean square position, i.e. $\langle f \rangle = \int \mathrm{d}x \psi^*(x,t) f(x) \psi(x,t)$. Quantum wave functions ψ are by nature complex quantities with a real part and an imaginary one. One then defines averages as where ψ^* is the complex conjugate of ψ. It is interesting to note here that the wave function is defined up to a phase $\exp(\mathrm{i}S)$ which will clearly not be measurable from quantities depending on the position. That nevertheless does not fully rule out phases from the game, as two wave functions may interfere, which will definitely concern their phases, and ultimately become visible for example in an interference or a diffraction figure. The phase explicitly appears in calculating expectation values of quantities $g(p)$ depending on the momentum. In that case, one has to calculate $\langle g \rangle = \int \mathrm{d}x \psi^*(x,t) g(-\mathrm{i}\hbar\partial/\partial x) \psi(x,t)$, where the differential operator acts on the function $\psi(x,t)$ which is on the right.

As the wave function is complex, one can always write it, without lack of generality, as:

$$\psi(x,t) = \sqrt{\varrho(x,t)} \mathrm{e}^{\mathrm{i}S(x,t)/\hbar} \tag{2.14}$$

where one recognizes the density ϱ defined in Eq. (2.13) and associated to the probability of presence. The phase $S(x,t)$ is related to the classical action, whose derivative provides the velocity information. This form of wave function is interesting as it allows us to understand in simple terms the meaning of the Schrödinger equation. Indeed when writing down the latter equation using ϱ and $S(x,t)$ one basically obtains the equations of fluid dynamics for a fluid of local density ϱ and local velocity field $v \propto \partial S(x,t)/\partial x$. The difference with classical fluid dynamics is that the equation involves a complementary term of purely quantum origin (directly proportional to $\hbar^2$) which modifies the local force field felt by the system. Still the analogy with fluid dynamics is extremely telling, retaining the essential wave character of a fluid.

2.4 QUANTIZATION AND QUANTUM NUMBERS

2.4.1 Confining potential and quantization

We have seen that the quantum energy of a *free* particle can take any value. In quantum terminology it means that it is not quantized. This results from the absence of a confining potential (and/or of interactions for a many particles system). To illustrate the point let us consider as a simple case a unidimensional (1D) box of size $2a$ with impenetrable walls. This is described by a potential which vanishes inside the box ($V(x) = 0$, for $-a < x < a$), being infinitely repulsive ($V(x) = +\infty$) otherwise. Because in the internal region the potential is strictly zero, the plane wave solution is still correct inside the box. In the region of infinite potential, the wave function can only be zero because the infinite repulsion hinders the possible presence of the particle. This means strictly zero probability of presence and consequently zero wave function. The major difference with the case without potential (see Section 2.3.1) is that one has to account for boundary conditions, namely the fact that the potential V jumps to infinity at $x = \pm a$. In order to preserve the interpretation of the wave function in terms of probability of presence, the wave function must be continuous at the boundary $x = \pm a$. This imposes a condition on the possible wave number k of the plane wave solution inside the potential well.

We can obtain the quantization conditions using an analogy with the stationary modes of the rope of Figure 2.1 discussed in Section 2.1.3. We showed there that stationary waves are possible only if the length of the rope is an integer multiple of half their wavelength. In the present case, as displayed in Figure 2.5, the analogous of the rope length is $2a$. The wavelength is given

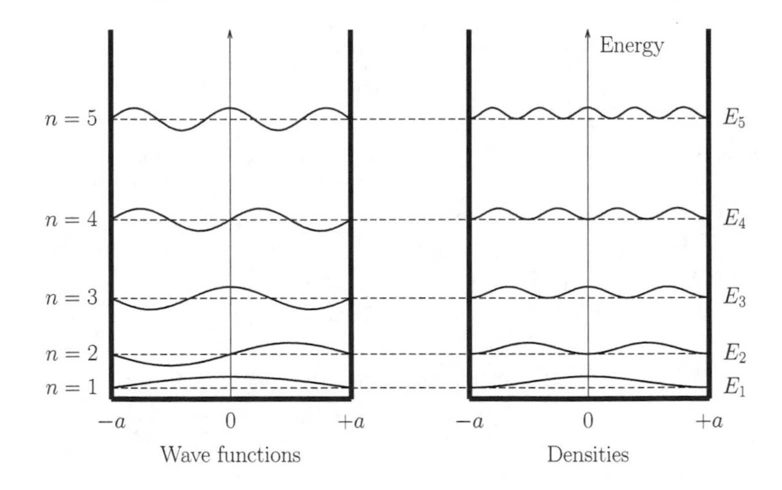

Figure 2.5: First five stationary wave functions (left) in an infinite 1D potential of width $2a$, and corresponding density (right). The curves have been upshifted at the value of the corresponding energy $E_n = n^2\hbar^2\pi^2/(8ma^2)$ to ease their visualization.

by the de Broglie relation $\lambda_B = h/p$ defined in Eq. (2.1), where we make the replacement $p = \hbar k$. The condition of having stationary waves reads then $k = n\pi/(2a)$, with n a non-negative integer (see Section 4.2.1 for more details). The energies $E = \hbar^2 k^2/(2m)$ of these states are therefore quantized

$E_n = n^2\hbar^2\pi^2/(8ma^2)$. Actually, the stationary solutions of the Schrödinger equation are linear combinations of the plane waves $\exp(\pm ikx)$, vanishing at $x = \pm a$. They may be written in turn as linear combinations of cos or sin functions with argument kx. Specifically, they take the form $\cos(n\pi x/(2a))$, for odd-n, and $\sin(n\pi x/(2a))$ for even-n.

The first stationary solutions of the infinite potential well are illustrated in Figure 2.5. They are ordered by increasing energies, starting at $n = 1$. Indeed, the value $n = 0$ has no physical interest, because it corresponds to a vanishing wave function. Notice that negative values of n only affect the sin form, changing it by an irrelevant sign. We shall therefore consider only positive n. The left panel provides the wave functions corresponding to the five lowest values of n, and the right panel gives the corresponding probabilities of presence. The latter shows that the particle is 'localized' at certain positions inside the potential well. This result on the quantization of wave numbers, apparently simple, is in fact essential for the forthcoming discussions and contains a lot of physics.

2.4.2 The general concept of quantum numbers

The first important point to realize is the fact that the boundary conditions imposed by a potential well have led to considerably restricting the number of acceptable possibilities for the stationary solutions. In terms of possible energies, the presence of an infinite potential well has reduced the infinity of plane wave solutions to a few possible energies $E = E_n = \hbar^2 k_n^2/(2m)$ with $k_n = n\pi/(2a)$ (n being an integer). This is the well known quantization of energy levels. In that particular case, the sequence of energy levels is $E_1 = \hbar^2\pi^2/(8ma^2)$, $E_2 = 4E_1$, $E_3 = 9E_1$, etc., which allows us to label these particular energy values by integers n (see Figure 2.5). We shall call n the energy quantum number. It suffices to fully specify the stationary levels.

Now imagine again an infinite potential well but in two dimensions, say x and y and with the same extension $2a$ along the two directions. The solution of the Schrödinger equation will proceed the same way as before but now involving the two spatial coordinates x and y. Quantization of energy levels again results from boundary conditions but now leads to possible energies of the form $E = E_{n,p} = \hbar^2(k_n^2 + k_p^2)/(2m)$ with $k_n = n\pi/(2a)$ and $k_p = p\pi/(2a)$. The two integers n and p are then the two quantum numbers which characterize the state. As in the 1D case, the stationary energies take simple values $E_{1,1} = 2E_1$, $E_{1,2} = E_{2,1} = 5E_1$, etc. The interesting point is that a given energy can be attained in several ways, for example with one quantum along x ($n = 1$) and two quanta along y (p=2) or vice versa. This phenomenon is quite general and is called the degeneracy of the level.

The phenomena described above, namely energy quantization and level degeneracy, are not specific to the infinite potential well we have used. It is in fact quite generic and appears for any potential. Let us come back for a moment to the 1D case. Even if the case of an infinite well is telling and provides analytical solutions, it constitutes a rather crude approximation of a confining potential in view of those stemming from interactions of a particle with other particles. Another very important potential is that of a harmonic oscillator. Indeed, any potential which possesses a minimum can be approximated by a harmonic oscillator in the vicinity of this minimum. This model potential also provides analytical solutions and is therefore very useful. We briefly describe here how it performs at the level of the quantization of energies. Consider a harmonic oscillator of frequency ω and denote by x the displacement of a particle of mass m with respect to the equilibrium position (the minimum of the potential). The stationary Schrödinger equation then reads:

$$\left[-\frac{\hbar^2}{2m}\frac{\mathrm{d}^2}{\mathrm{d}x^2}+\frac{1}{2}m\omega^2x^2\right]\psi(x)=E\psi(x) \quad . \tag{2.15}$$

This equation can be solved analytically. We simply give here the possible values of the energy E which are quantized and labeled by a non-negative integer n as:

$$E_n^{\mathrm{HO}}=\hbar\omega\left(n+\frac{1}{2}\right) \quad . \tag{2.16}$$

The case $n=0$ corresponds to the zero point energy $E_{\mathrm{kin}}^0=\hbar\omega/2$. In the general case (different from an infinite well or a harmonic oscillator), the expression of the accessible energies can only be obtained numerically.

We now proceed to the 3D case with the spherical shell harmonic oscillator. It can be seen as the product of three 1D harmonic oscillators, one in each spatial direction, and all with the same frequency ω. For a particle of mass m, it can be written as

$$V_{\mathrm{HO}}=\frac{1}{2}m\omega^2\mathbf{r}^2=\frac{1}{2}m\omega^2(x^2+y^2+z^2) \tag{2.17}$$

Each 1D harmonic oscillator has a quantized sequence of levels of energies $\varepsilon_n=\hbar\omega(n+1/2)$. At this point, no energy level is degenerate.

When combining the three 1-D oscillators we obtain again a simple spectrum whose levels can be written as

$$\begin{aligned} E_n^{\mathrm{HO}} &= E_{n_x}^{\mathrm{HO}}+E_{n_y}^{\mathrm{HO}}+E_{n_z}^{\mathrm{HO}} \\ &= \hbar\omega(n_x+n_y+n_z+3/2)=\hbar\omega(n+3/2) \end{aligned} \tag{2.18}$$

where $n=n_x+n_y+n_z$ fully characterizes each energy level. But the same energy can be reached in different ways, depending on which $n_{x,y,z}$ values are populated. This then leads to degeneracies. The ground state ($n=0$) can be filled with 1 particle only and is not degenerate. The second shell corresponds to $n=1$ which can only be attained by setting one of the three $n_{x,y,z}$ to 1, leading to a degeneracy equal to 3. The situation becomes a bit more complicated for the next shell $n=2$, the value 2 being attainable for two types of occupation: $n_x=2, n_y=n_z=0$ or any permutation of the indexes, or $n_x=$

$1, n_y = 1, n_z = 0$ and again any permutation of the indexes. This altogether makes 6 possibilities. And so on with higher values of n. The spherical or deformed harmonic oscillator model has been widely applied, sometimes with refinements, particularly to nuclei, as we will see in Section 7.2.1.

The quantization of energy levels is a well known basic feature of quantum mechanics. It is deeply rooted into the wave nature of the representation of a particle and also deeply connected to the existence of interactions. Indeed we have seen that a free particle may have any kinetic energy and is represented by a plane wave. As soon as interactions enter the game, either under the form of an external potential or as interactions between particles (whose result practically can often be represented by an 'external' field acting on *one* particle), quantization of energy levels occurs. This can as well be seen as an impact of boundary conditions on solutions of the Schrödinger equation. In the case of a realistic potential, the effect is not as dramatic as in the example of the infinite potential well we have considered above (where the potential basically played the role of boundary conditions, effectively), and wave functions have a certain extension 'beyond' the potential well (see the discussion on quantum tunneling in Section 2.5.1). But the basic mechanism analyzed in the infinite well remains qualitatively sound. Not surprisingly, we recover here the stationary wave solutions introduced in Section 2.1.3 as providing a basis of states on which to construct any solution of the Schrödinger equation. The ensemble of the thus identified energy levels is called the spectrum of the system. The states will be attributed quantum numbers following the mechanism outlined above, the quantum numbers providing a way to sort these states (whose numbers grow rapidly with energy, especially in a realistic 3D world). And we have seen that degeneracies occur, the effect growing in importance with energy.

2.4.3 Spin and the notion of quantum state

The quantization mechanism discussed above concerns energies and can be formally associated to the above introduced Hamiltonian, which is just one example of a quantity possibly evaluated to characterize a quantum state. In general, measurable quantities are quantized (with noticeable exception of the center of mass or position of a particle).

Another important example of quantized quantity is the angular momentum vector. In the quantum case, it covers two notions. The first one is the orbital angular momentum, which is just the quantum analog of the classical angular momentum that appears in any rotational motion which for a particle is written as $\mathbf{L} = \mathbf{r} \times \mathbf{p}$. In quantum mechanics, it is associated to the orbital quantum number usually denoted by the letter l, which appears in a non-trivial form as $\mathbf{L}^2 \to l(l+1)\hbar^2$. This number constitutes another characteristic of a quantum state, and can only take non-negative integer values 0, 1, 2 ... Since vector $\mathbf{L}$ possesses a spatial direction, one usually measures its projection along an arbitrary spatial direction (very often defined along the

z-axis). Quantum mechanics tells us that this projection is also quantized, and defines a new quantum number as $L_z \to m\hbar$. This third quantum number m can only take integer values between $-l$ and l. There are then $2l+1$ different possible values for m.

A quantum particle possesses another angular momentum, named its spin. It is a vector quantity $\mathbf{S}$ which has the dimension of an angular momentum, to which a spin quantum number s is associated in the same non-trivial form as for the orbital angular momentum, that is $\mathbf{S}^2 \to s(s+1)\hbar^2$. However, contrary to the orbital quantum number, the spin s can take non-negative values either integer (0, 1, 2 ...) or half-integer (1/2, 3/2, 5/2 ...). The projection of $\mathbf{S}$ along a spatial direction is also quantized, and defines a quantum number as $S_z \to m_s\hbar$, where m_s take $2s+1$ possible values between $-s$ and s in steps of one. A particular important example is the electron which is a particle of spin 1/2, with projections 1/2 and $-1/2$ which are usually referred to as spin-up and spin-down states.

Contrary to orbital angular momentum, spin has no classical analog. Sometimes it is compared to the classical equivalent to the rotation of a particle on itself. However, this image implies the particle has a spatial extension and/or a structure, and pushing the analogy to the limit, one can get equatorial velocities (assuming the particle to be a sphere) which are higher than the speed of light. One should rather consider spin as an intrinsic property of a particle, either structureless or composite, in the same way that it can possess a mass or an electrical charge. The spin of a particle is actually related to its magnetic moment and determines its behavior in a magnetic field. It is precisely that way that the spin was identified in the early 1920s by Stern and Gerlach.

Our former description of the quantum world in terms of wave functions as introduced in Section 2.3 does nowhere account for spin... and correlatively the simple form of the Schrödinger equation given in Section 2.3.1, that we here call the 'real-space Schrödinger equation', is unable to account for spin either. The problem comes from the fact that the representation of the wave function in real space, $\psi(x,t)$, cannot describe the spin degree of freedom because it is not a real-space quantity.

To account for spin, one has to take a broader view and characterize the system by a more general/abstract object called a *quantum state*, also called 'ket' and usually denoted by $|\psi\rangle$. This object integrates both the former real-space wave function representation and spin. The quantum state follows a generalized form of the real-space Schrödinger equation, simply called the Schrödinger equation, written in the abstract space of quantum states, the latter ones being abstract vectors. One can formally show that the quantum state $|\psi\rangle$ can be projected onto real space (in a similar sense as the geometrical projection of a vector along a spatial direction), producing the real-space wave function $\psi(x,t)$. If one proceeds the same way for the Schrödinger equation, one reduces it to the real-space version. One can either project the ket onto momentum space and get another kind of wave function, $\widetilde{\psi}(p,t)$, or even onto an abstract spin space, giving access to a spin wave function. $\psi(x,t)$ and $\widetilde{\psi}(p,t)$

are called respectively the real-space and the momentum-space representations of the quantum state $|\psi\rangle$. As for the spin, since it is intrinsically contained in the quantum state, it appears as a parameter of these representations. When considering situations involving spin, such as the coupling to a magnetic field for example, one explicitly enters by hand the spin value into the equation when necessary. In many situations one has to deal with the total angular momentum of a particle, which is the sum of the angular momentum and spin $\mathbf{J} = \mathbf{L} + \mathbf{S}$, to which a quantum number J is associated as $\mathbf{J}^2 \rightarrow J(J+1)\hbar^2$. The quantum rules to determine the possible values J states that it can take any value between the minimum $J_{\text{min}} = |L-S|$ and the maximun $J_{\text{max}} = L+S$ in steps of one.

This is not the place to further elaborate on these formal aspects. We simply close the discussion by mentioning that a true account of spin is actually beyond the natural scope of the Schrödinger equation. Instead, one has to refer to a relativistic approach, as proposed by Dirac in the late 1920s, to properly integrate spin in a global quantum description. This being far beyond the objectives of this book, for the sake of simplicity, let us admit that we can characterize the quantum state of a system by its (real-space) wave function and its spin.

2.4.4 From one particle to many particles

The hydrogen atom

Up to now, we have restricted ourselves to 1D situations, for the sake of simplicity. We discuss here a more realistic 3D case, namely the hydrogen atom. It will allow us to understand more complex atoms and in particular the structure of the periodic table of elements. One has to consider the Schrödinger equation for the relative motion between an electron and a proton, which interact through the Coulomb force. Since the ratio of the electron mass over the proton mass $m_e/m_p \sim 1/2000$, we can consider the proton as fixed and choose it as the origin of the spatial coordinates for the electron. As we are dealing with a central interaction, depending only on the relative electron-proton distance r, it is convenient to use spherical coordinates (r, θ, ϕ) and take advantage of the associated spherical symmetry. Mathematically this means that the wave function can be factorized in two terms, one depending on r and the other on the angular variables θ and ϕ. The second term corresponds to the so-called spherical harmonics $Y_{l,m}(\theta, \phi)$ which depends on the quantum numbers l, m associated to the orbital angular momentum. In fact, the general wave function of a particle in a central potential can be written as $\psi(r, \theta, \phi) = R_{n,l}(r)Y_{l,m}(\theta, \phi)$. For the hydrogen atom, the radial function $R_{n,l}(r)$ is solved analytically, and is given in any basic quantum mechanics textbook. We simply mention here that the quantum state is characterized with the so-called principal quantum number n, and the two orbital quantum numbers l, m. The energy levels are given by $E_n = -13.6/n^2$ eV, with $n = 1, 2\ldots$ and $l = 0, 1\ldots n-1$. The principal quantum number n thus defines 'major shells', while l defines 'subshells'. The energy is independent of the quantum number m and also of the electron spin, resulting in a degeneracy $2(2l+1)$.

The periodic table of chemical elements

This particular resolution of the Schrödinger equation actually exhibits a broader scope than the mere hydrogen atom. Indeed, the nomenclature provided by the quantum numbers (n, l, m) can be applied to classify atoms in the famous Mendeleev or periodic table of chemical elements displayed in Appendix B. Coupled to the Pauli exclusion principle which states that, for electrons of spin 1/2, an electronic state can be occupied by at most two electrons (one with spin up and one with spin down), we can deduce the maximal number of electrons in each electronic shell of an atom, that is its degeneracy, given by $2(2l+1)$.

The principal quantum number n defines the shell. The case $n = 1$ corresponds to the first row of the periodic table. In this case, $l = m = 0$. We can then fill this shell with 1 electron, and get the hydrogen atom, or with 2 electrons, forming the helium atom. The case $n = 2$ corresponds to the second row of the periodic table. We have either $l = m = 0$, filled with 1 electron (lithium) or 2 electrons (beryllium), or $l = 1$ and $m = -1, 0, 1$, that is 3 levels with the same energy and filled up with a maximum of 6 electrons (giving the atoms from bore to neon). We can go on with $n = 3, 4$, etc. and explain the shell filling of the successive atoms appearing in the periodic table, according to the Aufbau principle which states that electrons should fill the lowest available energy levels before filling levels of higher energy. One also has to consider the Hund's rule which states that if, at a given energy level, that is for a given couple (n, l), there are several different states (given by the different m), electrons first fill unoccupied states and only after occupying all the $2l+1$ states with one electron, these states are filled again with a second electron (which has an opposite spin compared to that of the first electron, to comply with Pauli exclusion principle).

The periodic table displays indeed periodicity in the properties of elements. The elements in a given column show remarkable similarities in chemical properties, which eventually have been understood in terms of the shell structure of the atoms. The last column corresponds to elements with completely filled shells. This happens when the atomic number reaches the values 2, 10, 18, 36, 54 or 86 (and presumably 118 also, which is the heaviest artificial element), which are the inert gases. They are chemically unreactive and therefore show little tendency to combine with other atoms. In contrast, the atoms in the neighbor columns, with one less (halogens) or one more (alkali metals) electron are highly reactive and readily combine with other atoms.

There are a number of other properties which exhibit also periodicities related to atomic shell structure. In Figure 2.6 are plotted three of them. The upper panel shows the ionization potential as a function of the atomic number Z. This is the energy required to remove one electron from a given atom, and is thus a microscopic property. It can be seen that the removal of an electron

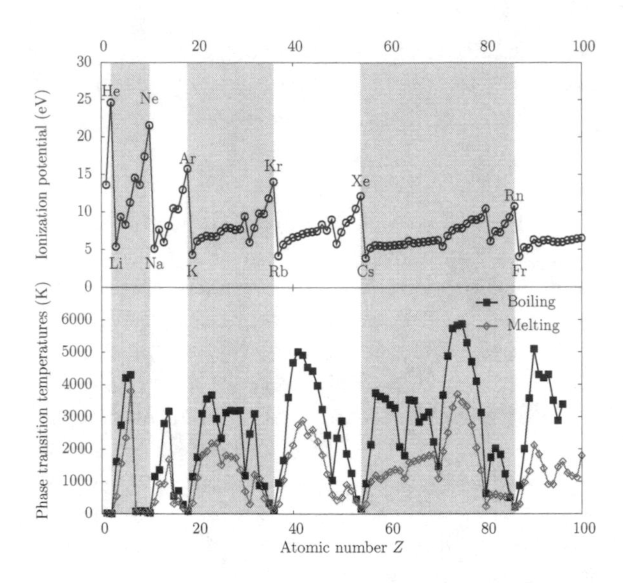

Figure 2.6: Microscopic and macroscopic periodicities of elements, as a function of the atomic number. Upper panel: ionization potential (in eV) of atoms. Lower panel: Boiling and melting temperatures (in K) of bulk elements. Data extracted from https://www.webelements.com/periodicity/

from an inert gas atom, with the electronic shells completely filled, requires the highest energy in each row of the periodic table. In contrast, the removal of an electron from an alkali atom requires the lowest energy because they have a less bound electron out of closed shells.

Interestingly, shell periodicities also manifest in macroscopic or bulk properties. The understanding of these properties requires the knowledge of the (microscopic) interatomic interactions. In Section 5.1.1, we will come back to the periodic table and show how the occupation of electronic levels in atoms helps to understand how atoms can bind together. Here, we show in the lower panel of Figure 2.6 the boiling and the melting points, that is the temperatures at which takes place a transition from liquid to gas or from solid to liquid, respectively. Such a phase transition means that interatomic bounds are broken with the help of thermal agitation (see Section 1.3.2). And the weaker the interatomic interactions, the easier to break them, and the lower the phase transition temperature. One clearly sees that such transitions in the inert gases occur only at relatively low temperatures, and always at the lowest temperatures in each row of the periodic table. On the contrary, the highest phase transition temperatures occur within an electronic shell, revealing a relative strong interatomic interaction in these cases. We therefore observe a direct connection between the filling of electronic shells in atoms and the strength of interactions in a solid or liquid phase. Incidentally, the chemical elements in the last column of the periodic table are called 'rare gases' because they mostly appear in the gaseous state, while elements with a partial filled electronic shell rather appear in the liquid or the solid state at ambient temperature.

The multiparticle case

We have seen that the quantum numbers of the hydrogen atom allows one to understand the electronic configuration of many-electron atoms. This however does not tell us what the wave function of an electron in an atom is. Actually, apart from the hydrogen atom, it is impossible to compute analytically these wave functions, even in the helium atom. Note however that there exists a good approximation in the case of ^{4}He, as we will see later in Section 2.5.4.

In the case of a N-particle system, the wave function depends on the coordinates of all particles $\Psi = \Psi(x_1, x_2, ..., x_N, t)$. If we normalize it such that $\int \mathrm{d}x_1 \ldots \mathrm{d}x_N \left|\Psi(x_1, \ldots, x_N, t)\right|^2 = 1$, then the modulus of Ψ provides a density of probability of presence, as before. But in that case, it means the probability to find at a given time t a particle at position x_1, another one at position x_2 and so on. This represents highly detailed information on the system. It is often sufficient, and certainly more intuitive, to consider the probability to find a single particle at a given point, which will then be the generalization of the probability of presence defined for one particle in Eq. (2.12). This is in fact nothing but the density of matter at this given position and it is readily attainable from the many-body wave function by integrating the modulus of Ψ over $N-1$ particles, as:

$$\varrho(x,t) = N \int \mathrm{d}x_2 \mathrm{d}x_3 \ldots \mathrm{d}x_N |\Psi(x, x_2, x_3, ..., x_N, t)|^2 \quad . \tag{2.19}$$

The forefactor N is the number of ways of choosing 1 particle out of N. $\varrho(x,t)$ gives the probability per unit volume of finding a particle at position x, irrespective of the positions of the $N-1$ other particles. We naturally recover $\int \mathrm{d}x \varrho(x,t) = N$ since we are in presence of N particles.

One can easily suspect that the N-particle case is much more involved than that with one particle. A major conceptual difficulty lies in the interpretation of the N-particle wave function. Indeed, the system is composed of N particles which all have their own 1-particle wave function. And if one can 'visualize' what a 1-particle wave function does represent, it is harder to envision what the N-particle wave function contains. This is in fact a very deep question which is reflected by a difficult formal and mathematical problem to represent $\Psi(x_1, x_2, x_3, \ldots, x_N, t)$ in terms of the individual $\psi(x_1,t), \psi(x_2,t), \psi(x_3,t), \ldots, \psi(x_N,t)$. One very often uses the spherical harmonics mentioned above as a starting point for these single-particle wave functions, in the case of an ensemble of fermions.

But this is by no means the end of the story. A short presentation of the various theoretical approaches of the many-body problem can be found in Section 4.4. Here we simply show as an illustration the wave functions of the

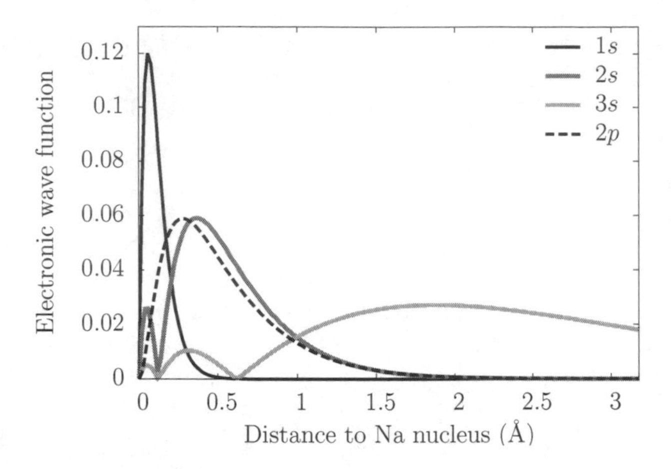

Figure 2.7: Electronic wave functions averaged over the angles for the $1s$, $2s$, $2p$ and $3s$ states, as a function of the distance to the central nucleus, for the sodium atom.

sodium atom averaged over the angles in Figure 2.7. Notice that the higher the angular momentum, the more extended the wave function and the larger the number of nodes (positions at which the wave function vanishes).

Let us finally mention that Ψ is not the simple product of the single-particle ψ's because the particles do interact with each other. This implies in particular that the probability of finding one particle somewhere, while knowing that the other ones lie somewhere else, is a nontrivial quantity. Such probabilities are indeed not simple products of the individual probabilities of presence of each particle. Particles are said to be *correlated.* These correlations can stem from interactions. But they already exist for non-interacting identical particles because of quantal statistics. We shall briefly discuss this point in Section 2.5.4, while more details are given in Section 4.3.2.

2.5 SOME UNEXPECTED QUANTUM FEATURES

2.5.1 Quantum tunneling

The wave-particle duality allows one to explain a purely quantal effect, namely the quantum tunneling of a particle. Consider a particle of energy E traveling towards a potential barrier of height V_0. Classically, the particle can surmount the barrier if and only if $E \geq V_0$. In other words, if $E < V_0$, the particle bounces back when it has hit the barrier. In quantum mechanics, we describe particles by wave functions which are, by nature, defined everywhere in space. By solving the Schrödinger equation with this potential barrier, one can therefore find a non-vanishing probability, even if very small, for the particle to be on the other side of the barrier.

This is illustrated in the left panel of Figure 2.8 in the case of a rectangular barrier of width a. In the case of a plane wave ψ_{inc} describing an incident

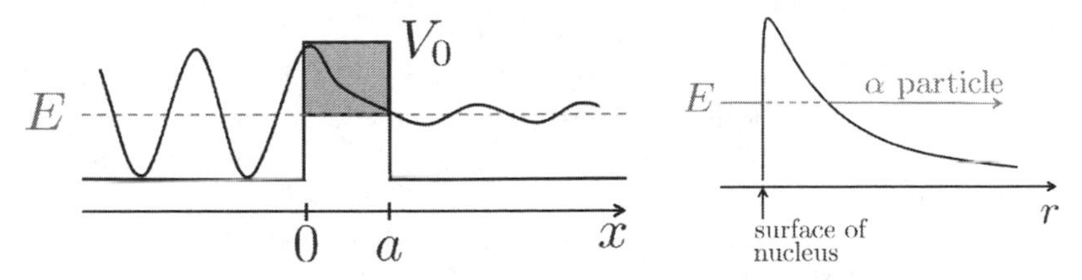

Figure 2.8: Left: Illustration of the quantum tunneling of a wave function of energy E through a potential rectangular barrier of height V_0 and width a. The shaded area is the region forbidden by classical mechanics. Right: case of an α particle tunneling through the potential barrier of a radioactive nucleus.

particle of mass m and energy $E < V_0$, the problem can be solved analytically. Before the barrier ($x \leq 0$), one has an incident wave e^{ik_1x} plus a reflected wave Ae^{-ik_1x}, where the wave vector is $k_1 = \sqrt{2mE}/\hbar$. Inside the barrier ($0 \leq x \leq a$), the solution reads $Be^{-k_2x} + Ce^{k_2x}$, with $k_2 = \sqrt{2m(V_0 - E)}/\hbar$. Finally, beyond the barrier ($x \geq 0$), the solution has the form of a plane wave De^{ik_1x}, analogous to the incident one. The constants A, B, C, and D are given by continuity conditions of the wave functions and their first derivative at $x = 0$ and $x = a$. We will not explicitly determine these constants.

We just note that inside the barrier, the wave function falls exponentially. Moreover it falls all the more quickly when the difference $V_0 - E$ is large. In other words, the higher the barrier, the less probably the particle can tunnel. One also notices that the amplitude of the transmitted wave function ψ_{out} depends on the barrier width a: the larger the barrier, the smaller this amplitude. Such a situation is impossible classically. This is emphasized in the figure by the shaded area above the horizontal line at E inside the barrier. In quantum mechanics, one says that the particle *tunnels* through the barrier.

Quantum tunneling is not a rare phenomenon. For instance, it is essential in the emission of α particles (that is helium nuclei) in radioactive decay. In an α-emitter nucleus, an α particle is trapped in the confining potential as depicted in the right panel of Figure 2.8. The shape of this potential is due on the one hand to the nuclear attraction from the other nucleons, and on the other hand from the Coulomb repulsion between protons. An α particle can escape by tunneling through the barrier and quantum calculations allow one to estimate the half-life of radionucleides with high precision.

In the core of our Sun, quantum tunneling is also crucial to allow fusion of hydrogen atoms. This thermonuclear reaction is needed to counterweight the gravitational collapse and therefore to stabilize the star (see Section 8.1.2 for more details on the stellar evolution). But hydrogen nuclei can fuse only if the Coulomb barrier between the two protons, of about 0.1 MeV, is overcome. This would correspond to a temperature of 1.5×10^9 K. However, at the core of the Sun, the temperature is 'only' 15×10^6 K. How can protons fuse in this context? By virtue of quantum tunneling. Even if the probability of tunneling is extremely small in that case (about 10^{-28}!), it suffices to allow the thermonuclear fusion and to stabilize our Sun.

Let us finally mention a widely used application of quantum tunneling, that is the Scanning Tunneling Microscope (STM). We have seen that the amplitude of the transmitted wave function through a barrier depends on the width of the barrier. The barrier involved in a STM is the space between the STM's tip and the metal surface to be scanned. By measuring the current of electrons that tunnel from the tip to the surface, one can deduce the distance between the tip and the surface. One can hence image the surface by positioning the tip at various places above the surface. Current STM have a resolution of about 1 % of atomic diameter.

2.5.2 Zero-point energy

The topic of this book is the description of quantum liquids and we shall see that liquids may indeed appear in unexpected situations, precisely because of subtle quantum effects. The basic mechanism responsible for the appearance of such a liquid behavior is called 'zero point energy'. In order to understand it in simple terms, let us come back to our favorite test case of *one* particle inside a 1D box of size $2a$ (see Section 2.4.1).

We have seen above that the energy levels of this particle were quantized with the spectrum $E_1 = \hbar^2\pi^2/(8ma^2)$, $E_2 = 4E_1$, $E_3 = 9E_1$, etc. Most interesting is the fact that there exists a minimum energy E_1 and that this minimum energy is not zero, the value which would correspond to the bottom of the potential well. This means that even in its lowest energy state, which corresponds to the *ground state*, the particle possesses a non-vanishing energy. This energy is called zero point energy and we shall denote it by E_{kin}^0. It can be seen as a kinetic energy in the sense that it corresponds to the minimum possible potential energy reachable by the particle. One also talks about 'zero point motion', in relation with this zero point (kinetic) energy. But mind that the value of E_{kin}^0 depends on the actual shape of the potential well and is then specific to it. It thus integrates both kinetic and potential effects, even in the case of an infinite well (in the latter case through the size of the well $2a$). Figure 2.9 precisely illustrates this effect in the case of the infinite potential well. We compare spectra and associated zero point energies of two such infinite potential wells, with different widths $2a$. As level energies scale with $1/a^2$, the wide well naturally leads to a smaller zero point energy and a smaller energy spacing than in the narrower one.

The appearance of a zero point energy is at complete variance with the classical picture: a classical particle in the present setup would have strictly zero energy as lowest energy (classical equivalent of the ground state). That would classically correspond to zero potential energy *and* zero kinetic energy ($E_{\text{kin}}^0 = 0$). The quantum non-vanishing ground state energy precisely stems

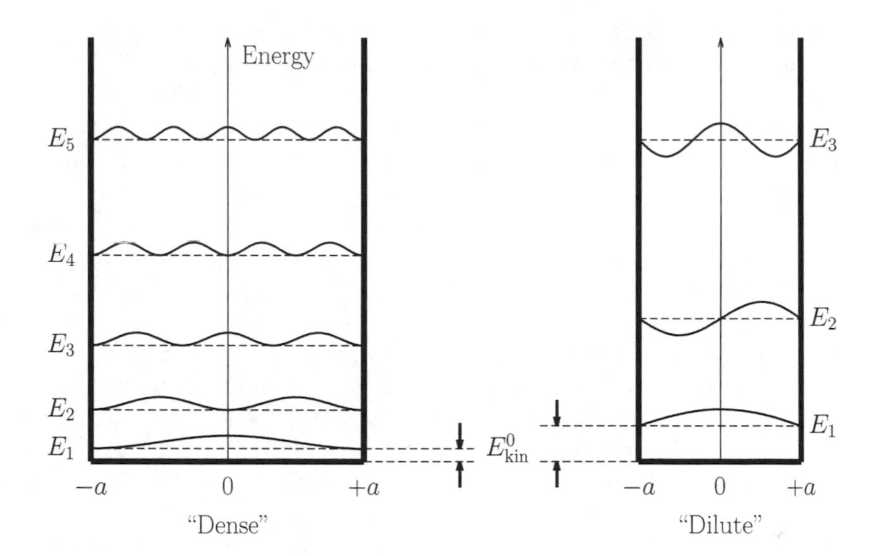

Figure 2.9: Stationary wave functions and corresponding energies E_n in an infinite potential well of width $2a$, for two different values of a. The zero point energy is indicated by E^0_{kin}.

from the wave behavior of the particle, whose space oscillations are bound to match box size, as discussed in Section 2.4.

This result, once digested, may seem surprising. Indeed it tells us that a confined quantum particle can never stay at rest! It is interesting here to come back to the notion of thermodynamic temperature introduced in Chapter 1 and more developed in Chapter 3 in connection with statistical physics. Indeed, temperature is intrinsically linked to the kinetic energy of the constituents of a system. And we have seen that when lowering the temperature *classical* particles tend to freeze so that at absolute zero temperature no motion becomes classically possible. But in the above quantum case we have just discussed, there is no (classical) temperature involved. We are, in classical terms, strictly at zero 'classical' temperature. And although we are considering only one particle and not an ensemble thereof, the situation is in fact quite similar. Indeed our particle is confined by the potential but we can as well envision this confining potential as the effect of interactions of the particle with its neighboring particles. These interactions can be translated in terms of potential energy. Therefore, they grossly 'allocate' a restricted volume to each constituent of an ensemble of particles. With everyday words we are in fact just discussing the notion of density ϱ (each particle occupies an elementary volume $v = 1/\varrho$). Again Figure 2.9 illustrates the point by comparing two systems of virtually different densities, a dilute one on the left and a denser one on the right.

We are then bound to admit that the strict rest does not exist in a quantum world, because of zero point motion. The existence of zero point motion has a further important conceptual consequence. It provides a new limit for which quantum physics becomes compulsory, beyond for example the criteria given by de Broglie wavelength or Heisenberg relation (see Section 2.2.3). We shall see in Chapter 3 that the temperature of a system of particles in equilibrium with an external environment provides a direct measure of the average kinetic energy of particles and vice versa. It is thus interesting to rephrase zero point motion in terms of a '1quantum' temperature T_Q which we shall define by assimilating E^0_{kin} to a thermal energy as $E^0_{\text{kin}} = k_{\text{B}}T_Q$. This quantum temperature therefore provides in a simple manner the temperature below which properties of a system are bound to be treated with quantum mechanics. Even more, because the existence of a non-vanishing quantum temperature leads to the existence of a minimal kinetic energy in a system, it *a priori* opens the possibility for a system to exist as a liquid, while classical temperature would not allow it. We are here at the core of our subject, at the true core of the mechanism which explains the very existence of quantum fluids.

It is interesting to evaluate the quantum temperature in a slightly more general way. Let us consider an ensemble of identical particles characterized by a density ϱ. As mentioned above, one can immediately relate ϱ to the volume v associated to each particle in the system. If we assume, for the sake of simplicity, that each particle occupies a cube of length $2a$, to comply with our interpretation of an infinite box potential, we can replace v by $8a^3$ and directly relate the density to a as $\varrho = 1/(8a^3)$ so that $a = 1/(8\varrho)^{1/3}$. Zero point motion can then be expressed in terms of the density. The obtained relation is very general if one omits geometrical factors (the cube) and can be summarized as

$$E^0_{\text{kin}} \propto \frac{\hbar^2}{2m}\varrho^{2/3}. \tag{2.20}$$

This relation has the great interest to relate zero point motion to the density of the system, which is a macroscopic characteristic. We shall use it at several places throughout this book.

A word of caution is in place here. In the above argument, we have simply attached sort of an 'exclusion' volume to each particle. The reasoning is a bit oversimplified. Indeed one should specify what covers the term 'identical' used to characterize the particles. Placing two identical quantum particles in the vicinity of each other imposes strong constraints (beyond possible interaction effects between the two particles) on the wave functions. This has to be analyzed in detail as we shall see that it leads to the definition of two classes of particles which behave differently when put together. It may even lead to introducing, in some cases, a new energy, the Fermi energy, characteristic of the system and which constitutes, to some extent, a complement of the zero point energy. More precisely, while the zero point energy provides a lower bound to energies, we shall see in the next chapter that the Fermi energy, in turn, provides an upper bound (see Section 3.4). Obviously, such an extra constraint on the nature of identical particles should heavily impact the notion of quantum fluid that is presently taking shape. We thus need to analyze this question in some detail.

2.5.3 Back to spin: bosons and fermions

As superficially outlined in the preceding section the notion of identical particles deserves some deeper comments in a quantum context. It is the goal of this section to address this question. We shall see in particular that depending on their spin, quantum particles do behave differently when put together. We shall identify two classes of particles depending on their spin and we shall see that assemblies of such particles exhibit specific behavior, even when no interaction is involved. Before considering this question let us come back to the crucial question of spin.

We have seen in Section 2.4.3 that spin is an intrinsic property of a particle, similar for example to its charge. And we have seen that spins can take either integer or half integer values (in $\hbar$ unit). Just as charge allows us to classify particles in neutral and (positively or negatively) charged particles, one can define two classes of particles, depending on their spin: the ones of integer spin are called bosons, the ones with half integer spin are called fermions. We shall see that bosons and fermions behave differently when put together.

Before discussing ensembles of fermions and bosons let us briefly analyze the nature of particles we shall consider in this book. Except electrons, which are elementary particles of spin 1/2, all the other particles discussed in this book are composite systems. Their spin then results from the composition of the spins of their constituents. Adding rigorously spins (or more generally angular momenta) in quantum mechanics is a bit technical. It is enough for our purpose to remember that the total spin is obtained by adding or subtracting the spins of each constituent, following some well defined rules. Practically this means that a system formed of bosons will always remain a boson. But a system of fermions may either be a fermion or a boson (adding or subtracting half integers lead either to integer or half integer values, depending on the even or odd number of values).

To illustrate the point let us consider a few examples of importance in forthcoming discussions. Let us start with neutrons and protons, themselves resulting from the association of 3 quarks. Quarks have all spin 1/2 and neutrons and protons as well. Associating neutrons and protons then leads to nuclei of various total spins, depending on the structure of each nucleus. We can recall here the historically interesting case of the ^{14}N nucleus whose structure was highly debated in the early years of the 20th century. Before the identification of the neutron, the structure of the atomic nucleus was considered as possibly being obtained by associating protons and electrons. In the case of ^{14}N, the nucleus would be constituted of 14 protons and 7 electrons, leading to an acceptable mass of about 14 proton mass m_p. But in this model, the nucleus would be formed of $14 + 7 = 21$ half-spin fermions and would thus possess a half integer spin. This was in full contradiction to nuclear spin measurements of that time which attributed (correctly) an integer spin to ^{14}N. The missing piece was of course the neutron which when introduced leads to a ^{14}N made of 7 protons and 7 neutrons, hence an ensemble of 14 fermions with therefore an integer spin.

The case of helium atoms also deserves some special comment. Indeed, there are two electrons in these atoms. However, there are two isotopes of helium, namely ^{4}He with 2 protons and 2 neutrons, and ^{3}He with only one neutron. There are then 6 half-spin fermions in ^{4}He atoms, which are bosons, and only 5 in ^{3}He atoms, which are fermions. Helium-4 atoms will be the single bosonic systems that we shall consider in the following. Still, as we shall see below, there will be some situations in which bosonic species will also appear and play a crucial role. The most important cases concern electrons and helium-3 atoms. Electrons and helium-3 atoms are spin-1/2 fermions but we shall be led to consider pairs thereof, in order to understand superconductivity in electronic systems and superfluidity of a liquid of helium-3 (see Chapter 9). In both cases, pairs of fermions leads to boson composites which exist under well defined temperature conditions.

2.5.4 Identical and indistinguishable

We are now in a better position to specify what we call identical particles. Particles will be called identical when they possess the same intrinsic physical characteristics, namely mass, charge, spin, the latter characteristics being a fully quantal one. This may not look so different from what we would define classically (up to the spin). But the quantum nature of identical particles actually adds a major feature to the picture, the fact that identical particles are indistinguishable. Let us take a simple example to illustrate the point by considering the collision along a horizontal direction of two particles, labeled '1' and '2', facing each other. After the collision, particle 1 may be deflected upwards or downwards and vice versa for particle 2. In a classical picture of such a collision process, say between two identical balls, the very existence of sharply definable trajectories permits us to distinguish between identical particles by simply following their trajectories. One can put an imaginary flag on particle 1 and explicitly follow its trajectory, so that one will know exactly where particle 1 goes.

Indistinguishability of two particles

In the quantal world, this simple notion disappears. As long as both particles are far enough from each other, the classical identification remains valid and one can decide which one comes from the left and which one from the right. On the contrary, when the two particles come closer, their undulatory character becomes dominant and what makes sense is the wave function of the pair of particles, not the individual wave functions. These wave functions superimpose to form that of the pair of particles and it becomes impossible to distinguish between the two. After the collision, one observes that one of the particles does leave upwards and the other one downwards, but it is impossible to decide which is which: they are *indistinguishable.* One is thus bound to consider both possible cases, as illustrated in Figure 2.10.

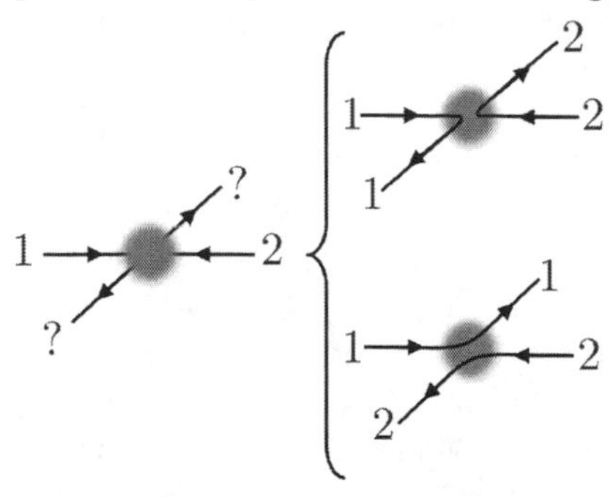

Figure 2.10: Indistinguishability of two identical particles in a collision.

In other words, if we exchange the role of the two particles, one should not see any experimental difference. The phenomenon is quite general and characteristic of any interaction process (labeled 'collision' in our simple test example) and one usually calls the two contributions the *direct* and the *exchange* contributions. Both have to be properly accounted for in order to describe an interaction process. This subtle effect is a clear sign of the wave behavior of the particles.

In quantum mechanical terms, the indistinguishable character of particles means that the wave function obtained by replacing one particle by the other should deliver the same physical values in a measurement. In other words, it can differ from the original wave function $\Psi(x_1, x_2)$ only by an irrelevant (because non-measurable in an experiment) phase $e^{\mathrm{i}\theta}$, that is $\Psi(x_2, x_1) = e^{\mathrm{i}\theta}\Psi(x_1, x_2)$. Now let us imagine that we exchange again particles 1 and 2. The newly obtained wave function will then differ from the second one by the phase $e^{\mathrm{i}\theta}$, and from the original one by the phase $e^{2\mathrm{i}\theta}$. However, with two successive exchanges, one should recover the original wave function: $\Psi(x_1, x_2) = e^{\mathrm{i}\theta}\Psi(x_2, x_1) = e^{2\mathrm{i}\theta}\Psi(x_1, x_2)$. This means that $e^{\mathrm{i}\theta} = \pm 1$. We thus have two cases: the wave function is either symmetric ($e^{\mathrm{i}\theta} = 1$) or antisymmetric ($e^{\mathrm{i}\theta} = -1$) with respect to the exchange of the two particles. The symmetric case corresponds to bosons, the antisymmetric case to fermions.

The wave function of a pair of identical non-interacting particles

Let us be more specific here and construct the wave function of the two-particle system from the single-particle wave functions. We consider that particle 1 is in a given quantum state described by a generic quantum number n. For the time being, we include the spin of the particle in this quantum number. We write the wave function of particle 1 as $\psi_n(x_1)$. Similarly, particle 2 is in state m (possibly different from n) and at position x_2. We write its wave function as $\psi_m(x_2)$. For a non-interacting system, the simplest ansatz for the 2-body wave function would be $\Psi(x_1, x_2) = \psi_n(x_1)\psi_m(x_2)$. However, $\left|\Psi(x_1, x_2)\right|^2$, which provides the probability of presence (see Section 2.4.4), is obviously not invariant under the exchange of 1 and 2. From the two single-particle wave functions, there are actually two different cases to be considered:

$$\Psi_{\mathrm{sym}}(x_1, x_2) = \psi_n(x_1)\psi_m(x_2) + \psi_n(x_2)\psi_m(x_1) \tag{2.21}$$

$$\Psi_{\mathrm{anti}}(x_1, x_2) = \psi_n(x_1)\psi_m(x_2) - \psi_n(x_2)\psi_m(x_1) \tag{2.22}$$

Note that these wave functions are not normalized. Now, the modulus square of these 2-particle wave functions are indeed invariant under the exchange of the particles. In addition, $\Psi_{\rm sym}$ is also invariant, that is, it is symmetric: it thus corresponds to a pair of identical independent bosons. On the contrary, a minus sign appears in $\Psi_{\rm anti}$ after one exchange: it is antisymmetric and thus describes a pair of identical non-interacting fermions.

More generally, there are some situations in which the space part of the wave function factorizes out from the spin part. More precisely, the 2-particle wave function, denoted by $\widetilde{\Psi}(1,2)$ to distinguish it from the previous notation $\Psi(x_1, x_2)$, is written as $\widetilde{\Psi}(1,2) = \Psi(x_1, x_2)\,\chi(1,2)$ where $\Psi(x_1, x_2)$ bears the spatial dependence only and $\chi(1,2)$ that from the spin of particles 1 and 2. In the case of two fermions, to have $\widetilde{\Psi}(1,2)$ antisymmetric under the permutation $1 \leftrightarrow 2$, one can either impose Ψ to be antisymmetric and χ symmetric, or the reverse. We also mention that $\Psi_{\rm anti}$ given in Eq. (2.22) is not constructed and considered as an abstract or formal object in pratice. Indeed, it is actually at first order a good (and simple) approximation for the wave function of the 2 electrons in ^{4}He.

It is finally worth mentioning that the consequence of exchange of pairs of particles on the wave function of the system turns out to be general: any multiparticle wave function made out of identical particles is symmetric with respect to the exchange of any pair of particles if the particles are bosons, or antisymmetric if the particles are fermions (see Section 4.3.1 for an example of the N-body wave function).

2.5.5 Correlated and independent particles

We thus have seen that, even if no interaction is present, or rather if the interaction is negligible, the nature of particles (bosons or fermions) immediately shows up when one considers the pair. The effect is deeply connected to the way two particles 'behave' with respect to each other. In other words, the question that we are addressing is whether the presence of a fermion/boson somewhere in space affects the presence of another fermion/boson somewhere else, and this without interaction effects. We have seen in Section 1.3.3 that the pair correlation function $g(r)$ brings useful information on the degree of correlation between pairs of particles. We have also related the notion of correlation with that of interaction: maxima and minima of g at given values of r reflect the range of the interactions between particles constituting the system (see Figure 1.6). Here, we do not consider any interaction. One is then tempted to believe that particles cannot be correlated without any interaction. However, in quantum physics, independent particles do not mean uncorrelated. We will see below that actually independent particles can be *correlated*. This depends on the bosonic or fermionic nature of the particles.

To compare the fermionic case with that of bosons, the natural system to test such effects is helium, which is known to be extremely inert so that interactions between helium atoms can safely be neglected as soon as the system is sufficiently dilute (see Chapter 6). Moreover, as seen in Section 2.5.3, helium can exist as a fermion (^{3}He) or a boson (^{4}He), so that the same experimental setup allows us to directly compare boson and fermion gases. The result of such an experiment is displayed in Figure 2.11 and exhibits an interesting bunching/antibunching effect at short separation. In the case of fermions (^{3}He), $g(r)$

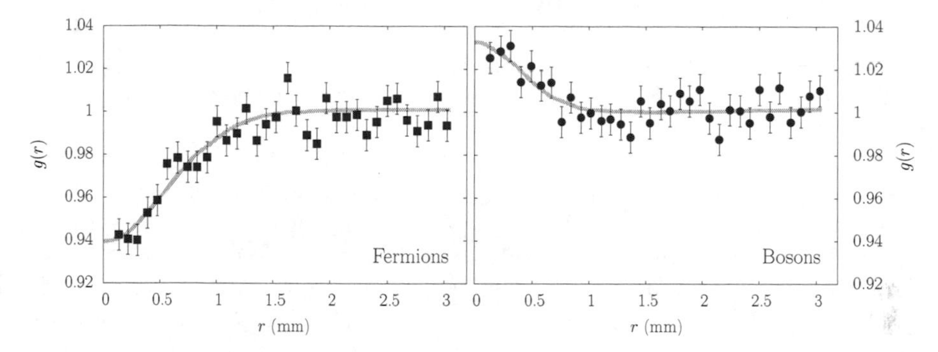

Figure 2.11: Antibunching of fermions (left) and bunching of bosons (right) observed in the values of the pair correlation function $g(r)$ at short distances, lower than 1 or greater than 1. The full lines are here to guide the eyes. Adapted with permission from W. Vassen and C. I. Westbrook, from T. Jeltes, et al., Nature 445 (2007) 402.

drops below 1 at short distance, reflecting the antibunching of fermions: there is a reduced probability (as compared to the uncorrelated classical case) of finding two fermions close to each other. In other words, two fermions have the tendency to 'repel' each other and this is the source of the Fermi pressure (see Section 3.4.1), essential to understand the stability of some classes of star, that is white dwarfs and neutron stars discussed in Chapter 8. But mind that the term 'repel' is not associated to an interaction. The effect is purely due to the fermionic nature of the particles. Two independent fermions, that is non-interacting, are *always correlated* through the exchange transformation. And actually, this constitutes the minimal amount of correlation in a gas of fermions. It is also worth noting that we have focused the fermionic antibunching on spatial effects (real space in quantum mechanics terms), ignoring the spin of each individual fermion. Actually, the antibunching effect applies to all degrees of freedom. If two fermions have opposite spins, they perfectly respect the antibunching effect (in terms of spin) and can then be located and remain at the same space point. This means that they can have the same representation in real space, and this without violating Pauli exclusion principle which is taken care of by the spin.

The effect is exactly the reverse for bosons which tend to 'attract' each other (or gather rather than separate). This creates the bunching effect and a conditional probability larger than 1 at short distance. This has also the counterintuitive consequence on the pressure of a boson gas: the latter is independent of the density. We will discuss this case in more detail in Section 3.3.3.

It is essential to remind readers here that these effects have nothing to do with a possible interaction (which in that case is safely negligible). The observed correlation effect is strictly due to the quantal nature of the pair of particles. In an idealized theoretical case, the correlation effect would lead to full exclusion for fermions, that is $g(r) \to 0$, as $r \to 0$. In contrast, full bunching for bosons implies $g(r) \to 2$, as $r \to 0$. In Figure 2.11, the amplitude of the observed effect looks limited and we are far from reaching the limiting values above. But this is due to the experimental difficulty in performing such a detailed analysis.

It is also important to stress the dependence of this bunching/antibunching effect on the relative distance between two particles. Although the correlation effect induced by quantum statistics on particles is global (symmetry/antisymmetry of the total wave function of the system with respect to exchange of 2 particles), it only extends over a finite range in practice. In other words, two sufficiently remote quantum particles will basically forget their fermionic/bosonic nature and, in that respect, behave as classical particles (and correlatively, $g(r) \overset{r\to\infty}{\longrightarrow} 1$). This in fact means that the wave functions of the two particles almost do not overlap because they are far away from one another. One can then identify separately each particle and the notion of indistinguishability loses its meaning.

The pair correlation function relies on the probability of joint detection of two particles. To talk about probabilities, we have implicitly considered an assembly of identical particles (and more precisely an ideal gas of either fermions or bosons). And dealing with an ensemble of particles opens the door to tools of statistical physics with some peculiarities at the side of ensembles of quantal particles. This is precisely the purpose of the next chapter to address such peculiarities.

2.6 CONCLUDING REMARKS

The notion of quantum state, replacing the classical description of particles in terms of position and momentum, is a central concept in quantum mechanics and will play a major role in this book. We have especially insisted on the simpler concept of wave function which somewhat 'materializes', in rather simple terms, the essential step from a localized picture (one particle, one position, one momentum) to the delocalized picture (with all its consequences) of the quantum description. But spin, for which only the abstract notion of state makes sense, has also been discussed in some detail because of its role in assemblies of quantum particles, an aspect which will turn essential throughout this book.

Before stepping to many-particle systems in the next chapter, we have seen that properties of quantum particles, are quantized: they can only take specific, well defined values. This quantization mechanism stems from the notion of quantum state, for example the boundary conditions imposed on the wave function by a potential well. It sets heavy constraints on the state of a particle, even more so when one considers ensembles of particles. Again such assemblies are heavily constrained by the nature, bosonic or fermionic (or equivalently the spin), of the considered particles.

As a final remark, let us recall the important notion of zero point motion, again a direct consequence of the above outlined notion of quantum state. Zero point motion sets a lower limit to the possible rest of a quantum particle (and thus directly impacts its localization). This is a key issue in liquids where motion competes with interactions to allow some relative disorder. And this precisely leads to the notion of quantum liquids. In the case of fermions, the fact that two particles cannot occupy the same energy state, because of quantum statistics, introduces another important mechanism tending to 'separate' the particles. In the case of bosons, on the contrary, particles tend to 'join' to form a conglomerate. We shall come back to these essential issues in the next chapter.

CHAPTER 3

Ensembles of particles

Thermodynamics draws its principles from a set of experimental laws, making no assumption about the constitution of matter. Historically, thermodynamics was developed before the general acceptance of the atomic structure of matter. Thermodynamics is thus a phenomenological *macroscopic* theory which relates different properties of matter, such as the differences between specific heats at constant pressure and volume, or the variations of these quantities with pressure. However, in spite of its numerous successes in explaining global properties of matter, the sole thermodynamic considerations cannot allow us to deduce the absolute values of say the specific heat or the equation of state of a given substance.

To overcome this limitation of thermodynamics, one needs to consider the structure and the constituents of matter. One objective of this chapter is to revisit some (macroscopic) aspects of thermodynamics in the light of microscopic considerations. Historically, the first attempt to link macroscopic properties of matter to the microscopic constituents is due to D. Bernoulli, who is considered as the founder of the so-called kinetic theory. His work was developed and extended later on by many other scientists throughout the 19th century. Among the most important contributors to the development of this field, we should mention R. Clausius, J. Clerk Maxwell, J.W. Gibbs, and L. Boltzmann. These many developments formed the basis of what is known today as statistical mechanics.

Kinetic theory and statistical mechanics are two different but related ways to predict the macroscopic properties of matter—in either gaseous, liquid or solid states—starting from those of its microscopic constituents, which we shall generically refer to as particles. We consider systems constituted of a very large number of particles, typically comparable to the Avogadro's number $\mathcal{N}_A = 6.02 \times 10^{23}$ mol^{-1}. We remind readers that this is the number of molecules contained in about 22.4 liters of a gas at standard values of pressure and temperature (1 atmosphere and 273 K). In principle, the properties of a system of particles could be deduced from the solutions of the equations of motion of all particles. To take a simple example, consider the characterization of the properties of a gas filling a room. Whereas the full microscopic description is characterized by a large number of mechanical variables ($6N$ in the classical description of a system of N particles), the macroscopic one is characterized by a small number of thermodynamic variables (pressure, temperature, viscosity, etc.). In practice, apart from the technical problem of solving these equations for a system containing a huge number of particles, this would provide us an enormous amount of information that would be useless anyway. Most of the physical quantities accessible experimentally are macroscopic and can only be related to microscopic properties through *averages* to discard meaningless (and unattainable) information. The term 'averages' calls for a statistical treatment of the physical properties at the microscopic level. And here, the usually huge number of particles involved in a macroscopic sample justifies such a statistical approach. Moreover, even in the case of relatively simple interactions between particles, the system very often displays collective properties which cannot be interpreted simply from the individual behavior of the particles. This is the case of phase transitions, critical points, fluctuations, etc. Averages are once again necessary here.

Such a probabilistic viewpoint to describe macroscopic variables of a system might look disturbing since thermodynamics in principle deals with physical quantities which are exactly known. For instance, in the first or the second law of thermodynamics, there is no room for any uncertainty on the internal energy or on the temperature. However, since these macroscopic quantities stem from microscopic constituents which are in very large number, we need to use a statistical reasoning to get average values of relevant quantities. The word 'relevant' is important here. There is indeed a selection of information as the result of the reduction of the (huge) number of variables. But this reduction is performed on, so to speak, an educated basis leading to select *only* the relevant variables. Note also that, as we have seen in the previous chapter, quantum mechanics is by essence probabilistic. However, this has nothing to do with the fact of manipulating a great number of particles: a single particle described by a wave function already implies uncertainties at the side of measurements performed on this particle.

In this chapter, we mainly focus on the properties of a system at equilibrium. We start with kinetic theory which applies Newtonian laws to a typical molecule of a gas. This allows us to deduce from averages over the total number of particles expressions for the pressure, the specific heat, etc., of a gas. We then move to statistical mechanics which gives a microscopic interpretation to a state of equilibrium in terms of accessible microscopic states. We build our reasoning on classical, distinguishable, particles. While historically, both kinetic theory and statistical mechanics were developed before the advent of quantum mechanics and therefore relied on classical mechanics, we extend the results to the case of quantal (thus indistinguishable) particles. This leads us to introduce quantum statistics to describe an ensemble of quantum particles. Since these statistics stem from the spin of the considered particles, we focus on the two statistics thereof by exploring some properties of the Bose-Einstein condensate and the Fermi gas.

3.1 KINETIC THEORY

By assuming that the pressure of a gas is due to the impacts of its particles on the container walls, D. Bernoulli, in the first half of thc 18th century, was able to deduce the empirical Boyle's law, namely that the product PV is a constant at fixed temperature. This was the starting point of the kinetic theory. More generally speaking, the aim of kinetic theory is to relate the motion of a typical molecule in a gas to macroscopic properties of the gas such as volume, pressure and temperature. Interactions are generally neglected. More precisely, particles do interact but only via elastic collisions, the latter being viewed as the result of a strong repulsive interaction when particles are close enough. One can ask if such an assumption is justified since interactions do play a fundamental role in explaining the structure of matter. It is nevertheless possible to explain, at least in a first approach, some properties of a given physical system even if interactions are ignored. The model of the ideal gas is based precisely on this idea. The volume is a result of the freedom the particles have to spread throughout the container, the pressure results from the collisions of the particles with the container's walls, and the temperature is related to the kinetic energy of the particles.

Frequently, we take the limit of infinite number of constituents ($N \to \infty$), implicitly assuming an equivalent macroscopic system. But at the same time, we should assume another limit. Indeed, if the number of particles in a box of volume V is multiplied by an arbitrary factor, then the volume must also be multiplied by the same factor so that both situations are macroscopically equivalent, that is, both must have the same density. Therefore, the previous limit $N \to \infty$ must be taken together with the limit $V \to \infty$, in such a way that the ratio N/V is a finite constant. This procedure is usually known as 'thermodynamic limit'.

In the ideal gas model, we consider a large number N of particles in a container which, for the sake of simplicity, will be chosen as a cubic box of side length L. The particles are in constant motion and, in the absence of external forces, they are uniformly distributed throughout the container. The number of particles per unit of volume is the same in any small volume within the box. All directions of the molecular velocities are equally probable, and the modulus of the velocities take all possible values. In a relatively dilute gas, the particles are most of the time separated by large distances, compared to their own dimensions. We assume that the range of the forces between particles is very short, comparable to the size of the particles. We can then reasonably accept that the particles interact with each other only when they collide. Collisions are considered as perfectly elastic, so that the kinetic energy and momentum are conserved. Because the duration of a collision is very small compared to the typical time between collisions, we can safely ignore the forces between particles, at least in a first step. The same assumption can be made to the collisions with the walls of the container, which have an enormous mass as compared to the molecular mass.

Once we have accepted these general considerations, we will consider our systems constituted of point-like particles (they occupy no volume), having no internal motion such as vibrations or rotations. For the time being, we will also assume that the constituents have no internal degrees of freedom such as spin. What makes the gas ideal is that its particles are assumed to be noninteracting: they are only subject to a motion resulting from the thermal energy. Experimentally, all gases behave in a universal way once sufficiently dilute. The ideal gas model thus corresponds to the case where interactions become negligible at thermal equilibrium, which is justified in the case of a very dilute gas.

3.1.1 The pressure of a gas

We consider N identical particles in the box moving in all directions and with various velocities. If there is no net motion of the gas, the average velocity is zero. To estimate the average speed of the particles, we then define their mean-square velocity by a simple arithmetic average: $\langle v^2 \rangle = \left(v_1^2 + v_2^2 + \cdots + v_N^2\right)/N$ where a quantity between brackets means its value averaged over the N particles. One can also define the mean-square velocity along any of the three spatial directions x, y, z and relate it to the total mean-square velocity. Indeed, for particle i, we have $v_i^2 = v_{i,x}^2 + v_{i,y}^2 + v_{i,z}^2$. Because of the huge number of particles and because these particles are all moving randomly in all directions, the three spatial averages are equivalent: $\langle v_x^2 \rangle = \langle v_y^2 \rangle = \langle v_z^2 \rangle = \langle v^2 \rangle/3$.

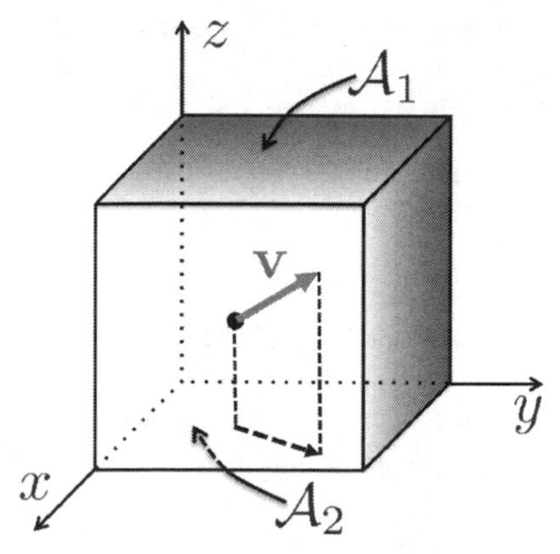

Figure 3.1: Point-like particle (black circle) of mass m and velocity $\mathbf{v}$(gray arrow with its projection in the xy plane as the dashed arrow) in a cubic box. The top and bottom walls are denoted by $\mathcal{A}_1$ and $\mathcal{A}_2$.

The average mean-square velocity of the particles is directly related to the pressure the particles exert on the walls of the box. To derive this relation, we ignore in a first step the collisions that particles may undergo with one another, and only consider elastic collisions of particles with the walls. Figure 3.1 shows a point-like particle of mass m and velocity $\mathbf{v}$ in a cubic box of volume $V = L^3$. Consider the motion along the vertical direction (z direction), with the particle bouncing back and forth from the walls $\mathcal{A}_1$ and $\mathcal{A}_2$. According to our assumption of elastic collisions, only the vertical component v_z of the velocity is varied and thus the kinetic energy is preserved.

After colliding on wall $\mathcal{A}_1$, the particle's momentum along the z-axis changes by $\Delta p_z = (-mv_z) - (mv_z) = -2mv_z$, and this is (in absolute value) equal to the momentum transferred to the wall $\mathcal{A}_1$. In other words, in the course of this elementary process, the particle, which has initially a momentum mv_z (necessarily pointing outwards to allow a collision) 'pushes' outwards the wall $\mathcal{A}_1$ by transferring it a momentum $2mv_z$.

After this collision, the particle travels back to the opposite wall $\mathcal{A}_2$, hits it with again a perfect momentum reversal $-mv_z \to mv_z$, and then comes back again towards wall $\mathcal{A}_1$, and so on. The time elapsed during the travel from $\mathcal{A}_1$ to $\mathcal{A}_2$ and back to $\mathcal{A}_1$ is given by $\Delta t = 2L/v_z$. The momentum transferred on $\mathcal{A}_1$ per unit of time $(\Delta p_z/\Delta t)$ is the force F_z that the particle exerts on wall $\mathcal{A}_1$ and is thus $F_z = (2mv_z)/(2L/v_z) = m{v_z}^2/L$. The total pressure on that wall is the force of all particles per unit of surface. The result is

$$\frac{F_z}{L^2} = \frac{m}{V}\sum_{i=1}^{N} v_{i,z}^2 = \frac{mN}{V}\langle v_z^2\rangle = \frac{mN}{3V}\langle v^2\rangle \quad .$$

We have assumed that in the travel between the walls $\mathcal{A}_1$ and $\mathcal{A}_2$, a particle does not collide with other particles. Although it is not an acceptable hypothesis, we can convince ourselves that the obtained result is correct, due to the fact that we are dealing with a huge number of particles. Because of the exchange rates in an elastic collision between identical particles, there will always be some particle colliding with the down wall with exactly the momentum of the particle coming off the up wall. In addition, the time of impact is very short compared with the time Δt. Therefore, ignoring collisions between particles is just a simple way to do the calculation.

We have calculated the pressure on the face $\mathcal{A}_1$s but it is obvious that the pressure is the same on all walls and in any point of the interior of the box, assuming in addition that the variation of pressure with height is negligible. Finally, the pressure can be written as:

$$P = \frac{1}{3}\, m\, \varrho\, \langle v^2\rangle \quad , \tag{3.1}$$

where $\varrho = N/V$ is the particle (or number) density of the system. This equation is fundamental as it shows how a macroscopic quantity as the pressure P is related to a microscopic quantity, namely the mean-square velocity of the particles.

3.1.2 Kinetic energy and temperature

The previous result on pressure allows us to establish another important relation between macroscopic and microscopic properties. As discussed in Chapter 1, the equation of state of an ideal gas can be written in terms of the density as $P = \varrho k_{\mathrm{B}} T$; see Eq. (1.2). Together with Eq. (3.1), this immediately leads to a relation between the average kinetic energy of a particle $\langle E_{\mathrm{kin}}\rangle$ and the temperature T

$$\langle E_{\text{kin}} \rangle = \frac{1}{2} m \langle v^2 \rangle = \frac{3}{2} k_{\text{B}} T \quad . \tag{3.2}$$

In other words, the thermal motion of the particles, measured by $\langle E_{\text{kin}} \rangle$, is directly related to a macroscopic quantity, namely the temperature.

Equation (3.2) is actually a particular case of a more general relation known as the theorem of equipartition of energy. This theorem states that in thermal equilibrium, the average energy of a particle is equally shared among the various degrees of freedom (translational, rotational, vibrational), each degree of freedom carrying $k_{\text{B}}T/2$. Here, we neglect vibrational and rotational motions of particles and only consider the three spatial directions for translational motion, whence the factor 3 in Eq. (3.2).

The mean-square velocity of particles in a gas thus only depends on its molar mass M (to speak in macroscopic terms, as the particle mass m is simply $m = M/\mathcal{N}_{\text{A}}$) and the temperature, and is independent of pressure or volume: $\langle v^2 \rangle = 3k_{\text{B}}T/m = 3RT/M$. For instance, the nitrogen molecules ($M = 28$ g/mol) in our atmosphere at 20°C have a root-mean-square velocity of 510 m/s. In the Sun's core, where the temperature is around 1.5×10^7 K, the average velocity of hydrogen molecules ($M = 2$ g/mol) is about 430 km/s.

The relation (3.2) is very interesting: the measure of the temperature of an ideal gas is also a measure of the average translational kinetic energy of its particles. In the macroscopic world, one can measure a temperature, and has thus access to a microscopic property of the system, the average kinetic energy. And conversely, the total microscopic kinetic energy provides a macroscopic and measurable characteristic, temperature. In an ideal gas, the internal energy U of a particle being only due to kinetic energy, we therefore have the specific heat capacity c_V of a particle equal to the factor of proportionality between $\langle E_{\text{kin}} \rangle$ and T, that is here $c_V = 3k_{\text{B}}/2$. We note that for a classical ideal gas, c_V is independent of temperature. The macroscopic version of this result is obtained by replacing k_{B} by the universal gas constant R. It can be related to the Dulong-Petit law which states that the molar heat capacity of a solid is $3R$. To obtain this law, we use once again the theorem of equipartition of energy mentioned above. The internal energy U of a solid is the sum of a kinetic energy and a potential energy, each of them carrying 3 degrees of freedom. We therefore end with $U = 2 \times 3RT/2$ and from that, $C_V = 3R$. This law is approximately observed at room temperature for many elements. However, it fails at low temperatures because quantum effects start to dominate. And we will indeed see in Sections 3.3.2 and 3.4.2 that the heat capacity of a quantal ideal gas does depend on temperature.

In fact, the Boltzmann's constant is just a factor to change units, and is the natural way to measure the temperature in units of energy or conversely. As a matter of fact, for systems at very low temperature energies are expressed in kelvin. We shall often employ the notation $\beta = 1/(k_{\text{B}}T)$ (some authors prefer to use the letter τ) to express temperature in units of energy.

3.1.3 Maxwell's distribution of velocities

Kinetic theory allows one to interpret microscopic quantities such as pressure and temperature, in terms of an arithmetic average of squared velocities. The average value of a quantity is the simplest information we can figure out from a system of particles. But it does not contain all the information one would like to know about it. For example, if we say that the average age of a population of 10,000 inhabitants is 40 years, the average age may be the same for a large number of possibilities. To mention only two extreme cases, it could happen that all members of the population are 40, or that half the population is 1 and the other half is 79. More information can be obtained if one knows the percentage of people within a given age interval.

From the mathematical point of view, this corresponds to access the distribution function of age. To be more general, we indicate by the letter X the variable considered, the total number of possible cases by N, and the number of cases with variable between X and $X + \Delta X$ by ΔN. The distribution function $f(X)$ is such that

$$\frac{\Delta N}{N} = f(X)\Delta X \tag{3.3}$$

The function $f(X)$ gives the probability that the variable X is between X and $X + \Delta X$. In the above example of a population, the distribution function is discrete and discontinuous, but if the number N of inhabitants is large enough, the ΔX can be considered as a differential element and the function $f(X)$ can be approximated by a continuous function. The same applies to the case of particles. As the distribution function is a probability function, it has to fulfill the condition

$$\int_0^\infty \mathrm{d}X f(X) = 1$$

which means that the total probability is equal to unity. We have put an infinite upper limit, although we know that there is a maximum age for a population or a maximum velocity for a particle above which $f(X)$ must vanish. But putting the upper limit to infinity is convenient to perform calculations, while it does not introduce errors in practice. Once f is known, any average value can be derived. For instance, if Q is a physical quantity which depends on X, its average value is calculated as:

$$\langle Q \rangle = \int_0^\infty \mathrm{d}X f(X) Q(X) \quad .$$

These ideas can be applied to the case of the velocities of gas particles. The velocity distribution function was obtained in 1859 by Maxwell. There are two assumptions for the establishment of this distribution. First, one assumes the molecular chaos: the three spatial components of the velocity are independent. This means that the number of particles such as the x component of their velocity is between v_x and $v_x + \Delta v_x$ is independent to the number of particles with their y component between v_y and $v_y + \Delta v_y$ (and similarly with the z component). The second assumption relies on the isotropy of the system: all spatial directions are equivalent. This has several consequences. First, the distribution function in each spatial direction should be the same, say f. Second, consider the x direction: f should be independent of the sign of v_x, that is f is a function of v_x^2. Third, all spatial directions being equivalent, at the end of the day, f should depend on v^2 only. Putting all things together, this means that the fraction of particles with a velocity $\mathbf{v}$ such as the x component is between v_x and $v_x + \Delta v_x$, the y component between v_y and $v_y + \Delta v_y$, and similarly for the z component, reads as:

$$f(v_x^2)f(v_y^2)f(v_z^2)\Delta v_x \Delta v_y \Delta v_z \quad .$$

Dividing the above quantity by $\Delta v_x \Delta v_y \Delta v_z$ provides the density of points in velocity space and the distribution function must fulfill the relation:

$$f(v_x^2)f(v_y^2)f(v_z^2) = f(v^2) \quad .$$

The only function satisfying this condition is an exponential function $f(v^2) = A\mathrm{e}^{\pm Bv^2}$. The plus sign is to be excluded to avoid an infinite probability if $v^2 \to \infty$, so we are left with the determination of the constants A and B. To this end, one uses spherical coordinates instead of Cartesian coordinates, and thus replace the velocity volume $\Delta v_x \Delta v_y \Delta v_z$ by $4\pi v^2 \Delta v$. In other words, the fraction of particles with velocity between v and $v + \Delta v$ is given by

$$\frac{\Delta N_v}{N} = A\mathrm{e}^{-Bv^2} 4\pi v^2 \Delta v \quad .$$

Integrating over velocities, one obtains the total probability of finding a particle with any velocity v, which must be equal to one. This normalization condition gives the relation $A = B^3/(\pi\sqrt{\pi})$. Therefore, there is a single free parameter left, that is B. It fixes the velocity v_{M} giving the velocity at the maximum of the distribution (see below), and we write it in terms of temperature. The final expression of the Maxwell's distribution is the following:

$$f(v) = \frac{4}{\sqrt{\pi}} \left(\frac{m}{2k_{\mathrm{B}}T} \right)^{3/2} v^2 \exp\left(-\frac{mv^2}{2k_{\mathrm{B}}T} \right) \quad . \tag{3.4}$$

This distribution can be further used as an input for more elaborate treatments in kinetic theory.

$f(v)$ is plotted in Figure 3.2. In the left panel, we have indicated by arrows three values of the velocity which are of particular interest: the most probable velocity v_{M}, which corresponds to the velocity of the maximum of $f(v)$, the mean average velocity $\langle v \rangle$, and the root-mean-square (rms) velocity $v_{\mathrm{rms}} = \sqrt{\langle v^2 \rangle}$. They are close to each other and are all proportional to the ratio $\sqrt{k_{\mathrm{B}}T/m}$. The maximum velocity v_m for example reads $v_{\mathrm{M}} = \sqrt{2k_{\mathrm{B}}T/m}$, $\langle v \rangle \simeq 1.128\, v_{\mathrm{M}}$, and $v_{\mathrm{rms}} \simeq 1.225\, v_{\mathrm{M}}$.

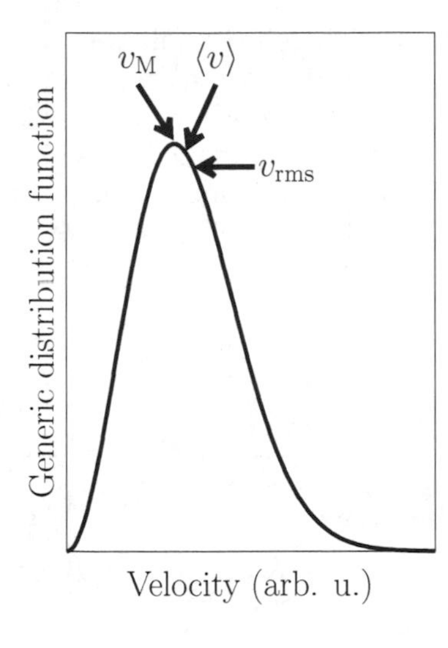

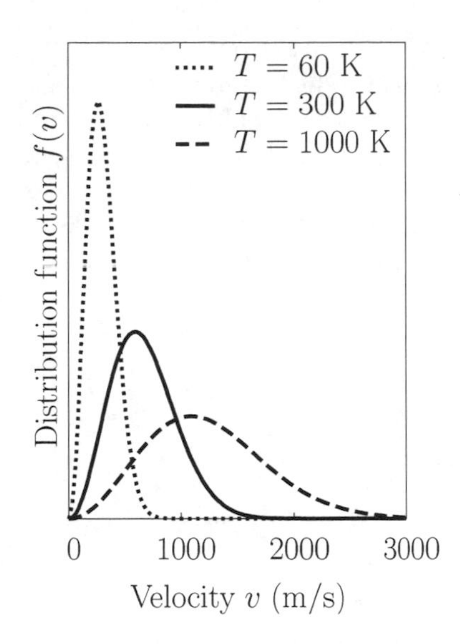

Figure 3.2: Maxwell's distribution function of molecular velocities in a gas in general (left), and in gaseous N_2 at three different temperatures as indicated (right).

In the right panel of Figure 3.2, the distribution function is plotted at different temperatures T for a gas of N_2 molecules as a typical example. Increasing T implies also increasing the number of molecules with higher velocities. A glance at these panels allows us to qualitatively understand some physical situations. For instance, the energy emitted by the Sun is originated in a nuclear fusion, where basically four protons form a helium nuclei (with the emission of two positrons and two neutrinos). The mutual Coulomb repulsion between protons hinders the process. But the high values of temperature inside the Sun provides enough kinetic energy to a great number of protons to overcome that repulsion and help the fusion reaction. As another example, consider the evaporation of water in a glass or in a pond. Evaporation is always taking place, even at low temperature, because there exist molecules with very high velocities (those in the tail of the distribution function), such that their kinetic energy allows them to escape from the global attraction due to the other molecules in the liquid.

3.1.4 Mean free path

Maxwell's distribution shows that in a gas, the particles can possess very different velocities, even if most of them move with a velocity close to v_M. The origin of such a broad distribution stems from elastic collisions between particles. There is actually an important physical quantity, the mean free path, directly connected to these collisions. It gives useful indications on the state of the considered fluid (gas or liquid). We establish its analytical expression in the following. Without any interaction, the particle follows a uniform motion along a straight line until it collides on another particle. Each encounter with another particle results in an abrupt change of velocity (direction and value). Following the motion of a particle along a long interval of time shows that in fact the particle follows a path which looks like a random motion. A typical example is displayed in the left panel of Figure 3.3, each solid dot representing the collision with another particle.

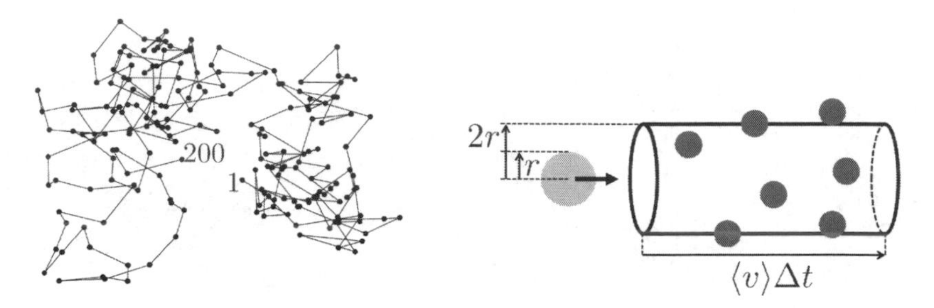

Figure 3.3: Left: random walk of a particle due to 200 collisions with other particles. Right: particle of radius r (light gray circle) with an average velocity $\langle v \rangle$, colliding with particles at rest (dark gray circles) overlapping the volume of a cylinder of transverse section $4\pi r^2$ and height $\langle v \rangle \Delta t$.

There is however the possibility of defining average quantities. For instance, a very useful one is the average distance traveled by a particle between collisions. This defines the mean free path, usually denoted by λ. It is reasonable to assume that the larger the density, the more collisions should occur, so that one can expect λ to be inversely proportional to the density. Now we should relax the hypothesis of point-like particles. Indeed, if the particles were dimensionless, the probability that they collide would be vanishingly small, while collisions do occur in real life. The spatial extension of the particles determines the mean free path λ, that is the typical distance a particle travels between two successive collisions: the larger the size of the particles, the smaller the mean free path.

To determine λ, we consider a hard sphere of radius r traveling at a mean velocity $\langle v \rangle$ during a time interval Δt in a medium of volume V containing N particles at rest, not necessarily of the same type of the traveling particle (see right panel of Figure 3.3). During the considered time interval, the particle travels the distance $\langle v \rangle \Delta t$. For the time being, we consider no interaction between particles and collisions are hence all elastic. The traveling particle will undergo collisions during Δt with the still particles which are in a cylinder of transverse section $\pi(2r)^2$ and height $\langle v \rangle \Delta t$. The number of collisions is then identical to the number of particles at rest in this cylinder, that is $4\pi r^2 \langle v \rangle \Delta t\, \varrho$. The mean free path of the traveling particle is thus the average distance covered during Δt divided by the number of collisions during the same time interval:

$$\frac{\langle v \rangle \Delta t}{4\pi r^2 \langle v \rangle \Delta t \varrho} \quad .$$

Now, in a gas, all particles are moving. Therefore, the previous reasoning is wrong in that the target particles do also move. To get a correct mean free path, one should actually distinguish the velocity appearing in the numerator to that in the denominator. Indeed, in the numerator, it is the speed relative to the container, that is $\langle v \rangle$. But in the denominator, it is the speed relative to the other (moving) particles, v_{relat}. A detailed calculation gives that $v_{\mathrm{relat}} = \sqrt{2}\langle v \rangle$. The final expression for the mean free path is:

$$\lambda = \frac{1}{4\pi r^2 \sqrt{2} \varrho} \quad . \tag{3.5}$$

λ is inversely proportional to ϱ. This dependency was actually expected since the denser the system, the more likely interactions between constituents and then the smaller the mean free path.

The second ingredient in λ is not simply the size r of the colliding particle but $4\pi r^2$ which is the transverse section of the cylinder used to estimate λ. This quantity is actually called the 'geometric cross section' and defines the transverse area within which a particle of radius r can scatter with another particle. This is the relevant quantity which enters the expression of λ since we have neglected any interaction between particles and only considered elastic collisions. The geometric cross section should be replaced by a cross section which accounts for interactions (for instance, electromagnetic repulsion or attraction if the particles are charged). Grossly speaking, a cross section is an 'equivalent surface of influence'. The larger the interaction (the larger the cross section), the more numerous the interactions, and the smaller the mean free path.

The typical cases that we consider in kinetic theory are situations where mean free paths are large as compared to typical distances between the system's constituents. This situation is typically reached in dilute systems with short range interactions, as in an atomic or molecular gas (air for example). On the contrary, when λ is of the order the distance between the constituents, the system is rather in a liquid state. In Section 1.3.2, by reasoning on a typical two-body molecular interaction, we have already related the distance between two constituent particles with the (solid, liquid or gaseous) phase of a system. The notion of mean free path delivers a complementary and more quantitative view to understand the nature of a gas or a liquid.

It should be noted that the above arguments hold for a classical gas. As soon as quantum effects come into play, the interaction process cannot be considered in such simple terms. For example, Pauli exclusion principle has a significant impact on interaction processes. Therefore, it can severely constrain values of mean free paths, even in strongly interacting systems such as atomic nuclei.

3.1.5 The path to equilibrium

In the previous discussion on the mean free path, we mentioned the potential role of interactions (at least the short-range repulsion which results macroscopically in a collision when two particles are sufficiently close to each other). The mean free path is also related to the path towards equilibrium. To characterize a gas by an implicitly uniform temperature and/or pressure does not appear immediately linked to molecular motion. Still, we have seen above that kinetic theory provides a way to link both microscopic and macroscopic descriptions with direct expressions of macroscopic quantities such as temperature and pressure in terms of averages over molecular velocities. Performing the average, though, does not necessarily imply that the system has reached a state in which the average is independent on the volume on which it is performed. This introduces a first subtle difference between global (thermodynamic) and local quantities. The term 'local' refers here to averages performed over a restricted volume of the system, but large enough to contain a sufficient number of constituents to validate an averaging strategy. It hence provides a description in terms of local characteristics of the system, in particular local number density $\varrho(\mathbf{r})$ and local velocity $\mathbf{v}(\mathbf{r})$, the latter directly providing expression for the local temperature and pressure following our previous discussion (see Sections 3.1.1 and 3.1.2). This corresponds to a fluid dynamical or hydrodynamical description of the system, somewhat half way between the macroscopic and microscopic worlds. Simple fluids are perfectly understood at such a level of description with a proper account of evolution of various local quantities such as density, pressure and temperature.

The phase space picture

Going one step further, one may nevertheless question how well such local velocity averages always provide a sound representation of the state of the gas. In classical physics, the microscopic description will reduce to the set of positions $\mathbf{r}$ and velocities $\mathbf{v}$ of the system's constituents. In such terms, local velocity averages amount to average over $\mathbf{v}$ for particles having about the same position $\mathbf{r}$ but with potentially very different $\mathbf{v}$. In particular, if the system has not reached even a local equilibrium, velocities are likely to exhibit very different values, so that representing the ensemble thereof by a mere average is clearly insufficient.

This thus brings us to a new (more microscopic) level of description in which one would like to account for this velocity distribution in some detail. Mind that we keep here the idea to consider averages. We do not mean here to come back down to a description in terms of the positions and velocities of each constituent. Rather, we envision a description in terms of a distribution in position and momentum (or velocity) space. As in the previous example of the distribution function of inhabitants' age in a population (see Eq. (3.3)) we consider here particles that we pile up in 'spatial boxes' if they have about the same position $\mathbf{r}$, and in 'momentum boxes' if they have about the same momentum $\mathbf{p}$ (or velocity $\mathbf{v}$). We therefore consider a description accounting for both spatial and momentum but still in coarse terms: particles are distributed in elementary 'cells', each cell being characterized by an average position and an average momentum. Such a picture is called a phase space picture. The resulting distribution function is usually denoted $f(\mathbf{r}, \mathbf{p}, t)$, and is related to the number of particles with position $\mathbf{r}$ and momentum $\mathbf{p}$ at time t. Indeed, the probability of finding a particle with position and momentum defined within the volumes $(\mathbf{r}, \mathbf{r} + \Delta\mathbf{r})$ and $(\mathbf{p}, \mathbf{p} + \Delta\mathbf{p})$ reads as:

$$f(\mathbf{r}, \mathbf{p}, t)\, \Delta\mathbf{r}\, \Delta\mathbf{p}$$

The spatial volume element $\Delta\mathbf{r}$ should be very small as compared to the physical dimensions of the system. But at the same time, it should contain a large number of particles to reasonably justify the use of statistical considerations. For instance, in a cubic micron, there are about 10^7 molecules of a gas, which is indeed a large number. The same should hold for the momentum volume element $\Delta\mathbf{p}$. Although from a strict mathematical viewpoint, neither $\Delta\mathbf{r}$ nor $\Delta\mathbf{p}$ are infinitesimal volumes, we shall assume it for practical calculations. For instance, the distribution function is normalized as $\int \mathrm{d}\mathbf{r}\, \mathrm{d}\mathbf{p}\, f(\mathbf{r}, \mathbf{p}, t) = N$ where N is the total number of particles in the integration phase space volume. By averaging over $\mathbf{p}$ only, one gets the number density (or the spatial distribution of particles) at time t: $\int \mathrm{d}\mathbf{p}\, f(\mathbf{r}, \mathbf{p}, t) = \varrho(\mathbf{r}, t)$.

The Boltzmann equation

The phase space level of description points towards nonequilibrium situations and correlatively provides a way to analyze the way the system proceeds towards equilibrium. It can for example describe how a gas thermalizes (first locally then globally) and reaches a temperature T by which it can be properly characterized. A similar argument would hold for pressure. This actually brings us back directly to the discussion of Section 3.1.4 on the mean free path. Indeed, the path to equilibrium directly results from interactions and the mean free path constitutes a measure of how much they enter the game. Starting from whatever distribution of particles exploring various velocities, interactions, even rare ones, will progressively bring the constituents of a system to a thermodynamical equilibrium through multiple successive interactions. This is exactly what occurs for example in winter when the outside temperature is very low (much lower than the inside temperature) and the door is opened at a given instant. Initially, both outside and inside spaces, at equilibrium, can be characterized by a temperature, say $T_{\text{out}} < T_{\text{ins}}$. Opening the door will cool down a fraction of the inside molecules near the door. Energy (or velocity) exchanges through interactions with the rest of the inside molecules will progressively restore a thermodynamical equilibrium at a temperature below the original T_{ins}, the amplitude of the temperature variation and the time it takes to restore equilibrium depending on the amplitude of the original perturbation, here the initial temperature difference.

To attain the time evolution of a system towards equilibrium, one refers to the time-dependent version of kinetic theory. The question is to write an equation for the time evolution of $f(\mathbf{r}, \mathbf{p}, t)$. This is a major issue and we shall not enter any technical discussion here. We will confine the point to introduce the famous Boltzmann equation which precisely provided the first sound description of such an evolution. The basic idea is to write the change of $f(\mathbf{r}, \mathbf{p}, t)$ in each $\Delta\mathbf{r}\Delta\mathbf{p}$ cell as a consequence of interactions between particles. In the typical system considered here, the interaction is short range and thus manifests itself by collisions between particles. This amounts to vary only the momentum component of $f(\mathbf{r}, \mathbf{p}, t)$ at each point $\mathbf{r}$ (within $\Delta\mathbf{r}$ of course). Let us consider a given momentum value $\mathbf{p}$. A particle with such a momentum acquires another momentum $\mathbf{p}'$ after undergoing a collision. This depopulates the $\mathbf{p}$ component of $f(\mathbf{r}, \mathbf{p}, t)$. When considering all possible $\mathbf{p}'$, these contributions are called the loss term I_{loss}. But at the same instant t and location $\mathbf{r}$, collisions can occur between particles with some momentum $\mathbf{p}''$ and attain the momentum $\mathbf{p}$. This correlatively populates the $\mathbf{p}$ component of $f(\mathbf{r}, \mathbf{p}, t)$. This contribution, when all possible $\mathbf{p}''$ are considered, is called the gain term I_{gain}. This leads to the Boltzmann equation which we can write as:

$$\frac{\partial}{\partial t} f(\mathbf{r}, \mathbf{p}, t) = I_{\text{gain}} - I_{\text{loss}} = I_{\text{coll}} \tag{3.6}$$

More sophisticated, including quantum ones, kinetic equations share the same structure, although with variations in details. A word of caution is necessary in Eq. (3.6) on the meaning of the time derivative of f: it encompasses a derivative with respect to t but also derivatives with respect to $\mathbf{r}$ and $\mathbf{p}$, the latters also varying in time.

An interesting point is that, because interactions are short range, collisions only affect local momenta. Still, because of these modifications of momentum, particles undergoing collisions at position $\mathbf{r}$ follow a trajectory different from the one they would have had otherwise. They hence explore other regions of phase space and in particular modify the distribution function at locations $\mathbf{r}'$ different from $\mathbf{r}$. The effect, rather than being instantaneous, is simply delayed. But such delays occur everywhere and at any time, so that the notion of delay does not mean anything physically here.

As a last but essential remark, we now discuss how equilibrium is actually reached and to which situation it does correspond. For qualitative discussions, in particular when close to equilibrium, one can show that the collision term I_{coll} can be replaced by a simplified expression:

$$I_{\text{coll}} = \frac{f_{\text{equil}}(\mathbf{r}, \mathbf{p}) - f(\mathbf{r}, \mathbf{p}, t)}{\tau} \tag{3.7}$$

where $f_{\text{equil}}(\mathbf{r}, \mathbf{p})$ is the equilibrium distribution (thus time-independent) and τ is a characteristic time. The time evolution is then governed by τ, itself expressed as a function of the equilibrium distribution (temperature in particular). There exist generalizations of this expression farther away from equilibrium. The interest of the simple form (3.7) is to provide a clear path towards equilibrium. In turn, equilibrium simply corresponds to a vanishing I_{coll}. From

Eq. (3.6), we see that the phase space distribution function does not vary anymore, reflecting an equilibrium situation. It is nevertheless interesting to note that $I_{\rm coll} = 0$ does not mean the absence of collisions between particles, by far. It rather means that the gain and loss terms exactly compensate. This is physically interesting: collisions between particles continue to occur (there are, as before reaching equilibrium, $I_{\rm gain} + I_{\rm loss}$ collisions per time step dt) but in such a way that they do not affect the momentum distribution. This brings us back to our earlier arguments on the fact that a system can be characterized by a temperature because of the 'rare' interactions between its constituents.

3.2 SOME ELEMENTS OF STATISTICAL PHYSICS

In the previous sections, some properties of a system at equilibrium were derived in kinetic theory of gas mainly by describing collisions of an individual particle with many other particles, and then replacing specific individual properties, as the velocity square, by the mean values obtained by an arithmetic averaging over all particles. Here, in statistical mechanics, we also derive properties of a system at equilibrium but in terms of probabilities. More precisely, we introduce the notion of macrostates, defined by thermodynamical or macroscopic quantities such as volume, pressure, temperature, and the notion of microstates which are all the accessible states of the microscopic constituents.

One in fact needs to refer to quantum mechanics to describe the microstates (or accessible states) of a system. Some basics of quantum mechanics were given in Chapter 2 but we have focused the discussion on the description of either a single particle (via a single wave function) or at most a pair of particles (in the pair correlation function for instance). Here we will concentrate on ensemble of particles, and in particular ensemble of fermions or of bosons. We will start our reasoning by manipulating classical, thus distinguishable, particles but also discrete values of the level energies. This will allow us to apply some results obtained with classical particles to quantum systems. Indeed, each quantum state has a definite energy. Quantum states possessing the same energy belong to the same energy level, and the degeneracy or multiplicity of an energy level is the number of different quantum states in that level. Such energy degeneracy effects already occur for one single particle (see Section 2.4.2). In this chapter, by dealing with ensembles of particles, we will be brought to consider *occupation numbers* associated to energy levels. These numbers are important quantities in quantum statistical mechanics. And one can easily imagine that with 10^{23} particles, the number of possible quantum states adding up to the same total energy can become astronomical. In the statistical description, we shall define a probability for each of these states to calculate the average value of physical properties of a quantum system.

Statistical mechanics is based on two postulates which are justified from the validity of the results deduced from them. The first one is that details of the system under study become irrelevant if we wait long enough. The multiplicity of the possible events will eventually wash out all memory of how the system had started. When a system reaches this condition, it is said to be in equilibrium. The second hypothesis is that for an isolated system in equilibrium, all accessible microscopic states are equally probable: there is no reason to prefer one accessible state to another. This second hypothesis is sometimes called the fundamental postulate of statistical mechanics. A state is accessible if its properties are compatible with the physical characterization of the system. For instance, the state of an ordered crystalline structure is accessible or not according to the value of temperature. States that are not accessible have a zero probability.

3.2.1 Microstates and probabilities

The aim of statistical mechanics is basically to count the number of equally probable accessible states of a system. We shall first count this number in the case of an isolated system, in which the volume V, the number of particles N and the internal energy E are fixed (mind that we have used here the notation E commonly used in statistical physics for the internal energy, instead of U used in thermodynamics). The macrostate is therefore characterized by the values (N, V, E). We consider the case of identical and distinguishable non-interacting particles, and assume that there is a certain number p of accessible states with energies, ε_0, ε_1, ..., ε_p. For the sake of simplicity, we assume discrete values for the energies, which is the current situation encountered in quantum mechanics. Notice also that we have followed the usual way of labeling states in quantum systems and started from 0 our counting of states. We are then left to distribute n_i particles in the boxes (or energy levels) labeled i in the following way: n_0 particles with energy ε_0, n_1 particles with energy ε_1, etc. with the following normalization restrictions:

$$\sum_{i=0}^{p} n_i = N \tag{3.8a}$$

$$\sum_{i=0}^{p} n_i \varepsilon_i = E \tag{3.8b}$$

We denote by D a distribution of the particles, that is a given set $n_0, \ldots, n_p$ regardless the identity of the particles, and a particular realization of this distribution (for which the distinction among the particles is now made) a microstate. The problem is to count the number of ways w of performing a given distribution of the particles D in the set of levels. The sum of all the different w's provides the total number of microstates. This number depends on both the number of particles and the total energy.

To illustrate the difficulty in the determination of the distributions and the associated number of microstates, we consider the simple case of $N = 6$ noninteracting particles, to be distributed within 7 accessible states. These states possess an energy 0, ε, 2ε, 3ε, 4ε, 5ε, 6ε respectively. We finally impose a total energy for the 6 particles of $E = 6\varepsilon$. The different ways of distributing the particles are represented in Figure 3.4 . There are 11 possible distributions in

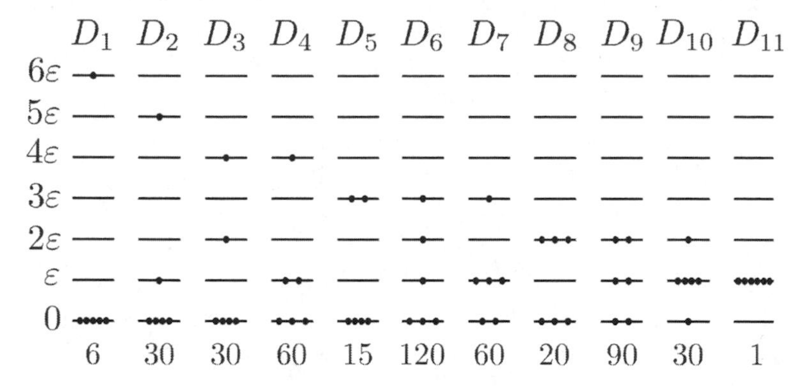

Figure 3.4: The eleven different ways (D_i) of distributing 6 identical and distinguishable particles (dots), with a total energy of 4ε, among 7 states of energy 0, ε, ..., 6ε, with the corresponding number of microstates at the bottom.

all, labeled D_j at the top of the figure. Each of these distributions corresponds to a number of microstates w_j indicated at the bottom. For instance, the distribution D_{11} contains only $w_{11} = 1$ microstate: all six particles with energy ε. The distribution D_1 contains $w_1 = 6$ microstates: there are 5 particles with zero energy, and a single one with energy 6ε, which accounts for the total energy E, but there are 6 possibilities for this single particle because the particles are distinguishable.

The total number of microstates is in this example $\sum_{j=1}^{11} w_j = 462$. They are all equally probable, according to the second postulate of statistical mechanics mentioned above. However, at the side of the distributions D_j's, we cannot consider them equally probable. For instance, our intuition tells us that D_6 is the most probable among the 11 distributions because it corresponds to the maximum number of microstates ($w_6 = 120$). More quantitatively, we can associate a probability P_j to the distribution D_j by considering the number of microstates in D_j divided by the total number of microstates: $P_j = w_j / \sum_k w_k$. For example, $P_6 \simeq 0.26$. The knowledge of the pairs (w_0, P_0), (w_1, P_1), ..., is called a probability distribution. With the definition taken above, we naturally have $\sum_j P_j = 1$, as it should be.

With only 6 particles and 7 energy levels, counting the total number of states is not difficult but may be tedious. In systems containing a large number of particles instead, the number of microstates becomes so incredibly large that counting them is simply unthinkable. Instead of literally counting them, it is better to find ways to get the total number of microstates without counting. We start again from a macrostate (N, V, E), to be distributed in p energy levels. The N particles are identical, distinguishable and noninteracting. Keep in mind that exchanging the momenta (or the energies) of two distinguishable particles having the same energy leads to a different microstate, although the set of n_i's remains unchanged. For a given distribution $D = \{n_0, \ldots, n_p\}$, we now calculate the number of accessible states $w(n_0, \ldots, n_p)$ in a direct way. Consider first the energy level at ε_0. The number of ways of choosing n_0 particles out of N is

$$\frac{N!}{n_0!(N - n_0)!} \quad .$$

Next, we have to choose n_1 particles with energy ε_1 out of $N - n_0$, for which the number of possible ways is:

$$\frac{(N - n_0)!}{n_0!(N - n_0 - n_1)!} \quad .$$

Proceeding along the same line, we choose the remaining particles for the other values of energy. $w(n_0, \ldots, n_p)$ is now given by the product of all these partial distributions:

$$w(n_0, \ldots, n_p) = \frac{N!}{n_0! \ldots n_p!} \quad . \tag{3.9}$$

Finally, the total number of microstates is the sum of the $w(n_0, \ldots, n_p)$'s for all pairs (n_i, ε_i) that satisfy the normalization conditions on N and E, given in Eqs. (3.8a-3.8b).

It should be noted that the classical nature of the particles (their distinguishability) makes that placing them on a given energy level is subject to no constraint. For example, the lowest energy configuration is attained by placing all particles on the lowest energy level, $n_0 = N$. And the next higher energy of the system will be reached by placing one particle in the first excited level, $n_0 = N - 1$ and $n_1 = 1$, this being possible in N different ways, depending on which particle is in level of energy ε_1.

3.2.2 Entropy and Boltzmann's distribution

The determination of number of microstates and the probabilities we can build thereof leads us to the description of the equilibrium of an isolated system. From the macroscopic viewpoint, the second law of thermodynamics tells us that the entropy S must reach a maximum value in such a state of equilibrium. We recall that, on the one hand, S has been initially introduced to quantify the notion of heat exchange (see Section 1.1.1). On the other hand, we have also mentioned that the entropy is a measure of the degree of disorder in a system, without any justification. Statistical physics will help us to understand the latter assertion and to revisit the second law of thermodynamics.

Let us consider once again the example depicted in Figure 3.4. The distribution D_{11} in which the 6 particles are all in level 1 is the most ordered distribution among the 11 possibilities. At the other extreme, we can claim that D_6 or D_9 are more disordered than D_{11}. And actually, D_6, with its $w_6 = 120$ related different microstates, is the most disordered distribution.

Now, let us imagine that we prepare the 6-particle system in D_{11} and let it evolve. If the system is isolated, it will spontaneously evolve towards the most probable distribution, that is D_6. Through this example, we intuitively anticipate that the entropy of an isolated system is related to the most probable distribution.

This relationship probably constitutes the foundation of statistical physics, brought by Boltzmann. For a given macrostate (N, E, V), he introduced a statistical entropy $S_{\rm B}$ by taking the definition $S_{\rm B} \equiv k_{\rm B} \ln \Omega$, where Ω is the total number of accessible states. Now, when the number N of particles gets larger and larger, the most probable distribution becomes increasingly dominant. And the associated number of microstates also becomes the only effective contribution to the total number of microstates Ω. We then make the approximation of replacing Ω by the number of states in the most probable configuration in the definition of the statistical entropy:

$$S_{\rm B} = k_{\rm B} \ln W \quad , \tag{3.10}$$

where W is the number of microstates in the most probable configuration. Note that W naturally depends on N and E. We will not comment further on the logarithmic dependence of $S_{\rm B}$ with W. Some elements of derivation are given in Section 4.1.4. We simply note that it complies with the fact that the more disordered the system, the higher W, and the larger $S_{\rm B}$. One interpretation of Eq. (3.10) is to reflect our degree of ignorance or uncertainty on the details of system. In the extreme case for which we know nothing about the system (apart from N, V and E), it is natural to assume that the system is in the most probable configuration.

Equation (3.10) also points out the notion of irreversibility in time, while at the microscopic level, the equations of motion, either classical or quantal, do not break the time symmetry. What happens is that, when a system evolves from a given initial microstate to a microstate compatible with the most probable configuration, there is a loss of information and this loss hinders the system to return to the initial state. For instance, going back to our 6 particles initially positioned in level 1, the fact that the system relaxes towards the most probable distribution D_6, it forgets that it was initially in D_{11}. This loss of information is the interpretation in microscopic terms of the second law of thermodynamics.

To determine the most probable distribution of particles (the n_i's) according to a given energy spectrum (the ε_i's), we start from the number of microstates $w(n_0, \ldots, n_p)$ for the distribution $D = \{n_0, \ldots, n_p\}$ (see Eq. (3.9)) and minimize it with respect to the various n_i's. The procedure is a bit involved and we will not detail it. The complete calculation can be found in Section 4.1.3. We only give here the final result, known as the *Boltzmann's distribution*:

$$n_i = N e^{-\beta(\varepsilon_i - \mu)} \quad . \tag{3.11}$$

We remind readers that $\beta = 1/(k_{\rm B}T)$. The chemical potential μ is determined by the normalization condition $N = \sum_i n_i$ at each value of T. Although similar to Maxwell's distribution of velocities (3.1.3), the Boltzmann's distribution is more general. It reduces to Maxwell's distribution if we only consider elastic collisions (short-range interactions) and if we assume that the energy ε_i consists in the kinetic energy only.

We now come back to the statistical entropy. W coincides with $w(n_0, n_1, \ldots)$ given in Eq. (3.9) provided that the n_i's follow the Boltzmann's distribution (3.11). By using Stirling's approximation, that is $p! \simeq p^p$ if $p \gg 1$, we end with:

$$S_{\mathrm{B}} = k_{\mathrm{B}} \left(N \ln N - \sum_i n_i \ln n_i \right) = k_{\mathrm{B}} \left(N \ln N - \beta\mu N + \beta E \right) \quad . \tag{3.12}$$

We finally recall that the thermodynamical entropy is related to the internal energy U by:

$$\left(\frac{\partial S}{\partial U} \right)_{N,V} = \frac{1}{T} \quad . \tag{3.13}$$

Since we have $U = E$, by calculating $(\partial S_{\mathrm{B}}/\partial E)_{N,V}$ from Eq. (3.12) and by comparing it with Eq. (3.13), we therefore deduce that the statistical entropy S_{B} coincides with the thermodynamical entropy S, thus reconciling the microscopic and the macroscopic viewpoints.

3.2.3 Ensemble averages

Microcanonical ensemble

In the previous sections, we have manipulated probabilities associated to distributions (or configurations) of a given macrostate (N, V, E), each distribution corresponding to a certain number of microstates. In practice, a way to avoid the knowledge of a probability distribution is to consider an ensemble of identical systems, in the sense that they all possess the same N, V and E. This constitutes a statistical ensemble. The elements of this ensemble describe the same real system but each element can be in any microstate of the real system. Then, for a given physical observable $\mathcal{O}$, its mean value $\langle \mathcal{O} \rangle$ is obtained by averaging over the statistical ensemble. If the ensemble is sufficiently large, $\langle \mathcal{O} \rangle$ corresponds to the value in the most probable configuration. For instance, the result of a coin tossing is either heads or tails. The probability $\mathcal{P}$ of getting heads is equal to that of getting tails, that is 0.5. Throwing the coin only once is not sufficient to observe $\mathcal{P} = 0.5$. One has to throw the coin many times and average over the successive experiences to obtain $\mathcal{P} = 0.5$. Instead of repeating the same experience many times, one can also consider a large number of coins, throw them together and compile the various results. This amounts to considering a statistical ensemble of the coin.

Up to now, we have dealt with isolated systems with a fixed number of particles N and a fixed energy E, in a fixed volume V. This corresponds to the *microcanonical ensemble*, also called, for obvious reasons, the NVE ensemble. In this ensemble, all microstates are equally probable, with a probability $p_{\rm micro} = 1/\Omega$ if Ω is the total number of microstates. The statistical entropy is $S = k_{\rm B} \ln \Omega \simeq k_{\rm B} \ln W$ with W the number of microstates compatible with the most probable distribution.

Canonical ensemble

In practice, a real system cannot be described in the microcanonical ensemble. There is always some uncertainty at the side of the perfect knowledge of E. Take for instance a gas with 10^{23} particles. It is of course impossible to know the velocity and the position of all these particles at each instant since no experimental device is able to measure this information. The latter information is anyway of little importance in practice, since one is more interested in the knowledge of macroscopic quantities as the temperature or the pressure of the gas. To account for the uncertainty on E, one uses the *canonical ensemble*. This ensemble is also called the NVT ensemble. Physically, this corresponds to the case of a system of N particles in a fixed volume V in thermal contact with a bath, the whole compound (system + bath) being isolated. The bath is usually considered much larger than the system. The thermal contact allows energy exchange between the system and the bath. In this ensemble, the microstates are not equally probable. More precisely, the microstate m of energy E_m has the probability $p_m = \exp(-\beta E_m)/Z_{\rm can}$ where $Z_{\rm can} = \sum_m \exp(-\beta E_m)$ is called the canonical partition function. Z guarantees that $\sum_m p_m = 1$ as it should be for a probability distribution. The energy of the system can fluctuate and one can show that its mean value is given by $\langle E \rangle \equiv \sum_m E_m p_m = -\partial \ln Z / \partial \beta$. We then identify the internal energy U of thermodynamics with $\langle E \rangle$. In the light of this identification, the first law of thermodynamics (see Section 4.1.1) is thus equivalent to the principle of the conservation of mechanical energy.

To deduce the expression of the entropy in the canonical ensemble, we start from that in the microcanonical ensemble: $S = k_{\rm B} \ln \Omega$. Using the fact that $1 = \Omega/\Omega = \sum_{1 \le m \le \Omega} 1/\Omega$, we rewrite this relation the following way:

$$S = -k_{\rm B} \ln \left(\frac{1}{\Omega} \right) = -k_{\rm B} \sum_{m=1}^{\Omega} \frac{1}{\Omega} \ln \left(\frac{1}{\Omega} \right) = -k_{\rm B} \sum_m p_{\rm micro} \ln p_{\rm micro} \quad ,$$

where m runs over the Ω microstates of the microcanonical ensemble. Considering now that the canonical ensemble and replacing $p_{\rm micro}$ by p_m, we obtain the general expression of the statistical entropy, applicable in any statistical ensemble:

$$S = -k_{\rm B} \sum_m p_m \ln p_m \quad . \tag{3.14}$$

Injecting the expression of p_m in Eq. (3.14) gives the entropy of the canonical ensemble:

$$S = k_{\rm B} \ln Z + \beta k_{\rm B} \langle E \rangle - k_{\rm B} \ln Z + \frac{\langle E \rangle}{T} \quad . \tag{3.15}$$

This allows us to link the macroscopic free energy $F = U - TS$ with the microscopic picture as $F = -k_{\rm B}T \ln Z$.

Grand-canonical ensemble

Finally, the situation in which the system can exchange particles with the bath (also called a reservoir) is possible, for instance if the walls separating the system with the bath are porous. This case corresponds to the *grand-canonical ensemble*, also called the μVT ensemble. As in the canonical ensemble, the microstates in the grand-canonical ensemble are not equiprobable. The microstate m of energy E_m and number of particles N_m has the probability $\mathcal{P}_m = \exp[-\beta E_m + \mu N_m)]/\mathcal{Z}$ where μ is the chemical potential and $\mathcal{Z} = \sum_m \exp[-\beta E_m + \mu N_m)]$ is the grand-canonical partition function. Both the energy and the number of particles can fluctuate. We have $\langle N\rangle \equiv \sum_m N\mathcal{P}_m = 1/\beta\, \partial \ln \mathcal{Z}/\partial\mu$ and $\langle E\rangle = -\partial \ln \mathcal{Z}/\partial\beta + \mu\langle N\rangle$. By injecting the expression of $\mathcal{P}_m$ in Eq. (3.14), we obtain the statistical entropy in the grand-canonical ensemble as:

$$S = k_{\rm B} \ln Z + \beta k_{\rm B}\langle E\rangle - \beta\mu k_{\rm B}\langle N\rangle \quad . \tag{3.16}$$

The grand-potential used in thermodynamics $\Phi \equiv F - \mu N$ is then related to the grand-canonical partition function as $\Phi = -k_{\rm B}T \ln \mathcal{Z}$.

To end this short review on the three different statistical ensembles, we mention that in the thermodynamical limit where $V \to \infty$ and $N \to \infty$, the grand-canonical and the canonical ensembles are very close. Indeed, with the probability $\mathcal{P}_m$, we can calculate the fluctuations of E and N (we recall that for instance, $\Delta E = \sqrt{\langle E^2\rangle - \langle E\rangle^2}$). We can show that $\Delta N/\langle N\rangle \simeq 1/\sqrt{V}$ and therefore, in the thermodynamical limit, the fluctuations of the number of particles vanish and N is approximately constant.

3.2.4 Quantum statistics

Up to now, we have considered that the identical particles can be distinguishable. In the quantum world, on the one hand, identical particles are indistinguishable. On the other hand, we have to take into account possible restrictions in the filling of energy levels, depending on the bosonic or fermionic nature of the considered particles. Indeed, as we have seen in Section 2.5.5, the possibility of exchanging identical quantal particles has a direct consequence on the way fermions or bosons do coexist in a many-particle system (antibunching or bunching effect respectively). This can be exemplified in a simple case of a system of noninteracting particles populating energy levels $\{\varepsilon_i, i = 0, 1, \dots\}$ created by an external field $V(\mathbf{r})$. The level spectrum itself is characteristic of the external field $V(\mathbf{r})$, for example a harmonic oscillator (see Section 2.4.2), from which it has been generated, including possible degeneracies. For the sake of simplicity, let us assume here that the energy spectrum is not degenerate. But because we consider either bosons or fermions, we have to account for spin. Still, the question is *how* the particles fill this spectrum in the ground state of the system, namely in order to reach the smallest possible total energy E. This also implies that we here consider a system at zero temperature. In the case of noninteracting particles, the total energy of the system is simply the sum of the energies of the occupied levels, that is $E = \sum_i \varepsilon_i$ where the summation runs over the number of particles. The single remaining question is hence how to arrange the particles amongst the levels. Contrary to the example of 6 distinguishable particles distributed among 7 energy levels, the situation is different in the quantal case of indistinguishable particles, because the way they are allowed to populate levels is specific to their nature (boson or fermion) or equivalenty their spin.

Let us start the discussion with fermions of spin number s. We have seen in Section 2.4.3 that each energy level has a degeneracy $g = 2s + 1$. For $s = 1/2$ (as electrons), this means that each energy level can be at most filled by 2 electrons (and this constitutes a particular case of the Pauli exclusion principle) provided that they have *opposite* spins (one with spin up and the other with spin down). Stated differently, for a given spin state (up or down), an energy level can be filled at most by *one and only one* fermion. Moreover, at the side of energy filling, the spin-up fermions are independent from the spin-down fermions. In the following, we will therefore rely our reasoning on one spin state only.

The number of particles in an energy level is called the *occupation number*. It is usually denoted by n_i where i is the label of an energy level. The ground state of N identical fermions at zero temperature is obtained by filling successively each energy level with one fermion, starting from the lowest level ε_0. We thus have $n_i = 1$ for $0 \leq i \leq N - 1$, and $n_i = 0$ for $i \geq N$ since we have exhausted the total number of fermions. This is illustrated in the left panel of Figure 3.5 for the case of 6 identical fermions. This means that the total

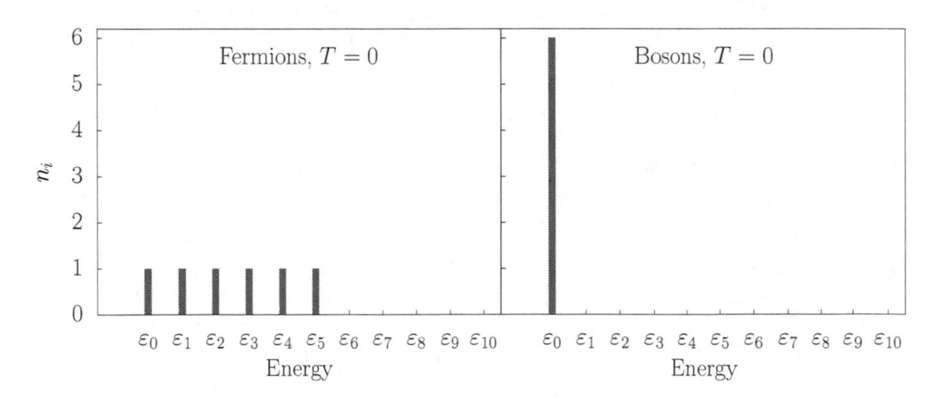

Figure 3.5: Occupation numbers n_i at zero temperature, appearing as boxes at the different energies ε_i, for a system of 6 identical particles, with the distinction between fermions (left) and bosons (right).

energy of the N fermions' system will be expressed as $E = \sum_{i=0}^{N-1} \varepsilon_i$. The case of bosons is completely different (right panel). We do not have here to care for spin. The bunching effect manifests itself by piling up *all* bosons in the lowest energy level. The total energy of the system then simply reduces to $E = N\varepsilon_0$.

3.2.5 Statistics at finite temperature

This picture at $T = 0$ is modified when the temperature is not strictly zero. One can imagine that the extra kinetic energy brought by $T > 0$ allows some particles to reach higher energy levels. In the example of Figure 3.5, this would correspond to energies $> \varepsilon_{N-1}$ in the case of 6 fermions, and above $> \varepsilon_0$ in the case of bosons. The calculation of n_i can be deduced for instance from the determination of the number of microstates W. This derivation is a bit involved and we will not detail it here. Some elements are given in Section 4.1.3 in the case of classical particles. The strategy is analogous in the case of quantal (fermionic or bosonic) particles. We simply give the final result for the occupation numbers in Table 3.1 below. In case of degeneracy,

Table 3.1: Different types of statistics written in terms of occupation number n_i, with $\beta = 1/(k_B T)$ and μ the chemical potential determined by the normalization condition $N = \sum_i n_i$ at each value of T.

Type of particles	Distingui-shables	Name of statistics	Analytical expression
Classical	✓	Boltzmann	$n_i = e^{-\beta(\varepsilon_i - \mu)}$
Bosons	×	Bose-Einstein	$n_i = \dfrac{1}{e^{\beta(\varepsilon_i - \mu)} - 1}$
Fermions	×	Fermi-Dirac	$n_i = \dfrac{1}{e^{\beta(\varepsilon_i - \mu)} + 1}$ $0 \leq n_i \leq 1$

due to spin, angular momentum or whichever, the distribution functions are multiplied by the degeneracy number.

The Fermi-Dirac and Bose-Einstein statistics are displayed in Figure 3.6 (see dashed curves and black boxes). For comparison, the occupation numbers at $T = 0$ are also shown as gray boxes. For bosons (right panel), the lowest energy state is depleted and higher levels are populated, according to the Bose-Einstein statistics. In the case of fermions (left), with $T > 0$, fermions initially filling levels from ε_0 to ε_{N-1} can be excited to higher levels, then leaving empty levels behind. For instance, we can have the fermion in level ε_{N-1} leaving that level and filling level ε_{N+1}, or the fermion in level ε_{N-2} going to level ε_{N+3}, etc. Actually, there are many different thermal excitations but they appear with different probabilities. This is in some sense similar to the distribution of 6 particles in 7 states discussed in Figure 3.4. Regarding all these different excitations, one has to average over them and one finally ends with the Fermi-Dirac statistics. This average explains why n_i is neither 0 nor 1 any longer, but can take a non-integer value in between. The same kind of average over different level occupations also holds for bosons.

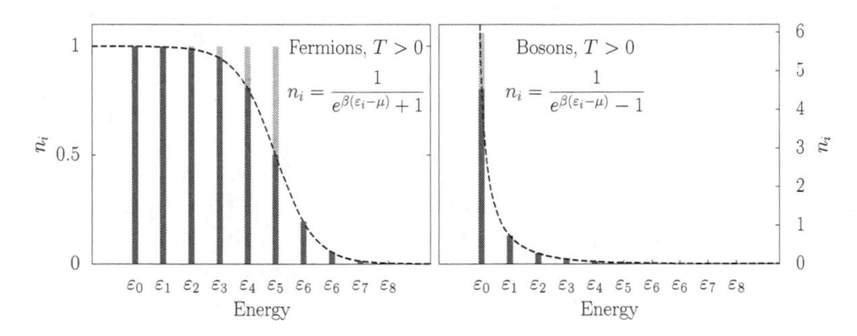

Figure 3.6: Occupation numbers n_i at finite temperature ($\beta = k_B T$) for a system of 6 identical particles, with the distinction between fermions (left) and bosons (right). The dashed curves emphasize the Fermi-Dirac (left) or the Bose-Einstein (right) distributions. The light gray boxes correspond to the occupation numbers at $T = 0$.

Both Bose-Einstein and Fermi-Dirac distributions have a common limit form when $e^{\beta(\varepsilon_i-\mu)} \gg 1$: they coincide with the Boltzmann distribution. In fact, the classical limit corresponds to average occupation numbers $\ll 1$, and it can be reached in two ways. The first option is, while keeping T constant, to reduce the particle density ϱ (either by increasing the volume or by reducing the total number of particles). Then the number of possible states per unit volume becomes larger than ϱ; or keeping constant ϱ and increasing T, the particles have access to high excited states, and can thus be distributed over a large number of states. In both cases, the average number of particles per state is much smaller than 1.

3.3 BOSON GAS AND BOSE-EINSTEIN CONDENSATION

A major result of the analysis of Section 3.2.4 is the fact that at strictly zero temperature, all bosons occupy the same lowest energy level of a spectrum. This effect, which we shall discuss in more detail in Chapter 6, is called Bose-Einstein condensation. In order to simplify the present discussion, we shall consider the case of an ideal Bose gas, namely without any interaction. This is obviously an idealization of a real experimental situation but it qualitatively (and to a large extent also quantitatively) does not alter the conclusions, while simplifying the picture. We also assume a given quantized energy spectrum, imposed by an external potential for instance. As stated above, in such an ideal Bose gas at zero temperature, quantum statistics imposes that all atoms do condensate in the lowest energy level, namely at zero energy ($\varepsilon_0 = 0$) in that specific case with no confining potential. This corresponds to a fully condensed Bose phase. It means in particular that all particles are associated to the same wave function which then represents a macroscopic object, namely the whole gas sample.

3.3.1 Boson gas at finite temperature

The natural question stemming up is to look at what occurs when temperature is not strictly zero. Intuitively one would bet that the perfect full occupation of lowest energy level will be progressively degraded to the benefit of higher energy levels. This will then gradually break the perfect homogeneity of the condensate. The remarkable point is that this progressive depopulation of the ground level will indeed proceed smoothly but only up to a finite temperature at which the condensate fully disappears.

Let us detail these general ideas. In the case of an ideal gas of bosons, the average occupation number $n_i(\mu, T)$ at temperature T and chemical potential μ, of particles with energy ε_i is given by (see Table 3.1):

$$n_i = \frac{1}{\exp\left[\beta(\varepsilon_i - \mu)\right] - 1} \quad . \tag{3.17}$$

It is determined as a function of N and T by the condition that the sum of the occupation numbers of all the energy levels be equal to the total number of particles N given in Eq. (3.8a). The minus sign in the denominator implies that the chemical potential must satisfy $\mu < \varepsilon_i$, otherwise negative values for the occupation numbers would result. Actually, the upper bound for μ is given by the lowest energy ε_0, as by definition ε_0 is a lower bound to all energies $\varepsilon_{i\geq 1}$. When μ is close to ε_0, $\exp\left[\beta(\varepsilon_0 - \mu)\right] \to 1$ and the occupation number of the lowest energy level

$$N_0 = \frac{1}{\exp\left[\beta(\varepsilon_0 - \mu)\right] - 1} \tag{3.18}$$

becomes increasingly large, and it is said that the bosons 'condensate' in the lowest energy state. This is the basis of the so-called Bose-Einstein Condensation (BEC).

3.3.2 Critical temperature and phase transition of a boson gas

It is convenient to write the total number of particles as

$$N = N_0(\mu, T) + N_T(\mu, T) = N_0 + \sum_{i\geq 1} n_i(\mu, T) \quad ,$$

where N_T is usually called the 'thermal component' of the gas (hence the subindex T), that is, the number of particles outside the condensate. For a fixed value of T, the function $N_T(\mu, T)$ increases smoothly as μ increases, reaching its maximum $N_{\rm max}(T) = N_T(\mu = \varepsilon_0, T)$ at $\mu = \varepsilon_0$, because $\varepsilon_0 < \varepsilon_{i\geq 1}$. For fixed values of μ, the function N_T increases with T and, for values of the temperature high enough, its maximum value $N_{\max}$ is greater than N, which is physically impossible. Therefore we define a critical temperature for Bose-Einstein condensation T_{BEC} by the relation

$$N_T(\mu = \varepsilon_0, T_{\mathrm{BEC}}) = N \quad .$$

The condensate component $N_0(\mu, T)$ in Eq. (3.18), behaves very differently: it is always of order of 1, except when μ tends to ε_0 where it diverges.

For temperature $T > T_{\mathrm{BEC}}$, one can neglect the value of N_0, and one has $N \simeq N_T$. However, for $T < T_{\mathrm{BEC}}$, the contribution of the condensate is crucial to fulfill the condition $N = N_0 + N_T$ because of the macroscopic occupation of the lowest energy state. Going down to $T \to 0$, the number of thermal particles tends to zero, and all the particles tend to occupy the lowest single-particle energy.

These functions and quantities can be evaluated in the case of bosons confined in a cubic box. The value of the critical BEC temperature is given by

$$k_B T_{\mathrm{BEC}} = \kappa \frac{\hbar^2}{m} \varrho^{2/3} \quad , \tag{3.19}$$

where $\kappa \simeq 3.31...$ is a numerical constant. The value of T_{BEC} is therefore fully determined by the density of the gas and the mass of its constituents. The numbers N_0 and N_T then take simple values as a function of T_{BEC}

$$N_T(T < T_{\mathrm{BEC}}) = \left(\frac{T}{T_{\mathrm{BEC}}}\right)^{3/2} N \quad , \tag{3.20a}$$

$$N_0(T < T_{\mathrm{BEC}}) = \left\{1 - \left(\frac{T}{T_{\mathrm{BEC}}}\right)^{3/2}\right\} N \quad , \tag{3.20b}$$

and of course $N_0(T > T_{\mathrm{BEC}}) = 0$ and $N_T(T > T_{\mathrm{BEC}}) = N$.

The above equations in particular show the specific behavior of the occupation of the lowest level when passing through the critical temperature T_{BEC}, as illustrated in the left panel of Figure 3.7. Another interesting quantity is the specific heat C_V, which gives an indication on the nature of the phase transition. In the cubic box case, it can be evaluated analytically. But since it implies special functions, we will not give the explicit expressions here and rather qualitatively discuss the temperature dependence of C_V. In the right panel of Figure 3.7, this dependence is displayed C_V in units of T_{BEC}. Starting from 0 at $T = 0$, it increases as $T^{3/2}$, until its maximum value $\simeq 1.927$ at T_{BEC}, where C_V exhibits a cusp. The latter reflects a discontinuity of the first derivative of the specific heat with respect to temperature, which is characteristic of a phase transition of first order (see Section 1.1.3). Note however that C_V is a continuous function at that point. Contrary to the classical ideal gas in which the heat capacity is independent of the temperature, C_V here strongly depends on T. For very large values of temperature, the specific heat tends to the classical value 3/2, as can be deduced from Eq. (3.2).

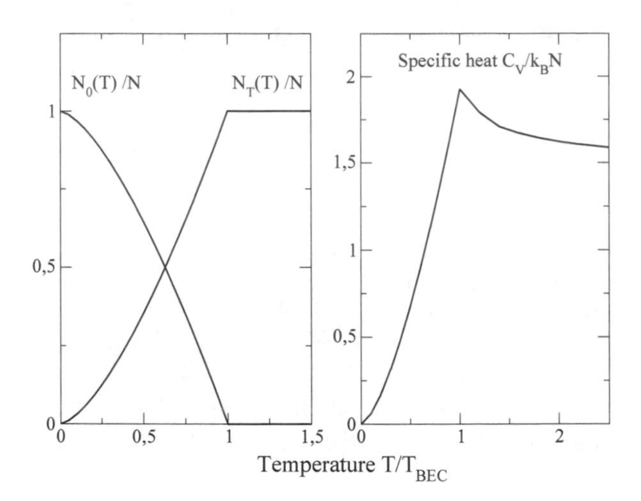

Figure 3.7: Occupation of lowest energy level (left) and specific heat (right) of an ideal Bose gas confined in a box.

The BEC transition is a purely quantum effect, neither existing in a classical ideal gas nor in an ideal gas composed of fermions. The term 'condensation' is used in analogy with a liquid-gas transition, in which at some critical temperature liquid droplets condense out of the gas to form a saturated vapor, eventually forming bulk liquid. There are however two main differences worth stressing. First, while the liquid-gas transition is driven by the interaction, the BEC transition is driven by quantum statistics of identical bosons, without need of an interaction (interactions actually tend to reduce the number of particles in the lowest energy state). Second, while in the liquid-gas transition the condensed molecules are separated in real space from the remaining ones in gas phase (spatial separation), the BEC bosons are separated in momentum space from the remaining ones in the 'normal' phase (momentum separation). The condensed bosons all occupy a single quantum state of zero momentum, while the non-condensed bosons all have a finite momentum.

3.3.3 Pressure of a boson gas

Another interesting aspect concerns the equation of state of such an ideal Bose gas. In Chapter 1, we have seen in Eq. (1.2) that, for a classical ideal gas, the equation of state can be written as $P = \varrho k_B T$. Mind that the assumption of an ideal gas means neglecting interactions, not absence thereof, as temperature can only be understood via interactions, even if they are rare. Henceforth, increasing the temperature means more thermal agitation, larger velocities and more interactions, and in turn a larger pressure. In this picture, the pressure then naturally vanishes at zero temperature. We observe the same property for the boson gas at $T = 0$. Indeed, it can be shown that this pressure reads:

$$P \propto m^{3/2} T^{5/2} \quad . \tag{3.21}$$

The pressure of a boson gas is thus strictly 0 at $T = 0$.

There is however an important difference between a classical ideal gas and a boson gas. In the first case, the pressure depends on the density. This expresses rather intuitive properties of a gas, as already recalled above. When the gas is compressed, it exerts a larger pressure on the walls of the container (increasing the density ϱ implies a larger pressure P) because the particles being closer to each other do interact more. The situation is completely different in a boson gas at $T \leq T_{\rm BEC}$, and somewhat counterintuitive. From Eq. (3.21), we see that the pressure depends on the temperature T and the mass m of the constitutive particles only. Since the incompressibility is defined as the derivative of pressure with respect to the volume, we then obtain the unphysical result of an infinitely compressible boson gas in the condensate phase. Actually, this is a trivial consequence of our assuming ideal point-like particles with no interaction, as nothing prevents such particles to (ideally) concentrate in a single point. A real Bose gas does not present that behavior.

3.4 FERMI GAS AND QUANTUM PRESSURE

The situation in a Fermi gas is totally different as compared to a Bose gas. The occupation number of particles of energy ε_i at temperature T is given by (see Table 3.1):

$$n_i = \frac{1}{\exp\left[\beta(\varepsilon_i - \mu)\right] + 1} \tag{3.22}$$

Analogously to the boson gas, μ is the chemical potential, which is determined as a function of N and T by the condition that the sum of the occupation numbers of all the energy levels be equal to the total number of particles $N = g \sum_i n_i(\mu, T)$ where g is the spin degeneracy. But the positive sign in the denominator excludes any type of divergence, contrary to the boson case, and no condensate can appear. We recall that in the boson case, n_i can take any positive real value. Here, the situation is very different, as can be immediately illustrated by considering $T = 0$.

In that case, $\beta \to +\infty$. While $\varepsilon_i - \mu$ has a finite value but can be either positive or negative, the argument of the exponential tends either to $-\infty$ if $\varepsilon_i < \mu$, and then $n_i = 1$, or to $+\infty$ if $\varepsilon_i > \mu$, and then $n_i = 0$. The distribution function n_i is therefore a step function. It is depicted as a gray line in Figure 3.8 and was already visible in the left panel of Figure 3.5. The interpretation of this distribution is that, owing to the Pauli exclusion principle, each available lowest energy level is occupied by a single fermion for a given spin state and this, up to a maximum energy fixed by the chemical potential. The chemical potential for fermions at $T = 0$ is referred to as the Fermi energy ε_F.

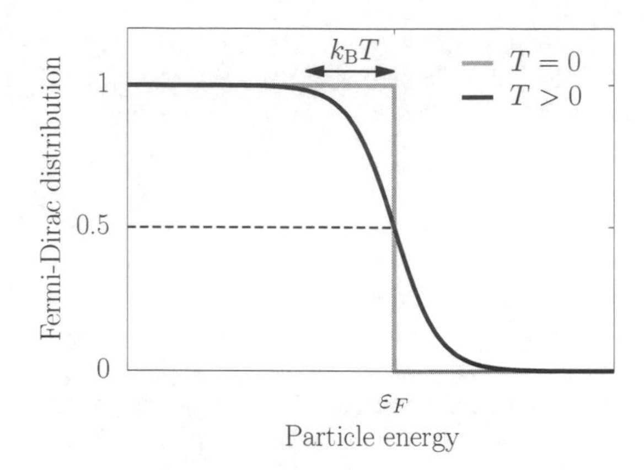

Figure 3.8: Distribution function for an ideal gas of fermions, see Eq. (3.22).

When T rises above zero, the particles are excited with an energy of order k_BT from their occupied state at $T = 0$. However, not all the particles can be excited. Only those particles at energies within about k_BT below the Fermi energy ε_F have a chance to be excited. The rest remain unaffected, because the possible excited states within k_BT from the initial energy are already occupied by other particles. As compared with the $T = 0$ case, the net result is that some states with energy below ε_F are emptied and other states above ε_F are filled (see black curve in Figure 3.8 and also dashed curve in Figure 3.6). The average occupation number of states with $\varepsilon_i = \varepsilon_F$ is equal to 1/2, and the chemical potential is always smaller than the Fermi energy. In fact, the relevant values of temperature are relatively small (as compared to Fermi energy, see below) in most physical systems of interest, apart from some specific cases such as the core of massive stars (see Sections 8.2 and 8.3).

At high temperatures, the Fermi gas becomes a classical gas. It is thus described by the ideal gas equation of state, provided that the gas is very dilute to safely neglect interactions. At very low values of temperature, the Fermi gas is often said to be a *degenerate gas*, being *fully degenerate* at $T = 0$. The word 'degenerate' is used here to stress that it differs from a classical ideal gas, owing to purely quantum effects. Notice that it has also been employed with a different meaning at some places, in relation to the spin degeneracy number. The study of degenerate gases has important applications to a variety of real physical systems, including conduction electron in metals, white dwarf stars, liquid ^{3}He and nuclear matter.

3.4.1 The fully degenerate Fermi gas

We discuss now the extreme case of zero temperature case illustrated in Figure 3.5. Although this seems to be an academic situation, it actually provides results applicable to different physical systems provided the temperature is small enough as compared to the characteristic Fermi temperature defined as $T_F = \varepsilon_F / k_B$. In those systems, thermal effects can be included as corrections to the fully degenerate case. This again justifies the interest of the zero temperature case. The latter in particular allows us to see immediately the connection between the Fermi energy and the size of the system.

In the Fermi gas model, one assumes that the system is composed of a very large number N of fermions in a very large volume V, and it is then convenient to take the thermodynamical limit, that is let them tend to infinity with the condition that the density of particles $\varrho = N/V$ remains finite. For such a macroscopic volume V, it is a good approximation to replace a sum over the possible states by an integral over momenta (see Section 4.2.3). In the present case, the momenta of occupied states are contained in the so-called Fermi sphere (also called Fermi sea), namely the sphere of momenta whose modulus $k = |\mathbf{k}|$ is bounded as $k \leq k_F$. This momentum k_F, called the Fermi momentum, is the maximum reachable value of momentum. Let us assume a system of spin-1/2 fermions, so that each momentum state is occupied by at most two particles: one with spin up and one with spin down. We also assume a saturated spin system, with the same number of spin-up and spin-down particles. By equating N to the number of occupied states in the Fermi sphere we obtain the relation

$$N = \sum_i n_i = 2 \sum_{k_i \leq k_F} 1 = 2 \frac{V}{(2\pi)^3} \int_{k \leq k_F} \mathrm{d}^3\mathbf{k} = 2 \frac{V}{(2\pi)^3} \frac{4\pi}{3} k_F^3 \,, \tag{3.23}$$

where the factor of 2 comes from the spin degeneracy; in general, for spin-s fermions the degeneracy prefactor is $2s + 1$. The factor $V/(2\pi)^3$ has a phase space origin (see Section 4.2.3). Both N and V quantities are big numbers, but the ratio $\varrho = N/V$ remains finite. From this equation we see that the Fermi momentum is related to the density of particles as

$$\varrho = \frac{k_F^3}{3\pi^2} \quad . \tag{3.24}$$

In some cases, instead of the number density, it is more convenient to deal with the radius of a sphere whose volume is equal to the volume per fermion: $r_0 = (3/4\pi\varrho)^{1/3}$. In the case of electrons in a solid or a cluster (see Chapter 5) it is referred to as the Wigner-Seitz radius, and the symbol r_s is used.

Equation (3.24) simply expresses in a different way that the chemical potential is fixed by the number of particles. At $T = 0$, the chemical potential is the energy of a level at the Fermi surface, called Fermi energy:

$$\varepsilon_F = \frac{\hbar^2 k_F^2}{2m} = \frac{\hbar^2}{2m} (3\pi^2 \varrho)^{2/3} \quad . \tag{3.25}$$

Another interesting quantity is the total energy of the system $E = \sum_i \varepsilon_i n_i$, which is purely kinetic energy, as no interaction nor any confining potential is present. It is obtained by summing up over Fermi sphere

$$E = \frac{V}{4\pi^3} \int_{k \leq k_F} \frac{\hbar^2 k^2}{2m} \mathrm{d}^3\mathbf{k} = \frac{V}{5\pi^2} \frac{\hbar^2}{2m} k_F^5 = \frac{1}{5\pi^2} \frac{\hbar^2}{2m} (3\pi^2)^{5/3} N^{5/3} V^{-2/3} \quad (3.26)$$

The proper quantity is the energy per particle, which can be written in terms of the Fermi energy:

$$\frac{E}{N} = \frac{3}{5} \frac{\hbar^2}{2m} k_F^2 = \frac{3}{5} \varepsilon_F \quad (3.27)$$

The total energy per particle is a fraction of the highest energy of an occupied state ε_F.

The integration in Eq. (3.23) was done for wavevectors up to k_F. One can also integrate from 0 to a given $k \leq k_F$ and get the number of electrons in a sphere of radius k. By using $\varepsilon = \hbar^2 k^2/(2m)$, we obtain the number of electrons with energy between 0 and ε. This allows us to define the number of electrons per energy unit, often denoted by $\mathcal{D}(\varepsilon)$ and also called the density of states, as (see also Section 4.2.3):

$$\mathcal{D}(\varepsilon) = \frac{\mathrm{d}N}{\mathrm{d}\varepsilon} = \frac{V}{2\pi^2} \left(\frac{2m}{\hbar^2} \right)^{3/2} \sqrt{\varepsilon} \quad .$$

At Fermi energy, namely for $\varepsilon = \varepsilon_F$, by injecting Eqs. (3.24) and (3.27) in the equation above, one obtains

$$\mathcal{D}(\varepsilon_F) = \frac{3N}{2\varepsilon_F} \quad . \quad (3.28)$$

In other words, up to the 3/2 factor, the number of electrons per energy unit at the Fermi level is the total number of electrons divided by the Fermi energy.

The energy expression Eq. (3.26) allows us to directly access the associated pressure using standard thermodynamics

$$P_F = -\left.\frac{\partial E}{\partial V}\right|_N = \frac{2}{15\pi^2} \frac{\hbar^2}{2m} (3\pi^2)^{5/3} \varrho^{5/3} = \frac{2}{5} \varrho \, \varepsilon_F \quad (3.29)$$

The key point is that one obtains a pressure which is a function of the density ϱ, but mind that this pressure is attained at strictly zero temperature, at variance with the standard pressure in a classical gas, which strictly vanishes at $T = 0$. The origin of this pressure, called the Fermi pressure, is of purely quantal nature. More specifically it is due to the fermionic nature of gas constituents, and reflects the Pauli exclusion principle operating for fermions.

The Fermi energy is also a key quantity to understand fermionic systems. We shall have numerous occasions to use it and realize its importance. For reasons which will become clear below, one also uses an alternative quantity, the Fermi temperature T_F, which possesses the same dimension as the usual classical temperature. This will allow us to compare usual classical temperatures of collisional origin to Fermi temperatures of quantum statistics origin.

We have discussed above the Fermi pressure which exists even at zero (classical) temperature T. But we have also seen previously in Section 3.2.5 that the Fermi gas may very well be at finite classical temperature T. And the 'classical' temperature will be the source of an associated 'classical' pressure P, which will compete/complement the Fermi pressure P_F. Actually both classical and quantal pressures join to create the actual pressure inside the system at $T > 0$. But pressure will be dominantly of quantum origin as long as $T \ll T_F$ and vice versa. As a final remark, we would like to stress again the fact that this quantum pressure at zero temperature is purely a fermionic effect. Indeed, as discussed in Section 3.3.3, the pressure of a boson gas at zero temperature is strictly zero.

We end the study of the fully degenerate Fermi gas by calculating the probability density $P(\mathbf{r}_1, \mathbf{r}_2)$ for the joint detection of a fermion at point $\mathbf{r}_1$ and another one at point $\mathbf{r}_2$. We have seen that this probability is related to the pair correlation function (see Sections 1.3.3 and 2.5.5). From the classical viewpoint, $P(\mathbf{r}_1, \mathbf{r}_2)$ reads:

$$P_{\text{class}}(\mathbf{r}_1, \mathbf{r}_2) = \sum_{i,j} \varrho_i(\mathbf{r}_1)\, \varrho_j(\mathbf{r}_2) = \sum_{i,j} |\psi_i(\mathbf{r}_1)|^2\, |\psi_j(\mathbf{r}_2)|^2 \tag{3.30}$$

where the sums extend over the occupied levels. We have added a subscript 'class' to emphasize that it corresponds to the classical case not affected by the exchange of any two particles. As we deal with fermions, the product of single particle wave functions has to be replaced with the antisymmetric wave function $\widetilde{\Psi}_{ij}(1,2) = \Psi(\mathbf{r}_1, \mathbf{r}_2)\chi(1,2)$ discussed in Section 2.5.4. If the spatial part $\Psi(\mathbf{r}_1, \mathbf{r}_2)$ is symmetric, the probability density coincides with the classical expression (3.30). But if the spatial part is antisymmetric (see Eq. (2.22)) we have to consider another probability density, simply denoted by $P(\mathbf{r}_1, \mathbf{r}_2)$:

$$\begin{aligned} P(\mathbf{r}_1, \mathbf{r}_2) &= \frac{1}{2} \sum_{i,j} \left| \psi_i(\mathbf{r}_1)\, \psi_j(\mathbf{r}_2) - \psi_i(\mathbf{r}_2)\, \psi_j(\mathbf{r}_1) \right|^2 \\ &= P_{\text{class}}(\mathbf{r}_1, \mathbf{r}_2) - \left| \sum_i \psi_i(\mathbf{r}_1)\, \psi_i^*(\mathbf{r}_2) \right|^2 . \end{aligned} \tag{3.31}$$

We observe the appearance of a new term which is somehow reminiscent of interferences between waves.

In the case of a homogeneous gas of noninteracting fermions, the single-particle wave functions are normalized plane waves; see Eq. (2.11). Moreover, the density ϱ is a constant. Summing up Eq. (3.31) over the Fermi sea gives:

$$P(r) = \varrho^2 \left\{ 1 - \left(\frac{3 j_1(k_F r)}{k_F r} \right)^2 \right\} , \tag{3.32}$$

where $r = |\mathbf{r}_1 - \mathbf{r}_2|$, owing to translational and rotational invariances, and j_1 is the first spherical Bessel function and reads $j_1(x) = \sin x / x^2 - \cos x / x$. This

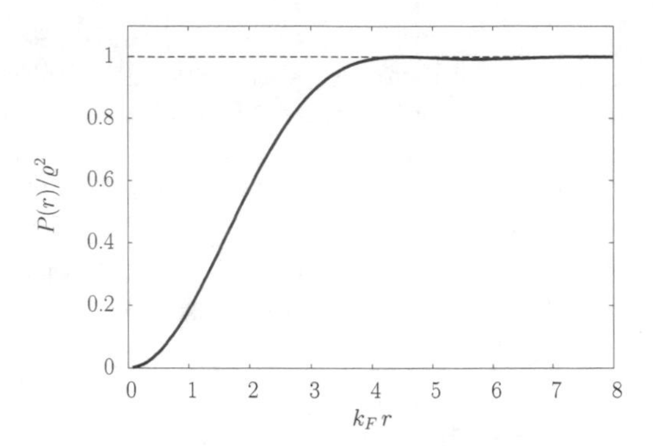

Figure 3.9: Probability $P(r)$ to find two fermions separated by a distance r in an ideal gas of fermions, divided by the number density square, ϱ^2.

two-body probability is plotted in Figure 3.9. We observe an antibunching effect of two fermions with respect to each other. This has to be compared to the antibunching observed in ^{3}He gas (see Figure 2.11). In particular, while in the experiment, we had $g(r) \to 0.94$ as $r \to 0$; in the ideal case explored here, we have $g(r) \to 0$.

3.4.2 The degenerate Fermi gas

We have defined the Fermi momentum k_F for a system at $T = 0$, and shown its relation to the particle density. Actually, this relation is also very useful for systems at finite T and/or subject to interactions, because it allows one to define several 'natural' scales related solely to the density. For instance, the Fermi energy ε_F gives a reference for energies, and also for temperatures, by defining the Fermi temperature as $T_F = \varepsilon_F / k_{\mathrm{B}}$.

Because of the scale naturally introduced by the Fermi energy, thermal effects in a Fermi gas are usually limited, as in most situations $T < T_F$ or even $T \ll T_F$, except maybe in very specific cases. Thermal effects can therefore be explored in the low temperature limit, which already covers numerous situations of physical interest. The key question then is to approximate Fermi factors by low temperature expressions. We have already noted that the strict zero temperature limit of a Fermi factor is a step function $\theta(\varepsilon_F - \varepsilon)$, which takes strictly 0 values for $\varepsilon > \varepsilon_F$ and which is strictly equal to 1 for $\varepsilon \leq \varepsilon_F$. The first order correction to the 0 temperature step function is a quantity peaked at Fermi energy and proportional to T^2. It turns out that its expression is mathematically involved. The simplest way to express temperature effects is to consider expressions involving integrals over Fermi factors $n(\varepsilon)$ times a function $f(\varepsilon)$.

The derivation is lengthy but straightforward, and leads, in the limit

$T/T_F \ll 1$, to:

$$\int_0^\infty \mathrm{d}\varepsilon \frac{f(\varepsilon)}{\mathrm{e}^{\beta(\varepsilon-\varepsilon_F)}+1} \simeq \int_0^{\varepsilon_F} \mathrm{d}\varepsilon f(\varepsilon) + \frac{\pi^2}{6}(k_\mathrm{B}T)^2 \frac{\mathrm{d}f}{\mathrm{d}\varepsilon}\bigg|_{\varepsilon=\varepsilon_F} + \ldots$$

Not surprisingly, this expression directly involves the ratio $k_\mathrm{B}T/\varepsilon_F$ which is precisely the quantity governing impact of temperature in a Fermi gas. This general expression allows for example to evaluate the temperature dependence of the Fermi energy $\varepsilon_F(T)$ as a function of the Fermi energy at zero temperature ε_F:

$$\varepsilon_F(T) = \varepsilon_F \left\{ 1 - \frac{\pi^2}{12}\left(\frac{k_B T}{\varepsilon_F}\right)^2 + \ldots \right\}.$$

Another interesting quantity is the thermal correction to the total energy of the fully degenerate Fermi gas, that is $(3/5)\varepsilon_F N$; see Eq. (3.27). We then write $E = (3/5)\varepsilon_F N + E^*(T)$ where the latter term is usually called thermal excitation energy. In the case of the simple Fermi gas, it reads:

$$E^*(T) = \frac{\pi^2}{4}\frac{N}{\varepsilon_F}(k_\mathrm{B}T)^2 \quad .$$

The latter expression in particular allows one to estimate the specific heat

$$C_V = \frac{\mathrm{d}E^*}{\mathrm{d}T} = \frac{\pi^2}{2}\frac{N}{\varepsilon_F}k_\mathrm{B}^2 T = \frac{\pi^2}{2} N k_\mathrm{B} \frac{T}{T_F} \quad . \tag{3.33}$$

At low temperatures, C_V therefore grows linearly with T. This substantially differs from the classical value $C_V = \frac{3}{2}Nk_\mathrm{B}$ (see Section 3.1.2), which is independent of T on the one hand, but also much larger than Eq. (3.33) on the other hand. Indeed, due to the Pauli exclusion principle, not all electrons in a Fermi gas can participate to the heat capacity since only those close enough to the Fermi level can be promoted to free levels by thermal excitations. This thus explains why the heat capacity of a Fermi gas is much smaller than the one of a classical gas. Equation (3.33) can also be interpreted the following way: at low temperatures, C_V is, up to some constant factors, the density of states at the Fermi surface times the thermal energy. In other words, the higher the density of states, the larger the number of electrons possibly excited by temperature, and the greater C_V.

3.5 QUANTUM FLUIDS

3.5.1 From gas to liquids

In classical physics, the simplest description of gases relies on the notion of the ideal gas, in which interactions between constituents are neglected. On the other hand, the presence of a temperature by nature reflects interactions between particles, interactions which are precisely overlooked in an ideal gas.

Still, at the simplest level, one can neglect these interactions and use instead a description in the simple terms of a macroscopic temperature. Of course the argument only holds for sufficiently dilute systems, typically for a rarefied gas inside a large container. In view of the above discussion, one can wonder how quantum particles will behave in such conditions.

Let us first consider such a dilute quantum gas in a large container. Because the system is sufficiently dilute, we can safely neglect interactions and we are then precisely in the framework of the cases discussed above. As we have seen in Section 2.5.2, the zero point energy, $E^0_{\rm kin} = k_{\rm B} T_Q$ where T_Q is the zero point temperature, scales with the density ϱ as $E^0_{\rm kin} \propto \varrho^{2/3}$. In a sufficiently dilute gas, $\varrho \to 0$ and the zero point motion therefore tends to disappear. In the case of fermions we should consider another quantum effect: Fermi statistics imposes a maximum energy ε_F which scales with density as $\varepsilon_F \propto \varrho^{2/3}$, while pressure scales as $P_F \propto \varrho^{5/3}$. It also tends to 0 for vanishing density so that Fermi statistics effects do become negligible in that case. A *very* dilute quantum gas (hence with no interaction) does finally behave as a *classical* ideal gas. The same holds, by the way, when the classical temperature of a system becomes large enough. Classical thermal agitation then takes the lead over zero point motion (see Section 3.2.5). Moreover, in the case of fermions, it also dominates Fermi motion, the latter related to the Fermi temperature T_F (see Section 3.4), for sufficiently large temperatures (mind that by construction $T_Q < T_F$). However, except for helium, it turns out that most systems we shall consider in the following are in a temperature range (typically room temperature), which is, on the contrary, far below their quantum temperature (T_Q and even more so T_F when relevant). The notion of 'high' temperatures leading to a classically justified description thus hardly holds in practice.

However, one may wonder whether the argument still holds when one cannot ignore interactions, as in a liquid or, even more so, in a solid. Indeed, in liquids and solids, interactions play a key role: this is what makes them liquid (or even solid), namely more (or much more, respectively) structured than (unstructured) gases. Interactions show up because the system is denser. We all intuitively associate liquids (and even more so, solids) to denser systems than gases. But if the density becomes larger, the quantum nature of particles will show up again. And the distinction between bosons and fermions will again start to play a role. This is all the more true since interactions do depend on quantum statistics, mostly because they often depend on spin. The equation of state (or, generally speaking, to avoid a too thermodynamically oriented term, the relation between energy and density) of quantum interacting systems will then depend on quantum statistics, just as in noninteracting gases (see Section 3.4). The pending question remains whether the system is liquid or solid and the answer will depend on how typical quantal agitation (in particular zero point energy) does compare to binding energy delivered by interactions.

Putting all pieces together, we see emerge a concept of quantum liquids, which we can now define more precisely. Quantum liquids, which constitute the core topic of this book, are systems in which quantum effects play a key role and where quantum driven motion (zero point motion in particular) remains large enough to compete with (and even dominate) interaction effects which tend to 'crystallize' the system if too intense. In turn, as we shall see below in more detail, it turns out that in quantum liquids, *classical* temperature effects can, to a large extent, be overlooked, because quantum properties induce larger effects than those induced by temperature (which, keep in mind, stem from interactions). This can be formalized by relating the actual temperature T to quantum temperatures T_Q and T_F : $T \ll T_Q, T_F$ in most cases of relevance (see Table 3.2 for a summary). In order to make the point more specific let us briefly present the various quantum liquids we shall discuss in the following.

3.5.2 Which quantum liquids?

Atoms

Let us start with the case of atoms and consider a system of atoms of mass number A. Their actual mass is about Am_p (where m_p is proton mass). Following the discussion of Section 2.5.2 and using the same notations, we introduce the typical distance L between two atoms as $L = 2a$ and we can then immediately express the zero point energy in this case as $E_k^0 = \hbar^2\pi^2/(2Am_pa^2) = h^2/(2m_pAL^2)$ which, when recast in temperature unit, becomes $T_Q = h^2/(2m_pk_\mathrm{B}AL^2) \simeq 100/(AL^2)$ (in Kelvin) when expressing L in Å, which is the natural unit of distance in such an atomic system. In a liquid, typical distances L are around 2-3 Å, which leads to $T_Q \simeq 10/A$, while it would only be about $T_Q \simeq 1/A$ in a gas. For average atomic masses ($A \sim 50-150$), which basically correspond to medium size chemical elements ($Z \sim 25-75$), we obtain $T_Q \sim 0.2-0.05$ K, therefore below 1 K. This corresponds to energies below 1 meV to be compared to typical interaction energies between eV and keV. Clearly, zero point motion is negligible as compared to interaction effects and cannot compete to impose a liquid behavior. At zero temperature, the system is thus a solid. It can only become a liquid when increasing the temperature and in this case the liquid nature stems from this *classical* temperature.

There are nevertheless noticeable exceptions to that simple result in the case of very light atoms, namely hydrogen and helium. Indeed atomic hydrogen ($Z = 1$, $A = 1$ or 2) when submitted to a very high pressure and high magnetic field is the unique chemical element which remains as a gas at very low temperatures. And helium ($Z = 2$, $A = 3$ or 4) is the unique element which remains as a liquid even at absolute zero of temperature. In this latter case, which is the single one we shall discuss in detail in the following, the origin for this liquid behavior is twofold. On the one hand, using the above estimate we find $T_Q \sim 2-3$ K which is a small value, but not that small. In addition, helium is the most inert chemical element in nature. Interactions between atoms in helium range are therefore very faint. Translated in terms of temperature, they attain values of the order of 1 K, namely fully comparable to T_Q. This is the basic reason why helium remains liquid at zero temperature.

To conclude on the case of atoms, we thus see that the simple general rule is that zero point motion is completely negligible as compared to interaction effects except for the lightest elements. The same argument holds for molecules, which by definition possess larger masses than atoms and hence even smaller zero point motion while interaction effects, but for exceptions, are much larger. The simplest solution to form other quantum liquids thus seems to consider lighter particles and we shall successively look at electrons and constituants of nuclei.

Electrons

Let us first consider the so-called *free electrons* of a metal, discussed at length in Section 5.2. We recall that electrons possess a mass m_e typically 2000 smaller than the one of the proton ($m_e/m_p \sim 1/1800$), which make them easily mobile. In a metal, a fraction of electrons (originally bound to their parent atom), named valence electrons, ensure binding between atoms. Specifically in metals, these 'binding' electrons remain only softly associated to their parent ion and then behave more or less freely (whence the term 'free').

The typical density in a metal ($\varrho \sim 10^{22} - 10^{23}$ cm^{-3}) provides distances between electrons of order $L \sim 2 - 5$ Å, from which one can easily evaluate the quantum temperature associated to zero point energy as $T_Q \sim 2 \times 10^5/L^2 \sim 10^4$ K. This value is notably larger than ambient temperatures of order 300 K. This clearly shows that in most situations involving metals the quantum temperature strongly dominates, so that quantum effects will play a key role to understand the properties of valence electrons in metals.

The role of interactions, still, deserves a further comment, as the question remains whether electrons behave as a gas or as a liquid. Indeed, we have mentioned the fact that electrons are not anymore strongly bound to their parent atom. The point should be specified a bit. This does not mean that interactions do not play a role. What we mean here is the fact that a given electron is not bound to a given atom anymore. The electron wave function is in fact spread over a sizeable number of ions. Binding therefore occurs 'globally' between the ions and the electrons, not 'individually' and the role of interactions between electrons and with ions hence remains crucial. Valence electrons in metals hence constitute a typical example of quantum liquid. Note also that ions, mostly because of their mass, remain basically at their equilibrium position (up to vibrational effects) and are certainly not constituting a quantum liquid.

Finally, a comment is in place here on the scale at which we are working. Standard metals belong to the human scale. In such cases, the metal is constituted of an enormous amount of ions/electrons (mind the above mentioned density) and is basically an infinite system. Now it turns out that one can also produce (and study) finite pieces of such metals, down to a few atoms/electrons. Such systems are called metal clusters and constitute essential objects to understand how matter builds up, not mentioning fascinating application from art to biology and medicine (see Section 5.3). In these systems one is facing a finite, possibly small, droplet of quantum liquid.

Plasmas also offer another a priori interesting case mixing electrons and ions, but this time with moving ions. Plasmas exist 'naturally' in stars and are used on Earth in fusion experiments. Again, because of high masses, ions do not constitute a quantum liquid. The case of electrons is different. Even if densities in massive stars may reach enormous values (5×10^{25} cm^{-3} in the core of our Sun), the typical distance between electrons is typically some fraction of angstroms. Therefore, the associated quantum temperature $T_Q \sim 2 \times 10^5/L^2$ will not go much beyond about 100 K which is vanishingly small as compared to typical temperatures in stars: some thousands of K at the surface of the Sun and about 10^7 K in its center. Very clearly, plasmas in massive stars are at such high temperatures that they can be treated classically (which is the usual approach in plasmas).

There is one noticeable exception for the behavior of electrons in stars. This is the case of white dwarfs (see Section 8.2). These stars look more similar to metals than to plasmas, with an ionic crystal and a 'sea' of electrons on top of it. The difference with 'normal' stars stems from the compacity of such stars. While having a mass of order the solar mass they possess a radius 100 times smaller which implies a typical distance between electron 1 million times smaller, indeed of order $L \sim 0.01$ Å. This leads to a quantal temperature $T_Q \sim 10^9$ K which is now much larger than the star temperature around $T \sim 10^7$ K. This means that electrons in a white dwarf do form a quantum liquid, and a quantum liquid at basically zero temperature ($T/T_Q \sim 0.01$!).

Nucleons

Nucleons (neutrons and protons) provide another interesting example of constituents of quantum systems, namely nuclei. One also often considers a fictitious system, called nuclear matter, which is formed of a (close to) infinite number of nucleons. Involved densities (in nuclei or nuclear matter) are enormous as well. The typical nucleon density in nuclei is $\varrho_0 \sim 0.17$ fm^{-3} (we recall that 1 fm=10^{-15} m). The associated distance between constituents is then $L \sim 2$ fm $= 2 \times 10^{-5}$ angstrom, which leads to a quantum temperature $T_Q \simeq 100/L^2 \sim 2.5 \times 10^{11}$ K. This is again an enormous quantity, in particular when compared to ambient temperature at which nuclei exist and are subject to experiments. This temperature (or the associated energy) has nevertheless to be compared to typical interaction energies between nucleons. The energy scale in nuclei is the MeV which in temperature terms amounts to 10^{10} K, only a bit smaller than T_Q. At that level of qualitative reasoning one should rather consider that zero point energy is of the order of magnitude of typical interactions between nucleons. It is therefore hard to conclude on the nature (gas, liquid, solid) of nuclei. There is fortunately another quantity which will allow us to conclude.

Since nucleons are fermions, we should also estimate the Fermi energy ε_F which provides the upper bound of nucleon kinetic energies. We have seen in

Section 3.4 that ε_F is related to the particle density as $\varepsilon_F = \frac{\hbar^2}{2m}(3\pi^2\varrho)^{2/3}$. Using the nucleon mass and the nucleon density ϱ_0, this leads to a value $\varepsilon_F \simeq 40$ MeV, that is, translated in temperature, $T_F \simeq 4 \times 10^{11}$ K. The latter is of the same order of magnitude as T_Q (although larger of course) and again of the same order of magnitude (probably larger) than typical interaction energies in nuclei. For example the average binding energy of a nucleon in a nucleus is about 8 MeV ($\sim 8 \times 10^{10}$ K) and this holds true for about all known nuclei (up to exceptions or 'exotic' situations). Since the average energy in a Fermi gas is $3\varepsilon_F/5$ (see Section 3.4), about 24 MeV, the typical kinetic energy of nucleons is clearly high enough to ensure sufficient motion of nucleons in nuclei. Nucleons in a nucleus therefore constitute a Fermi liquid. Because nuclei are finite, one can see them as droplets of liquid, more precisely as droplets of nuclear matter, nuclear matter being the 'macroscopic' associated liquid.

Although nuclear matter is primarily a theoretical concept (conceptually extremely rich), it also exists under a particular form in some stars, namely neutron stars. These stars are dominantly constituted of neutrons (see Section 8.3) at high densities, possibly several times ϱ_0. Increasing the density means increasing the Fermi energy (see Eq. (3.25)) which typically takes values in the 50-70 MeV ($T_F \sim 5 - 7 \times 10^{11}$ K) range. These stars are 'cold' for astrophysical standards with temperatures in the 10^9 K range. Again, neutrons in neutron stars constitute an archetypal quantum liquid, this time at macroscopic scale, at variance with finite nuclei.

3.5.3 A short summary

This rapid survey has allowed us to identify large classes of quantum liquids at various scales, from microscopic nuclei to macroscopic metals and astrophysics. In all situations we have seen that quantum effects play a key role and directly compare (in amplitude) to interaction effects. This is the basic reason why such systems are liquids, but *quantum* liquids because the agitation of quantum origin is enough to ensure a liquid behavior. We have also seen that such objects may show up either as finite systems, as for example in nuclei or metal clusters or as macroscopic phases, namely virtually infinite objects as in a metal or in stars.

Of course, the finite systems exhibit specific properties directly reflecting their finiteness but it turns out that they also exhibit generic trends, inherited from the bulk liquid, which usually allows us to understand them in simple terms. The point is interesting as in addition to the fact that quantum liquids of different constituents share similar pattern, the same also holds when considering finite or infinite phases thereof. We shall have numerous occasions to explore such effects in the following. Before actually looking at some of these systems in more detail we conclude this discussion by summarizing the key properties of the quantum liquids we shall discuss in the book. The summary

Table 3.2: Short summary of the various quantum liquids discussed in this book.

System	Metal Clusters	Liquid Helium	Nuclei	White Dwarfs	Neutron Stars
Constituents (considered)	Electrons (Ions)	Atoms ^{4}He, ^{3}He	Nucleons	Electrons	Neutrons
Spin	1/2	^{4}He: 0 ^{3}He: 1/2	1/2	1/2	1/2
Ambient Temperature	10–300 K	mK, K	300 K	10^7 K	10^9 K
Quantum Temperature T_Q	10^5 K	3–5 K	10^{11} K	10^9 K	10^{11} K
Interaction Energies	eV, keV $10^4 - 10^7$ K	μeV, meV $10^{-2} - 10$ K	MeV 10^{10} K	keV 10^7 K	MeV 10^{10} K
Typical Distances	2-5 Å	2-3 Å	2 fm	0.01 Å	1 fm
Density (per cm^3)	$10^{22} - 10^{23}$	10^{22}	10^{38}	$10^{28} - 10^{29}$	10^{39}

is presented in Table 3.2 and gathers conclusions drawn in previous discussions of the present section.

3.6 CONCLUDING REMARKS

The description of many particle systems is a key issue in physics and chemistry. We have addressed in this chapter some generic methods to deal with this crucial problem. We have first focused on classical aspects which provide a sound and accessible basis for understanding major questions in the field. We presented in particular kinetic theory which relates microscopic motion to global (thermodynamical) properties of a system. The statistical point of view has also been extensively discussed, again to relate microscopic and macroscopic worlds.

We have also spent some time to discuss specificities of statistical physics in a quantum context, especially in terms of ensembles of bosons and fermions. Focusing first on the ideal case of systems of non interacting particles we have been able to identify remarkable features of such systems. The formation of Bose Einstein condensates at low temperature is in particular an essential feature of bosons systems that we shall meet at various places throughout this book. We have also analyzed in some detail properties of (low temperature)

Fermi systems, emphasizing the key role played by the Fermi energy. These concepts will again be heavily used in many discussions in the following.

We concluded this chapter by stepping explicitly into the realm of quantum liquids. The step from the former Bose or Fermi systems essentially lies in the inclusion of interactions, in a way similar to classical liquids. The new aspect here precisely lies in the fact that liquid properties do not only stem from interaction but result from joint effects due to interaction and intrinsically quantum properties such as zero point motion or Fermi energy.

CHAPTER 4

Further theoretical aspects

This chapter aims at complementing the previous chapters, by giving a few technical details and presenting and discussing some theoretical developments which, for the sake of simplicity, have been either ignored or only treated in a superficial way.

4.1 THERMODYNAMICS AND STATISTICAL PHYSICS

4.1.1 The three laws of thermodynamics

We recall here the three basic laws of thermodynamics. As a first step, we consider only the case of closed homogeneous systems, that is those in which there is only one type of particles and the total number of particles N is constant.

First law. Consider an arbitrary thermodynamic transformation in a system from a given initial state to a given final state, and let Q denote the net amount of heat absorbed by the system, and W the net amount of work received by the system. There exists a physical quantity U, called internal energy, whose variation is given by $\Delta U = Q + W$. The internal energy is a 'state function' in the sense that its variation only depends on the initial and final states, whatever the path followed to go from one to the other. One consequence is that U is conserved if the initial state is identical to the final state while, in general, Q and W do depend on the chosen path.

The infinitesimal expression of the first law is $\mathrm{d}U = \delta Q + \delta W$. Note that the variation of U is described by the total differential $\mathrm{d}U$, while we consider infinitesimal variations of Q and W with the δ letter, to keep a trace that these variations depend on the path followed during the thermodynamic transformation, even if it is infinitesimal.

The first law simply expresses the conservation of energy. But it is not sufficient to describe the transformation of a system, as there are processes that never occur, even if the energy is conserved. Actually, in the first law, heat and work are considered on the same footing, while they are in practice not equivalent. Indeed, if work can be transformed into heat, the reverse process is often difficult to obtain, if not impossible. The second law expresses this experimental fact.

Second law. There exists no thermodynamic transformation whose sole effect is to extract a quantity of heat from a given heat reservoir and to convert it entirely into work (Kelvin's statement). There exists no thermodynamic transformation whose sole effect is to extract a quantity of heat from a colder reservoir and to deliver it to a hotter reservoir (Clausius's statement). In other words, even if a transformation is energetically possible, the second law determines if this transformation has a probability to happen.

There exists a state function S, called entropy, defined as $dS = \delta Q_{\text{rev}}/T$, where δQ_{rev} is the heat received by a system during a reversible transformation, which is conserved in any reversible cyclic transformation. For an arbitrary transformation, we have $dS \geq \delta Q/T$. One can relate S and U by considering a reversible transformation at constant volume (no work), for which we can write $\delta Q_{\text{rev}} = T\mathrm{d}S$ on the one hand, and $\mathrm{d}U = \delta Q_{\text{rev}}$ on the other hand. More generally, for a reversible transformation, we have $\mathrm{d}U = T\mathrm{d}S - P\mathrm{d}V$. This provides the thermodynamic definition of the temperature as $T = 1/(\partial S/\partial U)_V$.

Temperature drives the flow of energy between physical systems. There is also the case of systems that can exchange particles (they are called diffusive systems). In such a situation, one faces the change of the particle number N in the system. The corresponding intensive parameter is called the chemical potential. It is usually denoted by μ and it governs the flow of particles. If two systems (with a single type of particles) are at the same temperature and the same chemical potential, there is no net energy flow and no particle flow between them. If not, energy flows from the higher temperature to the lower one, and particles flow from the higher chemical potential to the lower one, until equilibrium is established between the systems. For a reversible transformation, the total differential of the internal energy is in this case written as $\mathrm{d}U = T\mathrm{d}S - P\mathrm{d}V + \mu\mathrm{d}N$. One can then define the chemical potential as $\mu = (\partial U/\partial N)_{S,V}$.

One can generalize the point to the case of a system consisting of a mixture of particles and denote N_i the total number of particles of type i. This situation is encountered when chemical reactions which absorb or create particles occur, or in phase transitions with the existence of the same chemical element but in different phases of matter. In such a case, we have $\mathrm{d}U = T\mathrm{d}S - P\mathrm{d}V + \sum_i \mu_i \mathrm{d}N_i$.

Third law. The second law defines the entropy of a system up to a constant, which is fixed by the third law. At absolute zero temperature the system must be in a state with the minimum possible energy. The third law has been formulated in several equivalent ways, two of which are the following. It is impossible to reduce the temperature of a system to absolute zero in a finite number of steps (Nernst's statement). The entropy of a system at absolute zero is a universal constant, which may be taken to be zero (Planck's statement).

Actually, the precise definition of entropy requires the description of the system in terms of its constituents. Both quantum mechanics and statistics are needed for this purpose, as has been discussed and derived in Chapter 3.

4.1.2 The analysis of phase transitions

We here complement the thermodynamics of phase transitions as discussed in Sec. 1.1.3. One can describe a phase transition through the Gibbs free energy defined as $G = U - TS + PV$. Its total differential is given by $\mathrm{d}G = S\mathrm{d}T + V\mathrm{d}P + \sum_i \mu_i \mathrm{d}N_i$. From this expression and according to the second law, one deduces that nature tends to minimize G. One consequence is that during a phase transition, say between the liquid and solid states, the Gibbs free energy of the liquid state must be equal to that of the solid state.

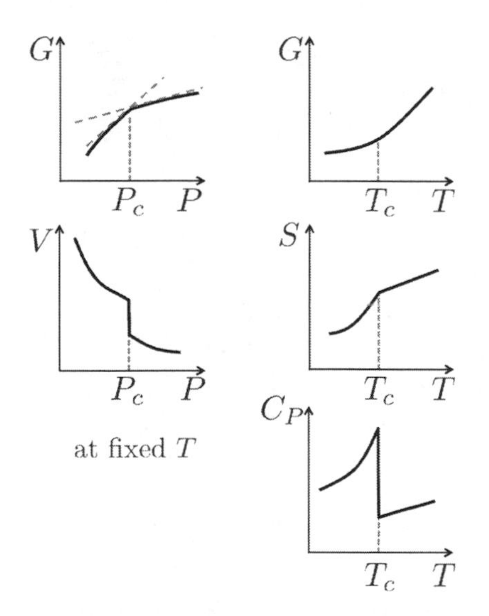

Figure 4.1: Left column: First order phase transition with a 'kink' visible in the Gibbs free energy (top) and a discontinuity of the volume V (bottom), both plotted as functions of pressure P. Right column: Second order phase transition with a kink in entropy S (middle) and a discontinuity of the heat capacity at constant pressure C_P, plotted as functions of temperature T (bottom).

Phase transitions can be classified according to the mathematical properties of G. The 'order' of the phase transition is given by Ehrenfest's criterium: a phase transition is said to be of n-order if the free energy has a discontinuous nth derivative. For instance, in the gas-liquid transition, the volume $V = (\partial G/\partial P)_{T,N}$ is discontinuous and we therefore have a first order transition, as depicted in the left part of Figure 4.1. The Ehrenfest's classification turned out to be incomplete in the sense that there exist phase transitions which exhibit a divergence, instead of a finite discontinuity, in a high order derivative of the free energy. For instance, the case presented in the right column of the figure, the heat capacity at constant pressure, given by $C_P = T\,(\partial S/\partial T)_{P,N} = T\left(\partial^2 G/\partial T^2\right)_{P,N}$, diverges at the transition temperature T_c. Such a phase transition is coined 'second-order' or 'continuous' phase transition. This occurs in some types of superconductors and in the superfluid transition (see Chapter 9).

Many fields of physics deal with continuous phase transitions near critical points. We simply mention here that an important issue consists in determining the related critical exponent. For instance, in the case of the divergence of the heat capacity C at the critical temperature T_c, C follows a typical power law as $C \propto |T_c - T|^\alpha$ where α is the critical exponent. Systems which exhibit the same critical exponent are said to belong to the same 'universality class'.

4.1.3 Boltzmann's distribution

We have deduced in Eq. (3.9) the number of ways of distributing N particles in p levels with n_i particles in level i. We want to determine the most probable distribution leading to the maximum number of microstates. It involves the factorial $N!$ of a large number N, resulting in a huge number. It is then more convenient to deal with logarithms.

We start with a given distribution $D = \{n_0, \ldots, n_p\}$ and the corresponding number of microstates $w(n_0, \ldots, n_p) = N!/(n_0! \ldots n_p!)$. The most probable distribution is obtained from the condition that, for any arbitrary variation δn_i of n_i, the quantity $\delta \ln w = \sum_i (\partial \ln w / \partial n_i) \delta n_i$ must vanish. The arbitrary variations of n_i must be consistent with a constant number of particles $N = \sum_{i=1}^{p} n_i$ and a constant total energy $E = \sum_{i=1}^{p} n_i \varepsilon_i$. This imposes some restrictions on the quantities $\sum_i \delta n_i$ and $\sum_i \varepsilon_i \delta n_i$. They are added to the equation on w, which becomes

$$\sum_{i=1}^{p} \left(\frac{\partial \ln w}{\partial n_i} - a + b\varepsilon_i \right) \delta n_i = 0 \,. \tag{4.1}$$

The constants a and b are called Lagrange multipliers, included here with a different sign for later convenience. Among the general solutions n_j of the previous equation, these Lagrange multipliers are chosen so that the constraints on N and E are fulfilled. As Eq. (4.1) must be satisfied for any arbitrary variation δn_i, we are left with:

$$\frac{\partial \ln w}{\partial n_i} - a + b\varepsilon_i = 0 \,.$$

In the expression Eq. (3.9) of $w(n_0, \ldots, n_p)$, we use Stirling's approximation, namely $\ln(p!) \simeq p \ln p - p$ if p is an integer much larger than 1. This gives $\ln w = N \ln N - \sum_i n_i \ln n_i$. The equation above becomes $\ln N - \ln n_i = -a + b\varepsilon_i$, which has the solution $n_i = N e^{a - b\varepsilon_i}$, known as the Boltzmann's distribution.

Although similar, it is more general than the Maxwell's distribution of velocities Eq. (3.4) which is the particular case of particles not interacting with each other, and where the ε_j are simply kinetic energies. This comparison can be used to convince ourselves that the parameter b must be related to the temperature. Actually, we have $b = \beta = 1/(k_\mathrm{B} T)$, resulting from the identification of the statistical and the thermodynamical entropies (see Sec. 3.2.2). It is also convenient to write $a = \beta\mu$, and deal with the new parameter μ known as the chemical potential. The Boltzmann's distribution is finally written as

$$n_i = Ne^{\beta(\mu-\varepsilon_i)} \quad . \tag{4.2}$$

The chemical potential is fixed by the normalization on the total number of particles (see Eq. (3.8a)):

$$N = \sum_{i=1}^{p} n_i = e^{\beta\mu} \sum_{i=1}^{p} e^{-\beta\varepsilon_i} \quad .$$

The sum in the right hand side is an important quantity, known as the canonical partition function:

$$Z = \sum_{i=1}^{p} e^{-\beta\varepsilon_i} \tag{4.3}$$

which can be directly related to thermodynamical potentials. We can then write $N = e^{\beta\mu} Z$ and

$$\frac{n_i}{N} = \frac{1}{Z} e^{\beta\varepsilon_i} \quad . \tag{4.4}$$

This gives the probability of having n_i particles in the energy level ε_i, in the most probable distribution.

4.1.4 Entropy and the number of microstates

We consider two different and isolated macrostates (E_1, N_1, V_1) and (E_2, N_2, V_2), both at equilibrium. We denote by $W_i(N_i, E_i)$ the number of microstates of the most probable distribution in system i. We now bring both systems in thermal contact, so that the energy and the number of particles can flow freely from one to the other. We have the total energy $E = E_1 + E_2$ and the total number of particles $N = N_1 + N_2$. What is $W(N, E)$ for the combined system?

Since E can be shared in many different ways between systems 1 and 2, we can as a first step fix the value of E_1 and look at the number of accessible states in the combined system. This number is the product $W_1(N_1, E_1)W_2(N - N_1, E - E_1)$. We now relax the value of E_1 and let it vary from 0 to E. Since the distributions at different E_1's are independent from each other, the total number of microstates is the sum of the previous products:

$$\Omega = \sum_{0 \leq E_1 \leq E} W_1(N_1, E_1) W_2(N_2, E - E_1) \quad .$$

This is not yet the number of microstates $W(N, E)$ of the most probable distribution of the combined system. Indeed, we have to minimize the above quantity with respect to E_1 with the restriction $E = E_1 + E_2$, written as $\mathrm{d}E_1 + \mathrm{d}E_2 = 0$. In other words, we have to solve

$$\mathrm{d}\Omega = \left(\frac{\partial W_1}{\partial E_1}\right)_{N_1} W_2 \, \mathrm{d}E_1 + W_1 \left(\frac{\partial W_2}{\partial E_2}\right)_{N_2} \mathrm{d}E_2 = 0 \quad ,$$

with the constraint $\mathrm{d}E_2 = -\mathrm{d}E_1$. We obtain the thermal equilibrium condition $\frac{1}{W_1}\left(\frac{\partial W_1}{\partial E_1}\right)_{N_1} = \frac{1}{W_2}\left(\frac{\partial W_2}{\partial E_2}\right)_{N_2}$, also recast as:

$$\left(\frac{\partial \ln W_1}{\partial E_1}\right)_{N_1} = \left(\frac{\partial \ln W_2}{\partial E_2}\right)_{N_2} . \tag{4.5}$$

In other words, at equilibrium, the quantity $\partial \ln W/\partial E$ of system 1 is equal to that in system 2. We can thus define the statistical entropy as:

$$S = k_\mathrm{B} \ln W \quad , \tag{4.6}$$

with W the number of microstates in the most probable distribution for the isolated macrostate (N, E, V). In macroscopic terms, thermal equilibrium means the same temperature for systems 1 and 2. This complies with Eq. (4.5) if we define a statistical temperature as:

$$\frac{1}{T} = \left(\frac{\partial S}{\partial E}\right)_{N,V} .$$

As we did in Sec. 3.2.2, this relation combined with the Boltzmann's distribution, and the fact that $\beta = 1/(k_\mathrm{B}T)$, allows us to identify the statistical temperature with the thermodynamical temperature, and also the statistical entropy (4.6) with the thermodynamical entropy.

4.2 QUANTUM DESCRIPTION OF MANY-BODY SYSTEMS

We consider here the case of a potential $V(x)$ independent of time t. The Schrödinger equation can be simplified by factorizing the spatial and time dependence of the wave function as in the case of a free particle: we write $\psi(x,t) = \varphi(x)\mathrm{e}^{-\mathrm{i}Et/\hbar}$, where $\varphi(x)$ satisfies the time-independent Schrödinger equation

$$H\varphi(x) = \left[-\frac{\hbar^2}{2m}\frac{\mathrm{d}^2}{\mathrm{d}x^2} + V(x)\right]\varphi(x) = E\varphi(x) . \tag{4.7}$$

In brackets appears the Hamiltonian operator which is the sum of kinetic and potential terms. This operator is related to the total energy of the particle. As in classical physics, if the potential is time-independent, the energy is a conserved quantity.

4.2.1 The role of boundary conditions

Boundary conditions play a major role in quantum systems. They in particular strongly impact the nature of solutions of the Schrödinger equation. In order to illustrate this point we consider, for the sake of simplicity, the case of no potential $V(x) = 0$, first for a system confined by bounded infinite walls at $x = \pm a$, as discussed in Section 2.4.1. The most general stationary solution of the Schrödinger equation can then be written as a linear combination of plane waves e^{ikx} and e^{-ikx}. A particle confined in such a box is subject to an infinite repulsion from the walls, preventing the particle to exit from the box. This means that the wave function must vanish at $x = \pm a$. The general solution is conveniently written in terms of the trigonometrical functions $\sin(kx)$ and $\cos(kx)$. The condition $\sin(ka) = 0$ implies ka to be an odd multiple of $\pi/2$, while $\cos(ka) = 0$ requires an even multiple of $\pi/2$. Both conditions are summarized in

$$k = n\frac{\pi}{2a}\,, \tag{4.8}$$

with $n > 0$ a positive integer. Indeed, the value $n = 0$ gives a vanishing solution, and solutions for negative of n are not linearly independent of those for positive n.

In the case of a uniform system, extended over a very large portion of space, it is convenient to consider periodic boundary conditions instead (see for instance Section 5.2.5). The entire (infinite) space is divided into adjacent cubic boxes of side length L, and one then requires each solution to be periodic with the period L for each of the three Cartesian coordinates, i.e., the solution is the same when one replaces $x \to x + L$, $y \to y + L$ or $z \to z + L$. The normalized solutions are

$$\frac{1}{\sqrt{V}}e^{i\mathbf{k}\cdot\mathbf{r}}\,, \tag{4.9}$$

where $V = L^3$ is the volume of the box, and the Cartesian components of the wave vector satisfy the quantization condition

$$k_i = n_i\frac{2\pi}{L}\,,\ i = x, y, z \tag{4.10}$$

with n_i an integer. The energies are then quantized as

$$E = \frac{\hbar^2}{2m}\left(\frac{2\pi}{L}\right)^2 n^2\,, \tag{4.11}$$

with $n^2 = n_x^2 + n_y^2 + n_z^2$.

Interestingly, both cases, that is one particle confined either in a finite well or in a periodic potential, lead to qualitatively similar behaviors, namely the quadratic quantization of energies (energies $\propto n^2$). This calls for two remarks. First, there is a direct impact of boundary conditions on quantization, a point which was already outlined in Section 2.4.1. Second, the similarity of quantization results suggests that a many-particle system in a periodic potential bears a strong resemblance with a set of 'similar' one-particle systems. This resemblance can actually be used both in solids (see Section 5.2.5) and, to some extent, in liquids (see Section 2.5.2).

4.2.2 The constructive role of stationary states

Once we obtained the function $\varphi(x)$, normalized to unity, the energy E can be calculated as the following integral

$$E = \int \mathrm{d}x \, \varphi^*(x) \left[-\frac{\hbar^2}{2m} \frac{\mathrm{d}^2\varphi}{\mathrm{d}x^2} + V(x)\varphi(x) \right] . \tag{4.12}$$

Integrating the first term by parts, one can write the particle energy E as

$$E = \int \mathrm{d}x \left[\frac{\hbar^2}{2m} \left(\frac{\mathrm{d}\varphi}{\mathrm{d}x} \right)^2 + V(x)|\varphi(x)|^2 \right] . \tag{4.13}$$

This is a particular example of the general rule of evaluation of average values of observables which has been mentioned in Section 2.3.3. The energy is the expectation value of the Hamiltonian $E = \langle H \rangle$.

It is interesting to note that the factorization $\psi(x,t) = \varphi(x)\mathrm{e}^{-\mathrm{i}Et/\hbar}$ of the wave function implies that the density $|\psi(x,t)|^2$ is independent of time. This reminds us of the well-known stationary solutions of standard wave equations in which spatial and temporal evolutions are separated and in which some points of space correspond to zero wave amplitude (nodes) and some other ones to maximum amplitude. In the present case, the stationary waves are the solutions of the time-independent Schrödinger equation. In general, these stationary waves provide a time-independent basis to expand the time-dependent wave function. The fact that the Schrödinger equation is indeed linear in ψ allows us to construct the expected (and observed) undulatory pattern such as interference and diffraction. This is by the way the standard strategy used in quantum mechanics to solve the time-dependent Schrödinger equation. One first obtains the stationary states $\varphi_E(x)$ with definite energy E, and then expresses the general solution of the time-dependent equation as a linear combination of stationary solutions:

$$\psi(x,t) = \sum_E c_E e^{-\mathrm{i}Et/\hbar} \varphi_E(x) \tag{4.14}$$

over all possible energies, c_E being constant coefficients. The general time dependence is therefore directly obtained from the trivial time dependence of the stationary solutions and the possibility for these stationary solutions to interfere with each other.

4.2.3 Density of states

An ideal gas may be modeled as N free particles confined in a cubic box. To calculate a quantity such as the total energy, one must take a sum over all the allowed states. In a large box, the momentum spacings are very small and it is a good approximation to replace the sum with an integral, in the following way. We consider the space of wave vectors $\mathbf{k} \equiv (k_x, k_y, k_z)$ and look for the number of states with momentum between $\mathbf{k}$ and $\mathbf{k} + \mathrm{d}\mathbf{k}$, which corresponds to the volume of the spherical shell. The quantization condition for momenta shows that a possible state occupies a volume $(2\pi/L)^3$ in momentum state. Therefore, the sum over states can be replaced with an integral over momentum, dividing by that volume occupied by a possible state:

$$\sum_{k_x,k_y,k_z} \rightarrow \frac{V}{(2\pi)^3} \int \mathrm{d}\mathbf{k}\,. \tag{4.15}$$

By integrating over all possible values of $\mathbf{k}$, one gets the total number of occupied states in the system. The right hand side should be multiplied by the degeneracy factor, if any, that is the number of particles which can occupy the same state. In the case of spin-s fermions, that degeneracy factor is equal to $2s + 1$.

By replacing the integral with the volume $4\pi k^3/3$, we obtain the number of states with wave vector less than or equal to k. Using the relation between the energy and the wave vector of a particle, we obtain the number of states with energy less than or equal to ε. Deriving the resulting expression with respect to ε, one obtains the so-called density of states:

$$D(\varepsilon) = \frac{V m^{3/2}}{\sqrt{2}\pi^2\hbar^3}\varepsilon^{1/2}\,. \tag{4.16}$$

The product $D(\varepsilon)\mathrm{d}\varepsilon$ gives the number of states with energy between ε and $\varepsilon + \mathrm{d}\varepsilon$.

4.3 THE MANY-BODY WAVE FUNCTION

4.3.1 Case of non-interacting particles

The idea here is to construct the wave function of a system consisting of N identical and independent particles. We use here the term 'wave function' as a generic term to avoid entering discussions around the more general (and more correct) formulation in terms of Dirac states. The account for spin degrees of freedom, which strictly speaking are not expressed as wave functions, is thus made practically by hand.

The case of two particles has been partially treated in Section 2.5.4. We recall that the wave function of the pair must be symmetric for bosons and antisymmetric for fermions. Denoting with $\psi_n(i)$ the wave function of particle i in state n, these pair wave functions are written as

$$\Psi_{\mathrm{sym}}(1,2) = \psi_n(1)\psi_m(2) + \psi_n(2)\psi_m(1) \tag{4.17}$$

$$\Psi_{\mathrm{anti}}(1,2) = \psi_n(1)\psi_m(2) - \psi_n(2)\psi_m(1) \tag{4.18}$$

apart from a normalization factor.

Notice that the antisymmetric wave function can be equivalently written in the form of a determinant

$$\Psi_{\text{anti}}(1,2) = \begin{vmatrix} \psi_n(1) & \psi_n(2) \\ \psi_m(1) & \psi_m(2) \end{vmatrix} \quad . \tag{4.19}$$

One can immediately guess the form of the antisymmetric wave function for a system of N particles. It can be written as a $N \times N$ determinant built up from the single-particle wave functions $\psi_i(j)$, where index i refers to the state and index j to the particle. It is commonly referred to as a Slater determinant, which we write as

$$\Psi_{\text{SD}}(1,\ldots,N) = \frac{1}{\sqrt{N!}} \begin{vmatrix} \psi_1(1) & \psi_1(2) & \ldots & \psi_1(N) \\ \psi_2(1) & \psi_2(2) & \ldots & \psi_2(N) \\ \vdots & \vdots & \ddots & \vdots \\ \psi_N(1) & \psi_N(2) & \ldots & \psi_N(N) \end{vmatrix} \quad . \tag{4.20}$$

The forefactor guarantees the right normalization of Ψ_{anti} assuming the single-particles normalized to unity.

Before considering the case of N bosons, it is convenient to write the Slater determinant in a different but equivalent way. We use all the possible permutations of N indices, generically denoted with the symbol P. A permutation is simply a rearrangement of the ordered set of integers $(1,\ldots,N)$, leading to the new set $(P(1),\ldots,P(N))$. There is a total number of $N!$ permutations of N different integers. The function (4.20) is also written as

$$\Psi_{\text{SD}}(1,\ldots,N) = \frac{1}{\sqrt{N!}} \sum_P (-)^P \psi_1(P(1))\ldots\psi_N(P(N)) \quad , \tag{4.21}$$

where the factor $(-)^P$ is ± 1, according to the odd-even symmetry of the permutation. Therefore, removing the factor $(-)^P$ in the previous expression, we suppress the change of sign associated to the permutation of particles or states. The result is a totally symmetric function, which is referred to as a 'permanent'. We write it as

$$\Psi_{\text{PER}}(1,\ldots,N) = \sqrt{\frac{n_1!\ldots n_N!}{\sqrt{N!}}} \sum_P \psi_1(P(1))\ldots\psi_N(P(N)) \quad , \tag{4.22}$$

where n_j is the number of particles in the jth state. Indeed, contrary to the case of fermions, there is no limitation for bosons to occupy a given state. The different normalization factor of Ψ_{PER} as compared to Eq. (4.20) is related to that fact. A permanent is the simplest symmetric wave function describing N independent bosons.

4.3.2 Many-body density matrices and correlation functions

Density matrices

Consider the stationary N-body wave function $\Psi(\mathbf{r}_1, \ldots, \mathbf{r}_N)$ and ignore, for the sake of simplicity, the possible spin degrees of freedom. Their inclusion and the generalization to the time-dependent case is straightforward.

The N-body density matrix is defined as:

$$\Gamma^{(N)}(\mathbf{r}_1, \ldots, \mathbf{r}_N; \mathbf{r}'_1, \ldots, \mathbf{r}'_N) = \Psi^*(\mathbf{r}'_1, \ldots, \mathbf{r}'_N)\Psi(\mathbf{r}_1, \ldots, \mathbf{r}_N) \quad . \tag{4.23}$$

Its diagonal part, $\varrho^{(N)}(\mathbf{r}_1, \ldots, \mathbf{r}_N) \equiv \Gamma^{(N)}(\mathbf{r}_1, \ldots, \mathbf{r}_N; \mathbf{r}_1, \ldots, \mathbf{r}_N)$, is simply the modulus square of the N-body wave function. With the normalization $\int \mathrm{d}^3\mathbf{r}_1 \ldots \mathrm{d}^3\mathbf{r}_N \, |\Psi(\mathbf{r}_1, \ldots, \mathbf{r}_N)|^2 = 1$, the diagonal N-body density matrix provides the density of probability to have a particle at position $\mathbf{r}_1$, another one at position $\mathbf{r}_2$ and so on.

This represents a highly detailed information on the system. In many situations one only requires the one- and two-density matrices ($N = 1, 2$, respectively). The diagonal one-body density matrix $\varrho^{(1)}(\mathbf{r}) = \Gamma^{(1)}(\mathbf{r}, \mathbf{r})$ gives the probability density of locating a particle at point $\mathbf{r}$ irrespective of the position of the remaining particles. This is nothing else but the particle density $\varrho(\mathbf{r})$, which is normalized to the number of particles $\int \mathrm{d}^3\mathbf{r}\varrho(\mathbf{r}) = N$. For a homogeneous system of N particles in a volume V, we have $\varrho(\mathbf{r}) = \varrho = N/V$.

Analogously, the diagonal two-body density matrix $\varrho^{(2)}(\mathbf{r}_1, \mathbf{r}_2) = \Gamma^{(2)}(\mathbf{r}_1, \mathbf{r}_2; \mathbf{r}_1, \mathbf{r}_2)$ gives the probability density of locating a particle at $\mathbf{r}_1$ and another at $\mathbf{r}_2$, irrespective of the positions of the remaining particles. It is normalized to twice the number of pairs $\int \mathrm{d}^2\mathbf{r}_1\mathrm{d}^3\mathbf{r}_2\varrho^{(2)}(\mathbf{r}_1, \mathbf{r}_2, t) = N(N-1)$.

Pair correlation function

Classically, the diagonal two-body density matrix for non-interacting particles is simply the product of two single-particle densities. Deviation from this behavior gives information about the 'correlations' of a given system, either due to interactions or to quantum statistics. One defines the pair correlation function as:

$$g^{(2)}(\mathbf{r}_1, \mathbf{r}_2) = \frac{\varrho^{(2)}(\mathbf{r}_1, \mathbf{r}_2)}{\varrho^{(1)}(\mathbf{r}_1)\varrho^{(1)}(\mathbf{r}_2)} \quad . \tag{4.24}$$

If there were no correlation between particles, $g^{(2)}(\mathbf{r}_1, \mathbf{r}_2) = 1$.

In a homogeneous system, the two-body distribution function only depends on the relative coordinate $\mathbf{r}_1 - \mathbf{r}_2$, since there cannot be any privileged point. Therefore, $g^{(2)}(\mathbf{r}_1, \mathbf{r}_2) = g^{(2)}(\mathbf{r}_1 - \mathbf{r}_2)$. In an isotropic system, there is no dependence on the angles of the vector $\mathbf{r}_1 - \mathbf{r}_2$ as well. We can then consider as a variable $r = |\mathbf{r}_1 - \mathbf{r}_2|$ instead. We hence simply write $g^{(2)}(|\mathbf{r}_1 - \mathbf{r}_2|) = g(r)$. From the normalization of $\varrho^{(2)}(\mathbf{r}_1, \mathbf{r}_2)$:

$$\begin{aligned} N(N-1) &= \int \mathrm{d}^3\mathbf{r}_1\mathrm{d}^3\mathbf{r}_2 \, \varrho_N^{(2)}(\mathbf{r}_2 - \mathbf{r}_1) = \varrho^2 \int \mathrm{d}^3\mathbf{r} \, \mathrm{d}^3\mathbf{r}_2 \, g^{(2)}(\mathbf{r}) \\ &= \varrho^2 \int 4\pi r^2 \mathrm{d}r \, \mathrm{d}^3\mathbf{r}_2 \, g(r) = \frac{N^2}{V} \int 4\pi r^2 \mathrm{d}r \, g(r) \quad . \end{aligned}$$

One deduces the normalization of the pair correlation function:

$$\varrho \int 4\pi r^2 \,\mathrm{d}r\, g(r) = N - 1 \quad . \tag{4.25}$$

Further interpretations of $g(r)$ in classical or quantal systems can be found in Sections 1.3.3 and 2.5.5 respectively.

Structure factor

In solids and liquids, the correlation function $g(r)$ is experimentally deduced from neutron scattering or X-ray diffraction. An incident beam, characterized by a given value of the momentum, is scattered by the constituents of the target system. The intensity of the deflected beam is analyzed in terms of its momentum and deflecting angle. Let us denote by $\mathbf{q}$ the difference between the incident and the scattered wavevectors, i.e. the momenta divided by $\hbar$. The measured intensity is proportional to the structure factor $S(\mathbf{q})$, defined as the following expectation value:

$$S(\mathbf{q}) = \frac{1}{N}\left\langle \left| \sum_i \exp(\mathrm{i}\mathbf{q}\cdot\mathbf{r}_i) \right|^2 \right\rangle = 1 + \frac{1}{N}\left\langle \sum_{i\neq j} \mathrm{e}^{i\mathbf{q}\cdot(\mathbf{r}_i - \mathbf{r}_j)} \right\rangle$$

The average above is calculated using a distribution probability (see Section 3.2.3). If the N-body system is described by a wave function $\Psi(\mathbf{r}_1, \ldots, \mathbf{r}_N)$, the structure factor can further be written as:

$$S(\mathbf{q}) = 1 + \frac{1}{N}\int \mathrm{d}\mathbf{r}_1 \mathrm{d}\mathbf{r}_2\, \mathrm{e}^{i\mathbf{q}\cdot(\mathbf{r}_1 - \mathbf{r}_2)}\, g^{(2)}(\mathbf{r}_1 - \mathbf{r}_2)$$

The second term above, apart from the $1/N$ factor, is nothing else than the Fourier transform of $g^{(2)}(\mathbf{r})$. One therefore accesses the pair correlation function from the structure factor by Fourier transform. For example, in a homogeneous system, we have:

$$S(\mathbf{q}) = 1 + \varrho \int \mathrm{d}\mathbf{r}\, e^{i\mathbf{q}\cdot\mathbf{r}}\, g(\mathbf{r})$$

This connection between measured structure factor and pair correlation is used in Section 9.2, in connection with the superfluidity of liquid helium.

4.4 HOW TO TREAT A QUANTUM LIQUID THEORETICALLY?

Answering this question would require a book of its own, if not a library. The problem of describing many-particle systems in quantum physics is a major issue which has been the focus of numerous investigations since the early days of quantum mechanics. The case of quantum liquids, which represents a large fraction of quantum many-body systems, plays a central role in this problem. Even more so, quantum liquids, for example helium or more recently cold atomic gases, are often considered as archetypal systems to test our understanding of many-body concepts and theories. We shall see many such examples in the second part of this book. For the time being, we only want to give a few keywords and research directions in the field. Obviously, we cannot be exhaustive here. Therefore we shall focus our discussion on a few general ideas, making possible the use of a few concepts and keywords for the following discussions.

The many-body problem

As most often in quantum mechanics, the starting point is the Schrödinger equation for the wave function Ψ_N describing the N-particle system one is interested in. We first consider the stationary version $H\Psi_N = E\Psi_N$ thereof to determine the ground state of the system or to provide a starting point for its time evolution (see Section 4.2.2). We recall that the Hamiltonian H contains the kinetic energy term and the potential term, containing the two-body (possibly three-body, etc.) interaction between particles. The interaction term is responsible for the technical difficulties encountered for solving the problem.

The wave function $\Psi_N(\mathbf{r}_1, \mathbf{r}_2, ..., \mathbf{r}_N)$ depends on $3N$ spatial variables, not accounting for spin or other degrees of freedom. This means that describing the N-body system implies accessing a space of dimension $3N$ and functions built in this space. The complexity is naturally enhanced by the fact that, in a real system, particles do interact with each other. In the case of a system of non-interacting particles, possibly subject to an external potential, the N-body wave function is a simple product of single-particle wave functions. This factorization strongly simplifies the problem. Unfortunately, such a situation is an idealization. A direct solution of the real interacting problem is beyond the possibility of any computer, even hard to imagine mentally, except for very simple systems containing a few number of particles. One then needs to develop systematic approximation schemes. We outline below a few such methods, only those which are mentioned throughout the book.

Mean-field based approaches: Hartree-Fock and beyond, Density Functional Theory

A first general class of methods consists in solving the problem by reducing its complexity, either by making some assumptions on the nature of the Ψ_N wave function and/or by simplifying the treatment of interactions. The simplest versions of such approaches are known as mean-field theories. Mean field corresponds to the assumption that the interaction of a given particle with all the others can be approximated by an average potential V_{mf}. Under such conditions, interactions can be 'hidden' in the mean field and the many-body wave function Ψ_N can be taken as Ψ_{SD} for fermions or Ψ_{PER} for bosons (see Section 4.3.1). The single-particle wave functions φ_i involved in Ψ_N feel the same 'external' potential V_{mf}, the latter being constructed by averaging the true interaction using precisely the set of functions $\{\varphi_i, i = 1 \ldots N\}$. Instead of solving the N-body Schrödinger equation, one solves a set of N *one*-body

Schrödinger equations

$$h\varphi_i(\mathbf{r}) = \varepsilon_i\varphi_i \quad \text{with} \quad h = -\frac{\hbar^2}{2m}\Delta + V_{\text{mf}}$$

where ε_i now labels the energy associated to the single particle i. With this setup at hand, the actual solution of the problem, say for example the ground state of the system, is obtained by minimizing the total energy of the system computed using Ψ_{SD} (in fermionic systems) as a test wave function. This reduces to solve the above N Schrödinger equations for each φ_i. This is an iterative process to be solved iteratively, as the potential depends on the set of solutions $\{\varphi_i, i = 1\ldots N\}$.

A typical simple version of mean-field was introduced by Hartree in the late 1920s for multi-electron atoms. Each electron, described by a wave function $\varphi_i(\mathbf{r}), (i = 1\ldots N)$, feels the attraction of the atomic nucleus (a true external field for the electrons) and an average of the interactions with the remaining electrons. In practice, the mean field is assumed to be the classical Coulomb potential created by the electron density $\varrho(\mathbf{r}) = \sum_i |\varphi_i(\mathbf{r})|^2$, thus neglecting Pauli's exclusion principle for electrons. This oversimplified mean field is known as Hartree approximation. It misses several key aspects. In particular, the wave function is simply a product of N single-particle functions $\varphi_i(\mathbf{r})$, instead of the expected Slater determinant (4.20). Instead, the Hartree-Fock approximation starts from a Slater determinant Ψ_{SD} as the many-body wave function to (self-consistently) calculate the mean field. This new mean field contains the above classical (Hartree) part and a complementing exchange term (see the discussion around Figure 2.10). The latter term is easy to express formally but very costly computationally speaking in the case of the Coulomb interaction, characteristic of electronic systems. The Hartree and Hartree-Fock approximations provide two powerful approximations which have served as basis for numerous formal and practical developments. Any improvement over them consists in reintroducing effects 'beyond mean field', that is beyond the mere factorization of the wave function, and therefore account for 'correlations', or interaction effects not considered in the mean field.

Another direction consists in keeping the simplicity of a mean-field approach, and develop 'simple' approximations for the contribution of exchange and correlations. The theory built on this strategy is called Density Functional Theory (DFT). It can be shown mathematically that such a strategy should work, provided one knows the exact correction to be inserted, and that this correction should be expressed in terms of the simple one-body density $\varrho(\mathbf{r})$. Unfortunately enough, the exact exchange correlation term is unknown and one has to refer to approximations. Such methods have been widely developed over the past half century and led to remarkable successes in particular in systems with a potentially sizable number of constituents (up to a few thousands).

Quantum Monte Carlo

The simple mean-field based approaches have met numerous successes but only provide approximate solutions. In general, the Schrödinger equation for a system of interacting particles can be exactly solved only in a few simple cases. Otherwise, the wave function is obtained numerically according to various approximate methods. There is a completely different path to address the problem which avoids the direct determination of the wave function. Instead, using statistical tools, they aim at obtaining stochastic estimates of expectation values of some operators, providing for them 'exact' results within controlled statistical errors. In general, this is true for boson systems but not for fermions, due to the antisymmetry of the wave function. Additional approximations and specific techniques are developed to overcome this difficulty. Stochastic methods are known under the generic name Quantum Monte Carlo, and come in several varieties.

We only give some hints about the Path Integral Monte Carlo method. It aims at describing a quantum system in thermal equilibrium, based on the partition function (see Section 3.2.3). All static properties of the system are obtained from the operator $\exp(-\beta H)$ (recall that $\beta = 1/k_{\mathrm{B}}T$), which underlies the partition function in the canonical ensemble. Accurate values for its N-body density matrix are obtained at high temperature. The basic idea to reach low temperatures is contained in the identity $\exp(-\beta H) = \exp(-\beta_1 H)\exp(-\beta_2 H)$, with $\beta = \beta_1 + \beta_2$. Starting from a high value T', this identity is used M times to reach the low temperature $T = T'/M$ one is interested in. In the Path Integral Monte Carlo approach, one obtains the density matrix at temperature T as an integral over all possible paths given by a product of M density matrices at temperature MT. At present it is the only numerical method capable of directly addressing finite-temperature and superfluid transitions in liquid helium.

Time-dependent approaches, quantum, semi-classical and classical dynamics

All the above discussed methods aimed at computing stationary solutions of the Schrödinger equation from which one can obtain either average quantities or, when knowing all stationary solutions (which in general is a formidable task), any time evolution of the system. Another option consists in trying to directly solve the time-dependent Schrödinger equation within an appropriate approximation scheme. Among the widely used methods, one can cite the Time-Dependent Hartree-Fock and the Time-Dependent Density Functional Theory approaches, both exploiting the simplifications of the many-body wave functions. There also exist time-dependent versions of correlated methods 'beyond mean-field'. However they are restricted to very small systems (a few particles at best). The solution of the time-dependent quantum many-body problem thus remains a challenge and only approximate solutions thereof are attainable.

Before closing this discussion, it is worth mentioning a complementing aspect which allows one to consider many applications otherwise unattainable, and this in a dynamical context. It relies on an approximation of the Schrödinger equation, but this time of a different nature. The idea consists in exploring how much the quantum system under consideration is quantal. For example, a highly excited quantum system is known to exhibit more and more classical features (see Section 2.2.3). Indeed, while electrons in the ground state of a solid are highly quantal, they become basically classical in high temperature plasmas in which quantum effects are washed out by temperature (see Section 1.5.2). The idea to approximate a quantum problem by a simpler, more or less classical one, is as old as quantum mechanics and we shall not review here all the achievements along that line. We only discuss two limiting cases which have been used in quantum liquids. The first idea is to explore whether one could rely on a simpler description of the system than the many-body wave function Ψ_N, typically in terms of a phase-space distribution function 'similar' to that used in Boltzmann equation (see Section 3.1.5). In fact, the connection can be established in a formal manner and leads to time-evolution equations which resemble the Boltzmann equation, although with a different meaning. These 'semi-classical' approaches are usually painful to solve. Except when the system is highly excited, they do not provide a clearly advantageous solution.

The next step consists in performing a direct classical treatment of the time evolution of the system. This amounts to associating one classical particle to any particle of the system. Each classical particle is then characterized by its position and momentum and follows Newton equations of motion. Motion is driven by the interactions between the particles. It corresponds to a more microscopic treatment than within a phase space or fluid dynamical picture, with a proper treatment of all interactions between particles. As such, all correlations are included... but only at a classical level. This in particular makes the description of truly quantum systems highly questionable. For example, electrons in a solid cannot be treated that way. On the other hand, such approaches, called Molecular Dynamics approaches, have been widely used to describe molecular motion inside a gas or a liquid (see the calculation of the pair correlation function in Section 1.3.3). They can reach a high degree of sophistication in the description of interactions between the molecules, even including some quantum effects. They have thus been used in many situations with a remarkable degree of success, as long as they are applicable.

The Games

CHAPTER 5

Electron fluids

The quantum liquid we are more familiar with, and probably without knowing it, is actually hidden in a solid. In our daily experience, we know that metals are good conductors of electricity. This is due to the fact that a huge number of electrons in metals can move almost freely like in a liquid. Their random motion however produces no net flow of charge, and an external electric field is required to force a collective motion of electrons and produce an electric current. Solids are composed of atoms, and their different physical and chemical properties are due to the ability of electrons to move... or not to move. Indeed electrons are the leading pieces that bind matter together at the nanoscale. Obviously, without the presence of atomic nuclei, the confinement of electrons in bulk matter would not be possible. But the fact remains that many properties of solids, especially metals, can be explained in terms of a fluid of electrons, a quantum fluid of course. In particular, the fact that they are fermions is essential to explain the origin of the chemical binding and the conductivity of a solid.

In this chapter, we first briefly review how atoms bind together in molecules and solids, and then we focus on the microscopic description of electric conductivity in terms of an electronic quantum fluid. We then move on and describe atomic and molecular clusters which are often viewed as systems interpolating atoms or molecules and bulk matter. We focus on the case of metal clusters which, to some extent, constitute droplets of electronic quantum liquid.

5.1 ELECTRONS IN ATOMS, MOLECULES AND SOLIDS

Electrons exist in relation to atoms (matter is basically neutral). They play a key role in the way atoms do bind together to form molecules or solids. But one cannot discuss electrons without reference to a confining agent, that is ionic cores or external fields which force them to stay together, simply because the electron-electron interaction has a repulsion of coulombic origin. The purpose of this section is to address this question and to better characterize the electrons in relation to their environment, taking as a starter their role of binding, both from the molecular and solid state physics point of view. There exists indeed a great variety of electronic binding, inducing for instance the conductor or insulator character of a given element. Before exploring how atoms bind together, let us review some properties of atoms.

5.1.1 Atoms

Electrons in an atom interact with each other and with the central nucleus through Coulomb interaction (gravitational effects can be safely neglected). Solving the Schrödinger equation for that system is a complicated problem. Fortunately there exists a useful approximation, the so-called mean-field approximation (see Section 4.4). An electron feels a strong Coulomb attraction due to the central nucleus, and interactions to the remaining electrons. The exact problem can be transformed into a simpler one, that is solving a Schrödinger equation for an electron moving in a mean-field generated by the nucleus and the other electrons. What matters for the present discussion are some general considerations made in Section 2.3, regarding the Schrödinger equation for a single particle. In particular, the stationary states are characterized by a set of quantum numbers that we have presented in Section 2.4.2.

We recall here that their energies depend on the principal quantum number n, which takes integer values starting at 1, and the orbital quantum number l, ranging from 0 to $n-1$. We have also briefly discussed in Section 2.4.4 the filling of the electronic levels according to some rules. And this has provided the periodic table of the chemical elements (see Appendix B). Let us here recall some important features. Conventionally, the various values of $l = 0, 1, 2, 3, 4\ldots$ are denoted by the letters $s, p, d, f, g\ldots$ respectively. Due to the Pauli exclusion principle, each energy level can be occupied by a maximum of $2(2l+1)$ electrons. The ground state of the atom is obtained by filling the lowest energy levels, which are grouped into degenerate 'sub-shells' (characterized by l) and 'shells' (labeled by n) formed by sub-shells with nearly equal values of energy. Rare gas atoms, which are in the last right column of the periodic table, contain closed shells, that is, all energy levels are occupied with the maximum number of electrons. The electronic structure of He is therefore denoted as $1s^2$, that of Ne is $1s^2 2s^2 2p^6$ or a 'core' of He plus $2s^2 2p^6$. This practically means that, around a helium 'core' constituted of 2 electrons in the $1s$ shell, are 8 electrons themselves shared between the $2s$ shell (1 electron with spin up and 1 electron with spin down) and 6 electrons in the 3 degenerate $2p$ shells. Note that the latter shells possess different spatial geometries and are usually denoted $2p_x, 2p_y, 2_z$ (see also Section 2.4.4). With the same notation, Ar is [Ne]$3s^2 3p^6$, Kr is [Ar]$3d^{10} 4s^2 4p^6$, etc. A consequence of the shell closure is that these atoms are particularly inert in chemical reactions. By contrast, alkaline and halogen atoms are highly chemically reactive species, since their electronic structure is that of a rare gas plus or minus an electron.

To be a bit more quantitative, we present in Figure 5.1 the electronic levels of two metallic atoms, namely copper and sodium. The height of the bars

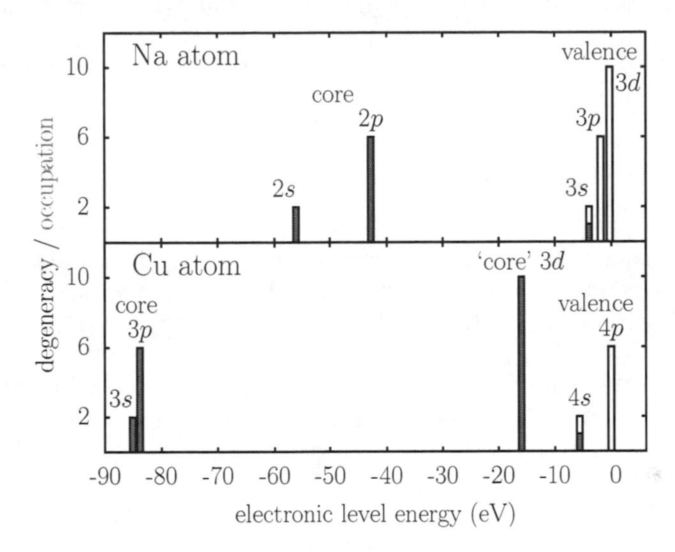

Figure 5.1: Comparison of electronic structure of two metal elements, namely copper (bottom) and sodium (top). Only levels with energy > 90 eV are shown. The height of the boxes indicate the degeneracy in each level and the shaded area the corresponding actual occupation.

represents the degeneracy of a given electronic level, and the shaded area in each bar shows the actual occupation in the neutral atom. Unoccupied states correspond to the available 'space' for additional electrons, if any. One can see an energy gap between the occupied $3s, 3p$ states and the $3d, 4s$ states in Cu, and between the occupied $2s, 2p$ states and the $3s$ state for Na. The electrons in the high-lying states are less bound, and consequently can interact with electrons of other atoms more easily than those in the low-lying levels, which form a core of less reactive electrons. The least bound electrons in an atom are called valence electrons. The distinction between core and valence electrons may differ under different circumstances. For instance, electrons in the $3d$ level of Cu can be considered as core electrons in low-energy situations involving interaction energies smaller than the energy gap of about 10 eV between $3d$ and $4s$ levels. In other dynamical processes, involving higher energies and/or frequencies, one might need to consider the $3d$ electrons of Cu amongst the active electrons.

It is also important to consider the Highest Occupied Molecular Orbital (HOMO) and the Lowest Unoccupied Molecular Orbital (LUMO) of a given atom. The $4s$ level in Cu ($3s$ in Na) is the HOMO, while the $4p$ level in Cu ($3p$ in Na) is the LUMO. Several chemical and physical properties of an atom rely on the energies of these levels. For instance, the difference between the energies of the LUMO and the HOMO corresponds to the lowest possible energy of an electronic transition from an occupied towards an empty level. The so-called ionization potential (IP), which is the HOMO energy with changed sign, corresponds to the needed energy to singly ionize the atom. The electron affinity (EA) is the minimal amount of energy released (if positive) or needed (if negative) to add an electron to the atom.

5.1.2 The molecular picture

When two or more atoms are close enough, the valence electrons of one atom feel more intensely the presence of the other atom. Actually, the wave functions of valence electrons superimpose with each other and, in some circumstances, they lose their identity. Strongly bound electrons tend to remain on their parent atom, while loosely bound ones have a tendency to leave it. Chemical bonds result from the 'share' or 'exchange' of these electrons between atoms. Let us consider what happens in the case of two atoms that form a dimer molecule. For simplicity, we assume the atoms to have a single LUMO and a single HOMO. Starting from simple considerations based on these electronic states of partner atoms, four major types of binding between atoms can be identified, directly linked to the degree of localization of the electrons involved in the bond, as schematically represented in Figure 5.2.

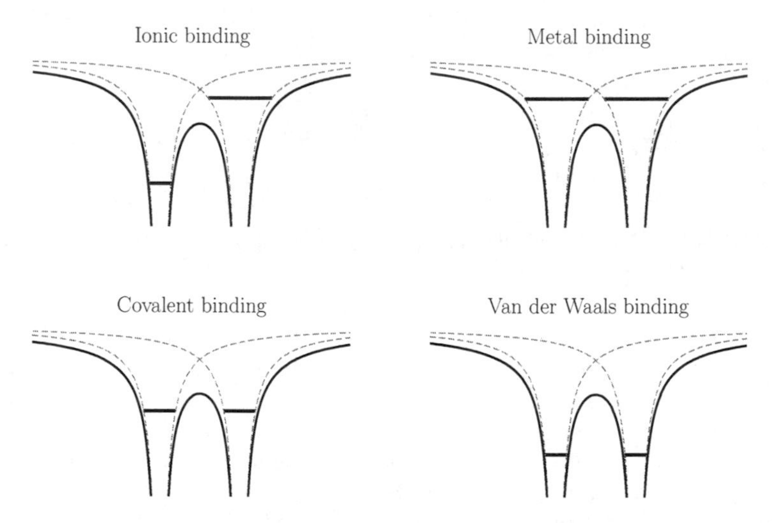

Figure 5.2: Four basic types of binding in a dimer with potentials originating from single atoms (dashes), the potential seen by an electron in the dimer (full curve) and valence levels (horizontal lines).

Ionic binding

We start with the case of a loosely bound electron, as for instance from an alkaline atom, transferred to an empty state of a hosting atom, as for example a halogen atom. The smaller the IP of the donor atom and the higher the EA of the acceptor atom, the easier the electron transfer with a localization of the electron mostly on the hosting atom. This leads to the formation of a cation strongly bound to an anion through this ionic bond, hence the naming ionic binding. NaCl, NaI or LiF are typical examples of molecules with an ionic bond and a binding energy of a few eV.

Metal binding

The extreme case of ionic binding occurs when the energy difference of the valence shells of the partner atoms is large. Otherwise, the probability of electron transfer becomes too low, due to too high an energy barrier between both atoms. This is often the case in homo-atomic dimers. Here, instead of a charge transfer, there is a charge sharing in which the valence electrons form a common electronic cloud extending over both atoms. Two sub-cases have to be distinguished here, depending once again on the electronic energies of the parent atoms. If the valence electrons are weakly bound and if the distance between both atoms is small enough, the electronic wave functions fully delocalize over both atoms. This constitutes the so-called metallic bond. The typical chemical elements which establish metallic bonds between each other are alkalines. They are coined simple metals because the metallic bond involves here only one loosely bound electron from the HOMO per atom. In transition metals instead, such as Cu, Ag, Au, or Pt, the shell below the HOMO also participates to the binding. For instance, in the case of Cu, the $3d$ shell is sufficiently close energetically to the $4s$ shell, to contribute to the metal binding, in particular through strong polarization effects.

Covalent binding

On the contrary, when electrons are initially more strongly bound to their parent atom, electrons do not delocalize and rather remain in the region of smallest potential energy, that is between the two partner atoms. This leads to the so-called covalent bond. C_2 and Si_2 are typical examples of covalent binding. It is hard in practice to distinguish the covalent and the metallic binding in molecules, as they correspond to ideal situations. For instance, in graphite, the carbon atoms are in principle covalently bound. Still, its electrical conductivity is high because electrons actually delocalize over the ionic structure. In general, there also exists a variety of situations in between the covalent and the metallic bindings, depending on the involved chemical element. Both types of bond exhibit binding energies of a few eV.

Theoretically, one can describe a formation of covalent bond in a dimer with the help of a linear combination of atomic orbitals. One starts with two atomic wave functions, the two HOMO for each atom, and constructs a molecular wave function as a linear combination of the atomic wave functions. According to the symmetry of the resulting wave function (see Section 2.5.4), a level of lower energy (the so-called bonding molecular orbital) and another one of higher energy (the antibonding molecular orbital) can show up when the two atoms are positioned close enough to form a dimer. This approach allows us to understand how this combination can give rise to a bonding and an anti-bonding.

Van der Waals binding
The last case of binding concerns closed-shell atoms, that is rare gas atoms. Here, neither charge transfer nor charge sharing are possible between both atoms, and electrons basically remain localized on their parent atom. Electronic binding is however still possible through mutual polarization of the electronic clouds of the atoms. This binding results from a dynamical correlation effect, with constantly fluctuating dipoles, rather than a mere static polarization. It can be understood via mutual dipole-dipole interactions between the atoms, which finally establishes a binding of the system. The resulting binding is much weaker (binding energy of a fraction of eV) than the ionic, metallic or covalent bonds. It however allows the build-up of possibly large compounds of rare gas atoms. This type of binding is called van der Waals binding. It can also describe the binding between small neutral molecules such as CO_2 which are compounds with closed shells. This is why this binding is also called intermolecular binding.

To end this section, we recall that the strength of interactions between atoms, stemming from the filling of electronic shells, can have macroscopic implications such as a particularly low (high) melting or boiling temperature if the interaction is very weak (strong); see Figure 2.6. Mind also that the different bonding mechanisms discussed above should be considered as ideal cases. Real cases often lie in between, except for particularly clean cases which precisely serve as prototypes. We shall come back to this aspect in Section 5.3.2 in relation with Figure 5.7.

Let us here be a bit more quantitative on the covalent and the ionic binding by presenting the ideas of the technique of linear combination of atomic orbitals (LCAO). It constitutes a simple and approximate method of quantum chemistry that yields a qualitative picture of the molecular orbitals in a molecule. In the case of a dimer formed by the sharing of a single electron between an atom A, located at $\mathbf{R}_A$, and another atom B, located at $\mathbf{R}_B$. We assume that we exactly know the so-called atomic orbital ψ_A (resp. ψ_B) for the atom A (B), possessing an energy E_A (E_B). To fix ideas, let us consider that $E_A < E_B$. The spirit of LCAO is to approximate the molecular orbitals (MO) of the dimer AB as :

$$\psi(\mathbf{r}) = c_A \psi_A(\mathbf{r} - \mathbf{R}_A) + c_B \psi_B(\mathbf{r} - \mathbf{R}_B) \quad , \tag{5.1}$$

where c_A and c_B are two constant parameters to be determined. This problem can be analytically solved. Here, we only discuss the final results. There are two independent solutions, denoted by $\psi_\pm$, of energy $E_\pm$ with the following hierarchy:

$$E_- \leq E_A < E_B \leq E_+ \quad . \tag{5.2}$$

The energies $E_\pm$ depend on $E_{A,B}$ but also on the interaction between atoms

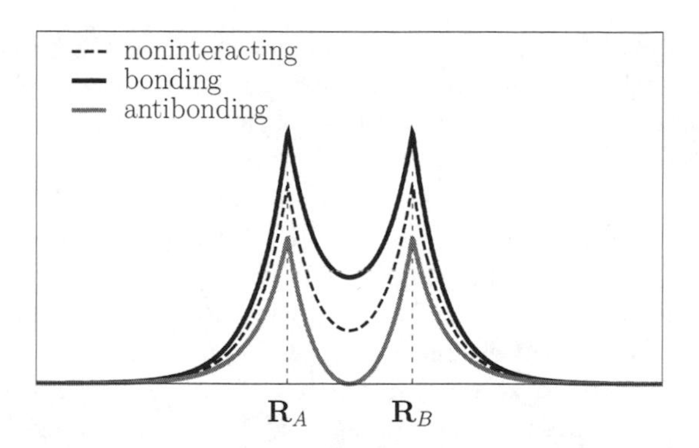

Figure 5.3: Electronic density of a monoatomic dimer when the two atoms do not interact (dashes), and for the bonding (black) and antibonding (gray) molecular orbitals.

A and B. ψ_- is said to be the 'bonding' molecular orbital, and ψ_+ the 'antibonding' one.

There are two extreme situations in which $\psi_\pm$ have a simple form. The first one concerns the case of a monoatomic dimer, that is A and B are identical atoms. We then have:

$$\psi_+ = \frac{1}{\sqrt{2}}(\psi_A + \psi_B) \quad \text{and} \quad \psi_- = \frac{1}{\sqrt{2}}(\psi_A - \psi_B) \tag{5.3}$$

Figure 5.3 shows the electronic density for two noninteracting atomic orbitals, the bonding MO and the antibonding MO. We now understand the naming 'antibonding' since there is no electronic density in between the two atoms. At the other extreme, the bonding MO provides more electronic density between both as compared to the noninteracting case. The bonding MO corresponds to the covalent binding.

The second extreme case is obtained when A and B, of different nature here, do not share the electron. The latter is rather almost completely localized on atom A, while atom B is empty. Then $E_- \simeq E_A$, $\psi_- \simeq \psi_A$, $E_+ \simeq E_B$ and $\psi_+ \simeq \psi_B$. This corresponds to the ionic binding.

5.1.3 The solid state picture

To describe the quantum liquid we are interested in, we must have a rough picture of solids. For our purpose, we will leave aside amorphous solids and glasses, and only consider crystalline solids, characterized by a regular arrangement of atoms, a periodic pattern which is called a crystal lattice. The solid can be regarded as a very large molecule, and the previous schematic classification in four types of binding also allows the understanding of bulk materials. There are thus ionic, metallic, covalent and van der Waals solids.

We can guess that neither ionic nor molecular solids are good electricity conductors, because there are no free electrons to carry a charge from one point of the solid to another. Covalent solids contain atoms which are bound by sharing valence electrons, while in metallic solids, valence electrons are shared by all ions in the crystal. Metals can be thought as a limiting case of covalent binding. Actually, about 3/4 of chemical elements present in nature are metals. Their atoms are located in the left-hand side of the periodic table of elements and have an electronic structure formed by an ionic rare-gas core plus a number of more loosely bound valence electrons.

As already mentioned above, the linear combination of the HOMO of two atoms leads to the formation of a (low-energy) bonding molecular orbital and a (high-energy) antibonding molecular orbital, explaining the formation of a dimer. A similar reasoning with the LUMO of each atom leads to two more higher levels for the dimer. We can push forward the argument by including more and more atoms to obtain a qualitative picture of the final structure of levels. More atoms means linear combinations of more orbitals—both occupied and unoccupied—resulting in an increase in the number of energy levels. Some of these levels are occupied by the valence electrons of the whole system. The other levels are unoccupied, and can be separated by an energy gap from the former. The result is schematically pictured in Figure 5.4. Increasing the

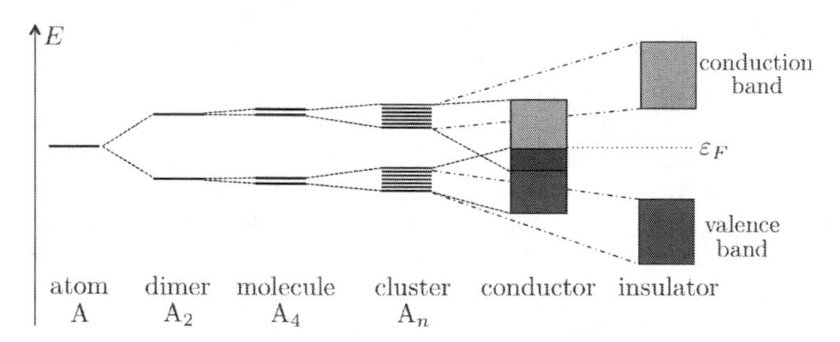

Figure 5.4: Schematic construction of band structure in a solid, starting from an atom, a dimer, a molecule then a cluster, and ending with two possible cases, that of a conductor, with overlap of the valence band (dark gray box) and the conduction band (light gray box), or that of an insulator, with a gap in between the valence and the conduction bands. ε_F is the Fermi energy.

number of atoms produces a smooth increase of the density of energy levels. Eventually, when the number of atoms is very large, it results in a continuous energy band structure. The valence band contains the lowest occupied electronic states, and its maximum energy defines the Fermi level. For reasons that will be explained immediately, the band of unoccupied electronic states is called the conduction band. The number density ϱ of electrons is very large. It can be estimated as the product of the number of valence electrons per atom Z_v by the number density of atoms in the sample. The latter is the product of

Avogadro's number $\mathcal{N}_A$ by the number of moles in the sample ρ_m/M_A, where ρ_m is the mass density and M_A is the molar mass. Finally, $\varrho = \mathcal{N}_A Z_v \rho_m / M_A$. Typical values of ϱ are of order $10^{22} - 10^{23}$ cm^{-3}.

In a crystalline solid, the band structure is characterized by the width of the bands, the gap between them and their relative location in the energy spectrum. Depending on the crystal composition and geometry, the bands may overlap or not, completely or partly. Among these possibilities, there are the two extreme cases of conductors and insulators. Figure 5.4 will help us understand the difference between them. In conductors, the valence band either contains many empty levels or overlaps with the conduction band. In any case, if one applies an electric field to the solid, the valence electrons can easily fill empty levels in the same valence band or in the conduction band at almost no energetic cost. As in the metal binding in molecules, in such a metal, electronic wave functions delocalize over the whole system.

In an insulator instead, the valence and the conduction bands are separated from each other by a finite energy gap. When it is large at ordinary temperatures, the thermal activation is not sufficient to allow electrons from the valence band to reach the conduction band. Note also that insulators are usually transparent, as diamond or glass. Indeed, photons in the visible spectrum (with energy between 1.5 and 3.1 eV) do not have enough energy to bridge the band gap (in diamond, this gap is equal to 5.33 eV at room temperature). In other words, electrons in the valence band which absorb photons from the visible light do not have enough energy to be promoted in the conduction band.

As in molecules, one can find a continuous path in between these two extremes depending on the size of the energy gap. The cases of small gaps, typically a few eV, are known as intrinsic semiconductors. Silicon (gap at 1.13 eV at room temperature) is a typical example used in many electronic devices nowadays. The Fermi level lies in between the valence and the conduction bands. Therefore, at finite temperatures, a certain amount of electrons can reach the conduction band and produce a small current. Electric conductivity therefore increases with temperature in semiconductors. However, the most interesting way to modify the electric properties of a semiconductor is by doping it, that is, by introducing a small amount of impurities which modify the properties of valence and conduction bands.

5.1.4 Quantum dots

Semiconductors are at the origin of interesting structures of various dimensions, in which electrons can be confined. These 'artificial' objects, the so-called quantum dots, exhibit interesting properties which can furthermore be more or less arbitrarily tuned. We will say a few words here on these specific systems.

Quantum dots are nanometer-sized dedicated structures of semiconductors, which contain a finite (possibly small) number of trapped electrons. They can somewhat be looked at as 'artificial atoms'. They have been studied in numerous devices, including transistors or diode lasers. It has even been imagined that they could become elementary pieces of quantum computers. Before discussing a bit their properties, it is interesting to briefly explain how they are constructed. Their existence stems from the mixing of solids with different band structures. As discussed above, in semiconductors, thermal agitation may be sufficient to overcome the gap. But the value of the gap varies from one semiconductor to the next. Conductivity can then become 'differential', at a given temperature, when mixing systems with different gaps. When combining semiconductors with different gaps, one then speaks of a semiconducting heterostructure. A typical setup merges a layer of one low-band gap material embedded into surrounding layers of higher band gap material, to form a 'quantum well'. In the low-band gap material, electrons remain more mobile and experience a barrier toward the high-gap bounds. When the layer is sufficiently thin, one performs a reduction of dimensionality, from a 3D to a 2D electron gas. By superimposing external electrical fields and/or using special etching techniques, dimensionality can be further reduced to 1D producing a quantum wire and even 0D, the latter object being called a quantum dot. Such a confinement may lead to very small structures containing a few electrons within a virtually 0D potential well. One should be cautious here with the word. Dot refers here to a 'small' structure, in the nanometer range, not to a true dot in the geometrical sense. But they can indeed contain very small numbers of electrons, whence the name.

As constituted of a finite number of confined electrons, quantum dots exhibit marked shell effects following the fact that electrons occupy discrete states. Interestingly, the number of electrons in quantum dots can be controlled within the construction procedure and certain numbers are strongly favored. This is a clear signature of shell closure, which has been confirmed by theoretical estimates. To a good approximation, the confining potential can be assimilated to a (possibly deformed) harmonic oscillator. But electron-electron interactions are also playing an important role in these systems which indeed exhibit strong correlation patterns as well. Quantum dots therefore appear as interesting and versatile electron systems. They indeed bear some similarities with atoms, in particular because of the dominant confining potential. But they also exhibit noticeable flexibility, for example in terms of electron number. The shape of the dot is also tunable, at least in two directions, by optimizing the confining fields. Furthermore magnetic fields can also play a key role in these systems as they can strongly affect electronic energies.

A simple description of a quantum dot is rather easily reachable. One can indeed reduce the system to a few interacting electrons of the semiconductor material, moving in a harmonic well. Because only low momenta contribute, the energy $\varepsilon(p)$ of an electron of momentum p can be expanded $\varepsilon(p) \simeq \varepsilon(0) + p^2/(2m^*)$ which defines an effective mass m^*. To a good approximation, m^* can be taken as that of the given semi-conductor material. This means in particular that the semiconductor environment is effectively accounted for in the description of the quantum dot electrons. The Coulomb interaction between electrons can be easily attained as well by using the corresponding dielectric constant of the material, which effectively amounts to modifying the Coulomb coupling constant e^2 to an effective one e^{*2}. Usually, one also rescales atomic units to effective ones, as the Bohr radius ($a_0 \to a_B$) or the Rydberg energy scale (1 Ry = 13.6 eV → Ry*), which delivers the so-called dot units. The latters then provide the correct scales to properly characterize relevant quantities in quantum dots. As an example, in a typical material such as GaAs, one has $m^*/m = 0.07$, $e^{*2} = e^2/\epsilon_0 = 0.08$, with $a_B = 9.8$ nm, 1 Ry* = 6 meV.

5.1.5 Towards quantum droplets of electrons

The above discussions in this section have incidentally raised an interesting question on the role of electrons in... 'electronic' systems. In all cases, electrons interact with each other of course. However, they feel in addition the confining potential provided by an external agent, usually ions. Still, among all the cases discussed, there are clearly two classes of systems which can be identified. In atoms or quantum dots, electrons do feel an external potential which is given. Electrons have no possible influence on this confining potential. It is important to realize here that this is not the case in molecules or solids (and in clusters; see Section 5.3). Indeed, the actual structure of a molecule, a cluster or bulk material (liquid or solid) stems from a subtle balance of interactions between all partners, basically electrons and ions. This in particular means that actual ionic positions are not given: they themselves result from all the interactions between all partners (electrons and ions). Therefore, electrons do influence ionic positions, just as ionic positions do influence electrons and the shape of the electron cloud (when it can be identified as such, in metallic systems in particular). This is obvious when looking at the variety of molecular structures. Even in simple molecules, very subtle effects are responsible for the actual shape of the molecule, for example to decide between a planar or a linear shape in triatomic molecules, such as water.

To make an analogy with a classical system/liquid, we could take the example of water. Imagine that water is plugged into a glass. The liquid fully adapts to the shape of the (rigid) glass. This exactly corresponds to the situation in an atom or a quantum dot. But now imagine that the container (the glass) is itself flexible, such as if we replaced the glass by a balloon. The actual shape of the balloon adapts itself to the content. The balloon, acts as an adaptable container whose shape depends on the quantity of water it contains. This somewhat corresponds to the molecular case.

This point on the 'dialectic' dialogue between electrons and ions represents an interesting issue for the purpose of this book as it raises the question of how and how much electrons do constitute a quantum fluid. More specifically, the question is whether one can speak of electron droplets in finite systems such as molecules and clusters. Indeed, even if electrons are key players to bind molecules or clusters, the latters exist because of ions and their interconnection with electrons. This sets the limits of the liquid nature of the system. Electrons do form a liquid (in particular in metals), but this is a liquid *around* the ions. The electron droplet is thus highly constrained by its environment. In other words, and at variance with helium droplets (see Section 6.3) or nuclei (see Section 7.1), electrons cannot constitute self-bound quantum liquid droplets.

To conclude this short analysis, we can say that electrons, in particular in metals, indeed constitute a unique quantum liquid. Such a system is not self-bound, so that electrons have to be considered with their environment. Still, as we shall see later, electrons remain major players in the game, leading to particularly stable structures (see Section 5.3.3) and being especially responsive to external perturbations (also because of their loose binding to ions), which in particular leads to a rich variety of situations in connection to light (see Sections 5.2.2 and 5.3.4).

5.2 PROPERTIES OF ELECTRONS IN A METAL

We have seen that, in a metal, electrons in the highest occupied states are shared by all the ions, and are then fully delocalized. It is in that sense that the electrons can be viewed as a fluid. We shall see in this section some characteristics of that fluid. As a starter, we shall consider the simplest picture of the electrons in a metal by considering the ions as a uniform positive charge background. We shall see that this oversimplified picture already provides interesting insights into the nature of metals. We shall continue the discussion by describing optical properties, the heat capacity and the electrical conductivity of a metal, which are mainly electronic effects, with respect to the ionic background. But ions are certainly not really forming a uniform structureless background and we shall finally discuss their impact on the band structure in a metal.

5.2.1 The jellium approximation

We present here a simple model to describe a metal sample. Its main hypothesis is that the positive ions, which are regularly distributed in a real metal, are replaced by a uniform background (the 'jellium'), which provides an external attractive potential for the electrons. Only the electrons contribute to the kinetic energy. The model only provides a qualitative account for real metals, but is reasonably adapted to metal clusters (see Section 5.3). One of the reasons in that case lies in the fact that, in experiments, metal clusters are produced at finite temperature (some hundreds of K). The temperature may induce sizable ionic motion so that experimental results are usually to be convoluted by this temperature blurring. A simple way to account for this thermal ionic motion (mind that such temperatures still remain negligible at electronic level, so that electrons remain basically degenerate; see Section 3.4.1) is to average it out in terms of a finite piece of jellium density. The jellium approximation allows one to overlook details of ionic effects and therefore emphasizes electronic aspects. In particular, it helps to understand in simple terms the nature of cluster/bulk binding.

We hence consider a system of N interacting electrons moving in the attractive background created by the jellium, characterized by the density $\varrho_I(\mathbf{r})$. We assume that electrons are described by a certain number of single-particle wave functions $\varphi_\alpha(\mathbf{r})$, α being an integer label. The electronic density can then be expressed as $\varrho_e(\mathbf{r}) = \sum_\alpha |\varphi_\alpha(\mathbf{r})|^2$. To keep the system neutral, both densities must be equal: $\varrho_e(\mathbf{r}) = \varrho_I(\mathbf{r})$ (and even equal to a strict constant ϱ_0 in the case of an infinite system).

The total energy of the system can be written as the sum:

$$E_{\text{tot}} = E_{\text{kin}} + E_{\text{II}} + E_{\text{eI}} + E_{\text{ee}} \tag{5.4}$$

where E_{kin} is the kinetic energy of electrons, and the other terms correspond to the Coulomb interaction between the various components of the system. From a classical viewpoint, the Coulomb interaction contributions exactly sum up to zero because the net charge density $e(\varrho_I(\mathbf{r}) - \varrho_e(\mathbf{r}))$ is by hypothesis set to zero (neutrality of the system)

$$E_{\text{II}} + E_{\text{eI}} + E_{\text{ee}}^{\text{dir}} = \frac{1}{2}\iint \text{d}^3\mathbf{r}\text{d}^3\mathbf{r}'(\varrho_I(\mathbf{r}) - \varrho_e(\mathbf{r}))\frac{e^2}{|\mathbf{r}-\mathbf{r}'|}(\varrho_I(\mathbf{r}') - \varrho_e(\mathbf{r}')) = 0.$$

But while ions can safely be treated as classical particles, electrons are to be treated quantally. This does not affect the electron-ion interaction but changes the electron-electron Coulomb interaction by complementing the classical (called direct part $E_{\text{ee}}^{\text{dir}}$ in the above relation) by the exchange contribution stemming from the indistinguishability of particles (see Section 2.5.4). The latter quantity can be written explicitly as

$$E_{\text{ee}}^{\text{ex}} = -\frac{1}{2}\sum_\alpha\sum_\beta \iint \text{d}^3\mathbf{r}\text{d}^3\mathbf{r}'\varphi_\alpha^*(\mathbf{r})\varphi_\beta^*(\mathbf{r}')\frac{e^2}{|\mathbf{r}-\mathbf{r}'|}\varphi_\alpha(\mathbf{r}')\varphi_\beta(\mathbf{r}) \tag{5.5}$$

and turns out to constitute the single non-vanishing contribution to the Coulomb contribution. The only potential energy contribution then comes from the exchange term, which is a purely quantal effect. To calculate this term, we describe the electrons by a Fermi gas, replace the single particle wave functions with plane waves and proceed as in Section 3.4.1. The computation is a bit involved and we simply give the final result in terms of the total energy per particle:

$$\frac{E_{\text{tot}}}{N} = \frac{\hbar^2}{2m}\frac{3}{5}(3\pi^2)^{2/3}\varrho_e^{2/3} - \frac{2e^2}{(2\pi)^3}(3\pi^2)^{1/3}\varrho_e^{1/3} \tag{5.6}$$

within having reintroduced the electron density, which, in a Fermi gas, is expressed from Fermi momentum $\varrho_e = k_F^3/(3\pi^2)$.

This result is interesting in many respects. First, it concerns the actual binding mechanism. Indeed, binding comes from the exchange term only. This is exactly true in this specific model, provided that we have made the above sequence of assumptions. But the result turns out to be almost true in any realistic neutral metallic system where bonding essentially stems from exchange. In other words, it is not so much the actual attraction between ions and electrons which ensures the binding, but rather an a priori tiny (finally not so tiny) effect of purely quantum nature reflecting the indistinguishability of electrons. Another more technical aspect concerns the actual form of the total energy obtained in the above model. It reduces to a simple expression as a function of density. One can then for example easily deduce the equation of state of this system and other thermodynamical properties, as we did for the degenerate fermion gas in Section 3.4.

5.2.2 Optical properties of metals

Under the application of an electric field, the valence electrons of a metal can be set to motion and then can create an electrical current. In some cases, the motion of these electrons can be collective and give rise to specific modes, also called plasma oscillations. And as any oscillation or wave propagation, such an excitation can be quantized and one then talks about plasmons. These modes are strongly related to the optical properties of the metal, in other words, its color. We here give some ingredients to understand these properties. For the sake of simplicity, the following discussion is done in 1D.

In general, the response of a metal to an electromagnetic field is described by the dielectric function $\epsilon(\omega, k)$ which relates the electrical displacement D to the electric field $E(x,t) = E(\omega)e^{\mathrm{i}(kx-\omega t)}$ and the polarization P as $D = \epsilon_0 E + P = \epsilon E = \epsilon_r \epsilon_0 E$. In the last expression, we have isolated the vacuum permittivity ϵ_0 and introduced the relative permittivity ϵ_r. The dielectric function is in general a complex function: $\epsilon = \epsilon_1 + \mathrm{i}\epsilon_2$ where $\epsilon_{1,2}$ are real functions. These functions have some important physical content: ϵ_1 is related to the reflectivity of the metal, while ϵ_2 describes its absorption properties.

In the following, we only consider the case $\epsilon_2 = 0$ and simply denote ϵ_1 by ϵ. We write the Newton equation of motion of a free electron at position x: $m\,\mathrm{d}^2x/\mathrm{d}t^2 = -eEe^{-\mathrm{i}\omega t}$. To solve this equation, we take $x = Xe^{-\mathrm{i}\omega t}$ and we find $X = eE/(m\omega^2)$. The polarization being $P = -\varrho e x$, we use the definition of the dielectric function to deduce its expression as:

$$\epsilon_r(\omega) = 1 - \frac{\omega_\mathrm{p}^2}{\omega^2} \quad \text{with} \quad \omega_\mathrm{p}^2 = \frac{\varrho e^2}{\epsilon_0 m} \quad . \tag{5.7}$$

ω_p is called the plasma frequency or the volume plasmon. It can also be deduced by considering a thin slab of metal, slightly displacing the electron gas as a whole in a direction normal to the slab and looking at the restoring force acting from this gas due to the positive background.

From Maxwell equations, we have in particular $\mu_0 \partial^2 D/\partial t^2 = \partial^2 E/\partial x^2$ with μ_0 the vacuum permeability or magnetic constant (in SI units, we have

$\mu_0\epsilon_0 c^2 = 1$). Using the definition of the dielectric function, we deduce the relation between ϵ, ω and k, also called dispersion relation: $\epsilon_r(\omega)\mu_0\epsilon_0\omega^2 = k^2$. We now discuss two cases. From Eq. (5.7), we see that, according to the value of ω with respect to ω_p, this changes the sign of ϵ_r. If $\omega < \omega_\mathrm{p}$, then $\epsilon_r < 0$, k is necessarily imaginary ($k = \pm i|k|$) and the electric field in the solid is damped (the amplified solution is not acceptable physically): $E(x,t) = E(\omega)e^{-|k|x}e^{-i\omega t}$. This corresponds to reflection only and explains why metals are 'shiny'. On the contrary, if $\omega > \omega_\mathrm{p}$, then $\epsilon_r > 0$ and k is real: an electromagnetic field can propagate inside the solid without damping. This also allows one to define the usual refractive index entering the Snell-Descartes laws, by $n = \sqrt{\epsilon_r}$.

Another collective mode appears at the surface of a metal. One can show that, by writing continuity conditions of the electric field at the interface between the metal and vacuum, the dielectric function must satisfy $\epsilon_r = -1$. By using Eq. (5.7) again, the latter condition defines another frequency $\omega_\mathrm{sp} = \omega_\mathrm{p}/\sqrt{2}$, naturally called *surface plasmon*. Volume and surface plasmon modes typically range between a few eV up to 20 eV. One can also encounter the term 'resonance' instead of 'mode' in the sense that the frequency of an incident electromagnetic wave, by matching a specific value, as ω_p or ω_sp, can excite a collective motion, analogously to the case of a classical resonant oscillator.

The volume and the surface plasmons of a material are usually measured by inelastic scattering of electrons through a thin film. An incident electron of a given energy (typically between 1 and 10 keV) transfers part of its energy to the film and one measures this loss of energy. In the obtained electron energy loss spectrum, peaks emerge, corresponding to a stronger absorption by the medium, as for instance by the excitation of a plasmon mode.

5.2.3 Specific heat capacity

The specific heat capacity of a metal can provide interesting information on the behavior of the degrees of freedom in a solid. We recall that the theorem of equipartition of energy mentioned in Section 3.1.2 allows one to estimate the specific heat capacity of a solid (known as Dulong-Petit law) as $C_V = 3N_\mathrm{ion}k_\mathrm{B}$, if N_ion is the number of ions in the lattice. Even if at room temperatures this law is approximately satisfied, when one decreases the temperature, quantum effects, and therefore departure from this law, show up.

The first effect comes from phonons which are lattice vibrations. Indeed, acoustic waves can propagate in a solid, inducing displacements of ions in the crystalline structure. And as any wave in quantum mechanics, such an acoustic wave is quantized and the quantum of energy is called a phonon. Statistics of phonons is identical to the distribution of photons from a black body at a given temperature T (see Section 2.1.4). If one assumes that all phonons have the same frequency ν (as Einstein did), the heat capacity reads as $C_V = 3N_\mathrm{ion}k_\mathrm{B}x^2e^x/(e^x - 1)^2$ with $x = h\nu/(k_\mathrm{B}T)$. When $T \to \infty$, one recovers Dulong-Petit law. At lower temperatures, C_V decreases as $e^{-h\nu/(k_\mathrm{B}T)}$, which is too fast compared to the experimental measurements. To solve this problem, Debye considered instead that different values of ν (hence, different values of x) should contribute to C_V. One should in principle sum up all contributions with ν varying between 0 and ∞. Debye however introduced a maximal allowed value, the Debye frequency ν_D. Indeed, on the one hand, the number of phonons is related to ν. On the other hand, the total number of phonons cannot exceed $3N_{\rm ion}$ (the factor 3 coming from the three spatial directions). This condition fixes ν_D. With the Debye temperature defined as $T_D = h\nu_D/k_{\rm B}$ (which is typically several hundreds of K), one can compute $C_V = 12\pi^4/5\, N_{\rm ion}k_{\rm B}(T/T_D)^3$.

This cubic temperature-dependence is well observed experimentally. However, at even lower temperature ($T \ll T_D$), another contribution stems from electronic excitations at finite T around the Fermi surface. More precisely, $C_V = \gamma T$ (see Section 3.4.2 for more details). This linear behavior then dominates the cubic one stemming from phonons. We recall here that $\gamma = k_{\rm B}^2/(\hbar k_F)^2 N\pi^2 m$ with N the total number of valence electrons, k_F the Fermi wavevector and m the mass of the electron. By measuring C_V/T at low T, one can access γ. Experimentally, one gets the right order of magnitude for γ but the measured value $\gamma_{\rm exp}$ often significantly differs from the theoretical $\gamma_{\rm theo}$. Conventionally, a thermal effective mass m^* is defined as $m^* = m\gamma_{\rm exp}/\gamma_{\rm theo}$. This effective mass can be as low as $0.11m$ for Sb, or as high as $2.18m$ for Li. The deviation from the mass of the free electron reflects various interactions in the solid: with the periodic lattice, with phonons and among valence electrons themselves. Although one often considers electrons in a metal as a free gas, as we shall do in the next section, one should keep in mind the underlying ionic structure and the fact that valence electrons are only 'nearly' free.

5.2.4 Electrical conduction

The extreme model of a classical electron gas, put forward by Drude in 1900, provides a phenomenological guide to get rough estimates on electrical and heat conduction. The idea is to apply kinetic theory to the conduction electrons in a metal. They are in thermal equilibrium with their environment through collisions. We assume that collisions instantaneously change the momentum of an electron, and this happens with a probability per unit time $1/\tau$, where τ is referred to with several names: collision time or mean free time. Between collisions, we neglect the interactions of an electron with the other electrons and with ions. We now estimate the electrical conductivity. Under the application of a constant electric field, an electron feels a force $-e\mathbf{E}$, which according to Newton's laws, is equal to the derivative of its momentum with respect to time, estimated as the ratio $m\langle\mathbf{v}\rangle/\tau$. Here, $\langle\mathbf{v}\rangle$ is an average velocity of electrons. We then have $\langle\mathbf{v}\rangle = -e\mathbf{E}\tau/m$. The current density is related to the average velocity by $\mathbf{j} = -\varrho e\langle\mathbf{v}\rangle$, where ϱ is the number of electrons per unit volume. From these two equations, one deduces the electrical conductivity σ

as the proportionality coefficient between the electric field and the current:

$$\mathbf{j} = \sigma \mathbf{E} \quad , \quad \sigma = \frac{\varrho e^2 \tau}{m} . \tag{5.8}$$

We have already seen that typical values of ϱ are $10^{22} - 10^{23}$ cm^{-3}. To estimate the collision time, we can write the mean free path of electrons as $\lambda = \langle v \rangle \tau$, with $\langle v \rangle$ the average electronic speed. Classical kinetic theory gives $\langle v \rangle = \sqrt{3k_{\mathrm{B}}T/m}$, that is about 10^7cm/s at room temperature. Assuming that the mean free path is given by the atomic interspacing, i.e. $\lambda \simeq 1 - 10$ Å, we obtain for the collision time the estimate $\tau \simeq 10^{-14} - 10^{-15}$ s. Using these values in Eq. (5.8), one gets the observed order of magnitude for σ. At Drude's time, this was good news. This was however achieved only by chance, as we shall see below. This model is based on classical ideas, not applicable actually here, and expressed as the ratio of the mean free path and the average speed which are wrongly estimated by 1-2 orders of magnitude.

Looking to the heat conduction, it is possible to get an expression independent of the collision time, which leads to a nice agreement with experiment. The observation that metals are also good conductors of heat led to the idea that valence electrons are also heat carriers. Drude also applied kinetic theory to calculate the thermal conductivity of metals κ which quantifies the flow of heat inside a sample under a difference of temperature of 1 K. The result reads:

$$\kappa = \frac{1}{3} \langle v \rangle^2 \tau \, \varrho c_V \quad , \tag{5.9}$$

where c_V is the specific heat capacity at constant volume for a single electron. From Eqs. (5.8) and (5.9), we get:

$$\frac{\kappa}{\sigma} = \frac{c_V m \langle v \rangle^2}{3e^2} \quad . \tag{5.10}$$

Since classical kinetic theory gives $\langle v \rangle = \sqrt{3k_B T/m}$ and $c_V = 3k_B/2$ (see Chapter 3) we get:

$$\frac{\kappa}{\sigma} = \frac{3k_B^2}{2e^2} T \quad . \tag{5.11}$$

This expression is in agreement with earlier observations by Wiedemann and Franz on the fact that, for a great number of metals, the ratio κ/σ is directly proportional to the temperature, with a proportionality constant roughly the same for all metals. From Eq. (5.11), this universal constant is equal to 1.11×10^{-8}W Ω K^{-2}. However, even if it has the right order of magnitude, it differs by a factor of 2 from the empirical value. Still, we should keep in mind that this is a classical model.

Sommerfeld was the first one to include quantum ideas in Drude's model. More precisely, he considered the free electron gas as a degenerate Fermi electron gas and used the Fermi-Dirac distribution instead of the classical Maxwell-Boltzmann distribution to calculate properties at thermal equilibrium. We apply here some of the results given in Section 3.4.1 and also complement them with some units commonly used. The number density of the electrons is related to the Fermi momentum $\varrho = k_F^3/3\pi^2$. We use the Wigner-Seitz radius $r_s = (3/4\pi\varrho)^{1/3}$, which is the radius of a sphere whose volume is equal to the volume per conduction electron. We also introduce the Bohr radius $a_0 = \hbar^2/me^2 = 0.529$ Å as a length unit, and the Rydberg Ry$= e^2/2a_0 = 13.6$ eV as an energy unit. With these units, the Fermi momentum, velocity, energy and temperature are written as

$$k_F = \frac{3.63}{r_s/a_0} \text{Å}^{-1} \quad , \quad \varepsilon_F = \frac{\hbar^2 k_F^2}{2m} = \text{Ry}\,(k_F a_0)^2 = \frac{50.1}{(r_s/a_0)^2}\,\text{eV}$$

$$v_F = \frac{\hbar k_F}{m} = \frac{4.20}{r_s/a_0} \times 10^8\,\text{cm/s} \quad , \quad T_F = \frac{\varepsilon_F}{k_B} = \frac{58.2}{(r_s/a_0)^2} \times 10^4\,\text{K}$$

Typical values are $r_s/a_0 \simeq 2-3$, $k_F \simeq 1/$Å, $v_F \simeq 10^8$ cm/s, $\varepsilon_F \simeq 1-15$ eV, $T_F \simeq 2_1 2$ eV$\simeq 2-14 \times 10^4$ K.

We now consider a gas of free valence electrons occupying the conduction band. We have seen in Section 3.4.2 that only electrons in states near the Fermi surface of that band substantially participate to the heat capacity. The first quantum modification to the Drude model is to replace the average velocity $\langle v \rangle$ entering Eq. (5.8), via τ, and Eq. (5.9) by the Fermi velocity v_F, which is independent of T and larger by a factor of 15 than $\langle v \rangle$ at room temperature. The second modification concerns the specific heat entering Eqs. (5.9) and (5.10). The quantal expression is $c_V = k_B(\pi^2/2)T/T_F$ (see Section 3.4.2). Contrary to the classical expression, that is $3k_\text{B}/2$, it depends on T. However, even at room temperature, it is 100 smaller than $3k_\text{B}/2$, indicating that the contribution of electrons to the specific heat of a metal is very small. In fact, heat conduction is rather related to the thermal motion of ions in the crystalline lattice. For example, diamond, which is not a good electrical conductor, is an excellent heat conductor. The Sommerfeld quantum free-electron model for metals corrects the Drude classical one, but only to point out several open questions. In particular, the collision time τ must be correctly interpreted. In any real metal, the electrons may interact with impurities and defects of the lattice, with phonons, and so on. The electrical conductivity will be affected by all these possibilities, leading to various temperature dependences.

5.2.5 Electrons in a periodical potential

In the previous sections, we derived some properties of metals assuming that electrons in a metal are free. This is however strictly speaking not true because of the presence of the positive ions of the lattice. Actually, describing electrons in a solid is in principle a complicated many-body problem, which calls for insightful approximations. One of these is the jellium approximation, discussed in Section 5.2.1, replacing the positive ions by a uniform background. Another approach relies on the 'nearly free electron' approximation. To detail it, we will assume in the following that the interaction from the lattice ions is considered to be small, that is a perturbation of a free propagation of electrons.

This will in particular illustrate how bands in a metal appear, as a quantitative complement of the discussion in Section 5.1.3. In the terms 'nearly free', one also assumes that interactions between electrons are ignored. This is a reasonable approximation because we can expect that the mutual repulsion of electrons is somehow shielded by the attraction with the positive ions. We thus consider that an electron is subject to an effective potential $V(\mathbf{r})$ (also called a pseudopotential), which must reflect the periodicity of the lattice.

Before proceeding, we should give here some basic elements of crystallography. For the sake of simplicity, we restrict the discussion to the monoatomic 1D case, that is a ring of N identical atoms, $N \gg 1$, separated by the lattice parameter a. Indeed, to represent an infinite system, one usually introduces periodic boundary conditions. This means that, while considering only a finite piece, one admits that the full system is just an infinite periodic sequence of copies of the finite piece. Doing such a trick in 1D amounts to replacing an infinite line by a segment whose extremities are the same, which practically means a closed ring. The length of this ring is therefore $L = Na$. There is obviously a translational symmetry: for any translation vector $T = na$, n being an integer, the potential created by the ions satisfies $V(x + T) = V(x)$. Any physical quantity originating from V must also possess this translational symmetry, as for instance, the electronic density $\varrho(x)$. From Fourier analysis, we know that any periodical function can be expanded as a Fourier series. The periodicity being here a, we can write $\varrho(x) = \sum_p \varrho_p e^{2i\pi px/a}$, with p integer and where the constants ϱ_p are the coefficients in this discrete Fourier expansion. We notice that the wavevector $2\pi p/a$ is a multiple of $b = 2\pi/a$ which defines a lattice parameter in Fourier space, also called the reciprocal space. Actually, a crystal structure is associated to the (direct) crystal lattice, defined by a, and the reciprocal lattice, defined by $b = 2\pi/a$. We will see below that the reciprocal lattice is strongly linked to measurable quantities.

Let us now consider an electron traveling in the lattice. We first describe it as a plane wave with wavevector k, $\psi_k(x) = e^{ikx}$. A scattering with an ion of the lattice will change its wavevector, say to k', and then $\psi_{k'}(x) = e^{ik'x}$. The change in wave function then reads as $e^{i(k-k')x}$. The scattering amplitude $\mathcal{S}(k, k')$ is obtained by integrating over the ring, taking into account the ionic potential $V(x)$. Because of the expression of the change of phase, the argument in $\mathcal{S}$ is $q = k' - k$, that is the relative change in wavevector. We can also Fourier expand V because it is periodical. More precisely, we have:

$$\mathcal{S}(q) = \int dx V(x) e^{-iqx} = \int dx \sum_p V_p e^{2i\pi x/a} e^{-iqx} = \sum_p V_p \int dx e^{i(2\pi p/a - q)x} \quad . \tag{5.12}$$

Note that $\mathcal{S}(q) = NS(q)$ where $S(q)$ is the structure factor (see Sections 1.3.3 and 4.3.2), which is related to the Fourier transform of the pair correlation function $g(r)$; see Eq. (1.5).

In the last expression of (5.12), the integral is non-zero if and only if $q = \pm 2\pi n/a$, n being a positive integer. This condition is called the Laue equation, here written in the 1D case. In other words, electrons are scattered by the lattice ions if and only if $k' = k \pm 2\pi n/a = k \pm nb$ where b is the parameter of the reciprocal lattice. We hence obtain the result that the scattering amplitude has non-vanishing values only if the scattered wavevector k' is separated from the incident wavevector k by a multiple of the reciprocal lattice parameter $b = 2\pi/a$. In other words, $\mathcal{S}$ provides a map of the reciprocal lattice.

We can go further by considering only elastic scattering, that is $k' = k$. For then, the Laue equation reads:

$$k = \pm\frac{\pi}{a}n \quad , \tag{5.13}$$

where n is a positive integer. The case $n = 1$ is of particular interest: the region $-\pi/a < k < \pi/a$ is called the first Brillouin zone. The physical interpretation of this result is the following: for a plane wave with wavevector in the first Brillouin zone, there is no scattering from the ions and the wave remains a plane wave. But if k approaches $\pm\pi/a$, the influence of the lattice ions becomes more and more important, with a constructive elastic scattering exactly for $k = \pm\pi/a$.

The existence of bands in the electronic structure of solids is intimately related to Eq. (5.13). To understand this, consider the plane wave $\psi_{\pi/a}(x) = e^{i\pi x/a}$. Because of the constructive scattering on the lattice, there is also the propagation of the plane wave $\psi_{-\pi/a}(x) = e^{-i\pi x/a}$. All in all, we have the superposition of both plane waves. Two possibilities arise:

$$\psi_+(x) = \sqrt{\frac{1}{2L}}\left[\psi_{\pi/a}(x) + \psi_{-\pi/a}(x)\right] = \sqrt{\frac{2}{L}}\cos\left(\frac{\pi}{a}x\right) \quad , \tag{5.14}$$

$$\psi_-(x) = \frac{1}{i}\sqrt{\frac{1}{2L}}\left[\psi_{\pi/a}(x) - \psi_{-\pi/a}(x)\right] = \sqrt{\frac{2}{L}}\sin\left(\frac{\pi}{a}x\right) \quad . \tag{5.15}$$

The forefactor in $\psi_\pm$ takes care about the normalization over the ring (we recall that $L = Na$). These wave functions are standing waves and the corresponding densities are depicted in the left part of Figure 5.5, together with the ionic potential $V(x)$. We observe that $|\psi_+|^2$ is mainly located on top of the ions and exhibits nodes in between, while we have the reverse density probability for $|\psi_-|^2$. This reminds us of the bonding and antibonding states when one forms a dimer by approaching two atoms (see Section 5.1.2).

Let us now calculate the energy of those standing waves. To this end, one starts with the Schrödinger equation $\left(-\frac{\hbar^2}{2m}\frac{d^2}{dx^2} + V(x)\right)\psi_\pm = \varepsilon_\pm\psi_\pm(x)$, multiplies it by $\psi_\pm(x)$ and integrates over x. The calculations are lengthy but straightforward. We give here only the results. However, before doing so, let us consider a specific case for $V(x)$. Indeed, we have already exploited the Fourier expansion of $V(x)$ in Eq. (5.12). Without a loss of generality, we can consider only the even part of this expansion, that is $V(x) = \sum_p V_p \cos(2\pi px/a)$, with

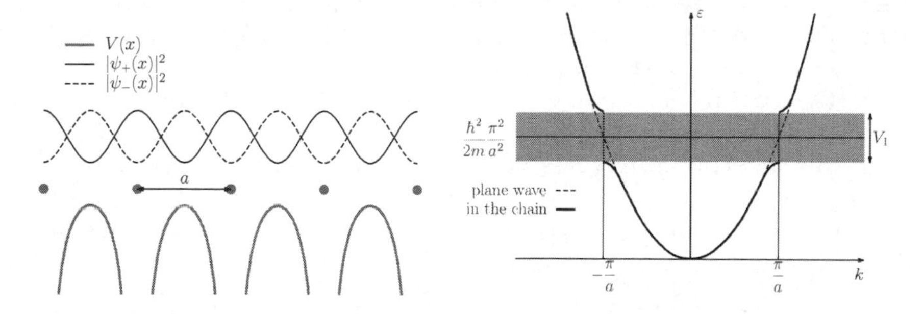

Figure 5.5: Left: 1D monoatomic lattice of parameter a, made of ions (black circles) creating the ionic potential $V(x)$ (thick gray curve), the electronic densities $|\psi_{\pm}|^2$ (thin dashed and full curves). Right: energy ε of an electron in the lattice, as a function of the wavevector k. V_1 denotes the first order coefficient in the Fourier series of $V(x)$.

index p running over all integers. We will even reduce it to the first term: $V(x) = V_1 \cos(2\pi x/a)$. In this case, the energy of $\psi_{\pm}$ reads:

$$\varepsilon_{\pm} = \frac{\hbar^2}{2m}\left(\frac{\pi}{a}\right)^2 \pm \frac{V_1}{2} \quad . \tag{5.16}$$

There is therefore an energy gap at $k = \pm\pi/a$ given by $\varepsilon_+ - \varepsilon_- = V_1$, which is the amplitude of the first order in the Fourier expansion of $V(x)$. In other words, there is no solution of the Schrödinger equation at $k = \pm\pi/a$ with an energy ε such that $\hbar^2\pi^2/(2ma^2) - V_1/2 < \varepsilon < \hbar^2\pi^2/(2ma^2) + V_1/2$. This is what one calls a *forbidden (energy) band.* One can convince oneself that each order in the Fourier expansion will produce the opening of gap at $k = \pm\pi p/a$ equal to V_p. The first energy gap is represented in the right part of Figure 5.5.

We are now in conditions to specify a bit more the qualitative description of energy bands given in Section 5.1.3. The lowest energy bands are filled with electrons which are tied to the individual ions. The highest occupied energy band is called the valence band. Depending on the specific atoms and lattice, the valence band may be completely or partially filled with electrons. Possible energy bands with higher energies are called allowed bands, each of them being separated by a forbidden band.

The space periodicity of the lattice has also an unexpected consequence: a perfect crystal would possess an infinite conductivity. This is at variance with Drude's model of conductivity (even in the corrected version by Sommerfeld) which describes it via collisions of electrons with ions of the lattice (see Section 5.2.4). We remind readers however that this model is classical and considers electrons as point-like particles. Here, to explain the band structure in a metal, we have treated electrons as wave functions, with a delocalization through the whole metal. A major theorem in solid state physics is Bloch's theorem which states that the solutions of the Schrödinger equation for a periodic lattice are the product of a plane wave e^{ikx} and a function $u_k(x)$ which possesses the same periodicity as the lattice. This product is called a Bloch wave. The wavevector k is here a quantum number in the sense that it is quantized, exactly as for a particle confined in a box (see Section 2.4). For instance, in the case of the ring considered above, we have $\psi(x) = u_{k_n}(x)e^{\mathrm{i}k_n x}$ with $k_n = 2\pi n/(Na)$, n being an integer. A Bloch wave defined by a wavevector k being a stationary solution of the Schrödinger equation, which accounts for the lattice through V, it will keep the same wavevector k forever if this lattice is infinite and perfectly periodic. In other words, a Bloch wave is not scattered by a perfect lattice. Electric conductivity σ being due to scattering of electrons, that is electrons which undergo a change of wavevector, is infinite in an infinite perfect crystal. We have indeed seen in the particular case of wave functions at the boundary of the first Brillouin zone that they result from coherent and constructive interferences of the scattered waves. What Bloch's theorem tells us is that this result is completely general. In practice however, measures of σ always give finite values because there always exist disturbances to perfect periodicity, either due to defects or impurities in the crystal, to phonons at finite temperature or at the boundaries of the lattice.

5.3 CLUSTERS

Atoms can bind to form molecules, when only few atoms are involved, or bulk material when a virtually infinite number of atoms bind together (see Figure 5.4). Our understanding of both kinds of systems is complementary, the various scientific fields involved (chemistry, molecular physics, solid state physics, ...) having developed their own, best adapted, tools of investigation/analysis. In all cases, electrons play a crucial role by providing the proper 'fluid' leading to actual binding between atoms. We have seen that covalent bonding leads to electrons localized in between actual bound atoms while metallic bonding allows electrons to spread over the whole system, possibly far away from the atoms themselves. This holds true in bulk as well as in small molecules, although often expressed in different terms. A deeper understanding of such effects requires exploring systems between atoms/molecules and bulk in a more systematic way. This is achieved through a new kind of system called clusters which are assemblies of atoms (or molecules) with sizes between a few constituents to a virtually infinite number thereof.

Cluster physics is both a new and old science. Indeed, clusters, not called that way though, have been used/studied since very long ago. Early use of clusters goes back to Roman times with the inclusion of small metal particles in glass, which delivers fascinating reflection pattern from light. Clusters also played the major role in the development of traditional photography. Image forming on a photographic plate or film is the direct result of coupling of small silver bromide clusters to light. The size of the cluster directly controls its rate of reaction to light and, technically speaking, the sensibility of the film. In both issues (glass, photography), clusters are always linked to their environment and this was the rule for many decades with studies concentrating on clusters embedded into a medium or deposited onto a surface. The study of clusters as free objects really developed only during the last quarter of the 20th century, with the ongoing availability of cluster sources allowing to produce clusters of various sizes, in a controlled way. As we shall see below, the study of clusters, as specific objects, provides a key understanding on how atoms bind together in a systematic way and how, simply speaking, actual matter forms. Furthermore, the specific role of electrons, in particular in the case of metal clusters, will directly bring us back to our topic of quantum liquids.

5.3.1 From atom to bulk via quantum liquids

Sizes

As stated above, clusters interpolate between atoms/molecules and bulk. A key issue thus concerns the way this interpolation takes place, in other words, how bulk matter emerges from single atom constituents. One can first think about the mechanisms of creation of clusters. Indeed, in cluster sources, one aggregates single atoms in a more or less controlled way, allowing the production of clusters constituted of a few atoms ($3-5$), rather similar to usual molecules, up to huge numbers of them (10^5-10^7), usually very close to finite pieces of bulk.

To give a rough estimate of cluster sizes, let us take 0.5 nm as a typical bond length in molecules, attributing an average volume of 0.125 nm^3 to each atom. A huge cluster of 10^7 atoms will then typically represent a volume of 10^6 nm^3 equivalent to a cube of 100 nm side. A moderate size cluster of 100 atoms corresponds to a nm side cube. We immediately see that clusters exactly cover the size range of today's numerous investigations on the so-called 'nano'-systems.

There is also a specificity that distinguishes bulk matter and clusters: the number of atoms at the surface as compared to the total numbers of atoms. Let us consider again a cubic cluster made of 100 atoms. We have roughly $100^{2/3} \simeq 20$ atoms at its surface, so a non-negligible amount of the total number of atoms. In a 1 mm side cube, with a lattice parameter of 0.5 nm, we roughly have $\left(10^{-3}/(5\times10^{-10})\right)^2 = 4\times10^{12}$ atoms at the surface, out of a total number of about 8×10^{18}. Therefore, the ratio surface to volume is about 5×10^{-7}, a very small value. Even in a 1 μm side cube, this ratio is still as small as 5×10^{-4}. In other words, in bulk, the effect of a finite surface, related to a finite volume of course, can be safely neglected (apart from cases where one is interested in the physics at the interfaces of a solid). This is obviously not the case at all in a cluster.

The role of interactions

The particularly high surface/volume ratio in a cluster has a strong impact on the way one can theoretically address such an object, especially at the side of the interactions. We have seen in Section 1.4.1 that surface tension in a classical liquid droplet results from the fact that particles at the surface undergo fewer interactions than particles in the bulk of the liquid. This produces locally an inward force and globally a spherical shape, in the absence of other forces as gravity.

The same kind of mechanism is at work for the electrons in a metal cluster. Note however that treating these electrons alone is nonsense since, as already mentioned in a bulk metal: electrons form a liquid around the ions and do not constitute a self-bound system on their own (see Section 5.1.5). These are actually the resulting forces between electrons and ions that bind the system together. The same holds for metal clusters. The example of Na_8 in Figure 5.7 shows a compact 'cubic-like' ionic structure (actually two squares rotated with respect to each other) and an almost spherical electronic shape.

Metal clusters can be also viewed as electron quantum liquid droplets that are highly constrained by their environment, at variance with nuclei and helium droplets. To illustrate this point, we consider again the Na_8 cluster, but now deposited on an insulator surface, for example that of the ionic crystal NaCl. As in the case of a classical droplet deposited on a surface, the competition of interactions inside the cluster and between the cluster and the surface will ultimately decide the shape of the cluster. The surface, being virtually infinite, may locally rearrange but only to a small extent, especially in a strongly bound ionic insulator as in NaCl. At the side of the deposited cluster, the actual result is a typical wetting behavior with a deposited Na_8 reduced to one layer of ions plus the surrounding electron cloud having a pancake shape around the ions.

Scalability

Interpolation between atomic and bulk scale is not nevertheless a trivial issue, and as we shall see, clusters are not finite pieces of bulk nor simply large molecules. To illustrate this point, one can look at a given observable X as a function of cluster size N. In most of the cases, the radius scales as $N^{-1/3}$ (see Section 3.4.1). Then X can be expanded as $X(N) \approx X_0 + X_1 N^{-1/3} + X_2 N^{-2/3} + \ldots$ with X_0 the bulk value of this observable, reached for $N^{-1/3} \to 0$ that is $N \to \infty$. The other correction terms stem from planar surface effects (X_1) and from curvature (X_2). Similar expansions are also very often used in nuclei (see for instance Section 7.1.2). On top of this general behavior, one can observe some fluctuations due to quantum shell effects (see Section 5.3.3) especially for small cluster sizes. This apparently irregular pattern gets smoother and smoother as N increases and ultimately converges towards the bulk value X_0 for very large sizes. One should however mention here that there is no strict rule on the convergence rate or on the amplitude of fluctuations along the path towards bulk values. Both can be large or small, depending on the nature of the observable and of the cluster material. But the qualitative behavior, up to these quantitative details, is always the same. This behavior is called scalability which one can define as a measure of how much a given physical quantity scales with system size.

Scalability of observables basically reflects the evolution of cluster properties with size and thus directly points towards the evolution of binding properties. It is interesting to link observables to this binding behavior, as directly as possible. To this end, we show in Figure 5.6 the cohesive energy per atom for cationic alkaline clusters as a function of cluster size. The cohesive energy

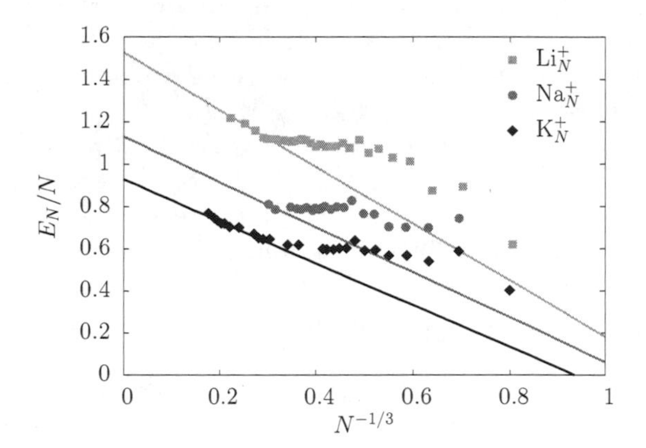

Figure 5.6: Cohesive energy per atom, E_N/N, for alkaline cationic clusters of size N, as a function of $N^{-1/3}$. Data reproduced from C. Brechignac, H. Busch, Ph. Cahuzac, and J. Leygnier, J. Chem Phys. 101 (1994), 6992, with the permission of AIP Publishing.

of an atomic system is the energy required to break all the bonds between the constituting atoms, to obtain isolated atomic species. It can be calculated in a solid, denoted by E_{coh}, or in a cluster of finite size N, then denoted by E_N/N. The straight lines in the figure correspond to the first two terms in the expansion of the cohesive energy per atom: $E_N/N \approx E_{\text{coh}} + E_1 N^{-1/3}$. Convergence to this line very much depends on the alkaline and occurs for $N \gtrsim 40$. Note that the observation of convergence to bulk also depends on the resolution of a measurement.

Clusters as objects of study

Linking atomic pieces to bulk is an important issue, especially from a fundamental point of view. But one can also consider the problem the reverse way. And the motivation then rather comes from the applications viewpoint. Indeed the industrial evolution over the past decades tends to miniaturize many devices. This is typically true for electronic systems, which size has collapsed over the last half century, allowing the development of unexpected portable devices, such as cellular phones or laptops, to cite emblematic examples. Electronic component sizes have thus been constantly decreasing reaching nowadays the tens of nanometers scale which brings them in the same size range as large clusters. Correlatively, a proper understanding of the behavior of such devices will require a deep understanding relying on cluster science and requiring the account of size with the associated specific effects. And, as we shall see, cluster properties directly reflect quantum properties of electrons and this implies an increasing role of quantum effects to understand behaviors of future portable electronic devices.

As a last remark we would like to point out that the previous analysis on size interpolation can also be considered from another viewpoint. Indeed, we have seen from Figure 5.6 that clusters exhibit specific properties (because the latters strongly depend on size). This therefore means that they do constitute systems on their own with their own interesting properties. Cluster science actually built up over the past half century precisely to study these properties for systems identified as a specific class of objects. And we shall see that this fundamental understanding played, and will most probably play, a key role in many scientific domains, such as biology, material science or even astrophysics, where clusters are playing an increasing role.

5.3.2 Clusters and metal(lic) clusters

The opportunities offered by cluster science are numerous and this is not the place to discuss them extensively. We shall only consider some illustrative cases and connect them to our fundamental topic, namely quantum liquids. Although electrons play a key role in whatever kind of clusters, the relation to the notion of quantum liquid does depend on the actual nature of the cluster. For example, the degree of scalability of observables, discussed in the previous section, actually strongly relies on the type of binding. Not surprisingly, cluster binding types basically match the various ones we have identified in the case of simple molecules (see Section 5.1.2) or, with some different terminology, in solids. The point is illustrated in Figure 5.7 where four typical examples of clusters are shown: the Na_8 metal cluster (upper left), an Ar_{13} molecular (or van der Waals) cluster, a Na_4F_4 ionic cluster and the famous C_{60} fullerene. The latter one can be considered as a typical example of covalent bonding in clusters (as usually observed in bulk carbon or silicium). This cluster exhibits a well defined atomic structure resembling a soccer ball shape. Electrons are mostly localized along the various links between the atoms. The case of Ar_{13} is typical of a van der Waals cluster, as in all rare gas clusters. We note in this case that each atom carries its own electron cloud with little overlap with the electrons from other sites. This is characteristic of the weak binding between rare gas atoms. Ionic binding is illustrated in the Na_4F_4 cluster. It represents an 'ideal' mixture (alkali + halogen) with basically four electrons exchanged between Na and F, yielding four 'naked' Na^+ cations (no more valence electron) and four 'dressed' F^- anions with rather well localized electron charge. Binding (actually strong binding) mostly results here from the Coulomb interaction between positive and negative charges. Finally, a typical metallic bonding is illustrated in the case of the Na_8 cluster. Here the electronic cloud

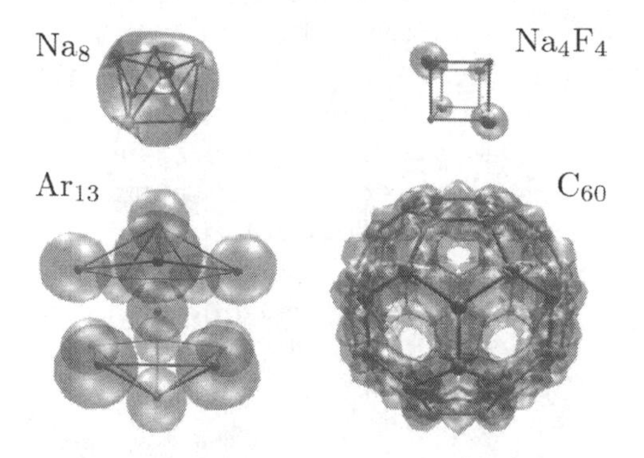

Figure 5.7: Examples of cluster exhibiting one of the four different types of binding, with ions denoted by small balls and electron clouds by transparent isodensity surfaces: metallic binding for Na_8, ionic binding for Na_4F_4 (large spheres are F ions), van der Waals binding for Ar_{13}, and covalent binding for C_{60}. Binding between nearest atoms is materialized to guide the eyes only. The four clusters are plotted with the same scale for better size comparison.

extends more or less smoothly over the whole system. Valence electrons do not belong anymore to a single ion, but rather to the system as a whole. To some extent, detail of the ionic positions becomes a secondary issue (although the connection is of course deep; see Section 5.1.5) and many cluster properties are primarily driven by the electron cloud itself.

Let us finally stress the fact that the same scale has been used in all four examples. This carries an important message on relative sizes, both at the side of the actual ionic positions and in terms of the spatial extension and shape of the electron cloud. This also illustrates the actual nature of binding between the atoms. The weak binding in rare gas clusters for example leads to soft extended clusters. To a large extent, molecular clusters can finally be considered as a collection of weakly interacting atoms. The other extreme is provided by ionic clusters with a strong binding associated to compact structures. Covalent binding also leads to compact structures (mind the actual size of C_{60} with 60 atoms as compared the 3 other examples with about 10 atoms each) and leads to almost as strong bonds as ionic ones. Metal clusters are generally softer bound than covalent ones, constituting an intermediate class between the almost unbound molecular clusters and the more tightly bound ionic or covalent clusters.

A word of caution is necessary before closing the discussion here. The sorting scheme presented here relies on idealizations. As in molecules or in solids, real cases may fall in between these categories. There are even examples where the type of binding depends on the cluster size, as in $Hg_N{}^-$. Indeed, there are strong experimental evidences which indicate that these clusters are rather covalent for $N \leq 13$. For intermediate sizes, they exhibit semi-conductor-like densities of states. And finally a closure of the band gap is observed for $N \geq 300$, leading to a metallic behavior. Without considering such extreme cases, there are numerous situations where properties of a system are representative of a bonding type, while other ones are rather associated to another type. Although all kinds of clusters do play an important role (we illustrate the point on two cases below), metal clusters, because of the specific properties of their delocalized electron clouds will play an important role for the topic of this book. This point is discussed from a formal point of view in Section 5.1.5 and will be largely illustrated on some of their specific properties in Sections 5.3.3 and 5.3.4.

5.3.3 Shells and electronic shells

Whatever the level of detail at which ions are described, metal clusters can, to a large extent, be pictured as sets of little interacting electrons inside an external confining potential. This immediately leads us to the idea that electrons will gather in shells of increasing energy, following simple principles of quantum mechanics (see Section 2.4.2). This is effectively what is observed experimentally. As in the case of atoms, a simple indicator of such an effect is provided by abundances with the simple rule: the more abundant a species, the more stable it is. This enhanced stability can be associated to an electronic shell closure. In the early days of cluster physics, measurements on small alkali clusters have been rapidly associated to enhanced stability/abundances leading to a sequence of 'magic' numbers 2, 8, 20, 40, 58, 92, ... By analogy, the same nomenclature ($1s$, $2s$, $2p$, etc.; see Section 5.1.1) used in atoms has been exported to metal clusters. Note however that some electronic shells, as $1p$ or $2d$, are nonexistent in atoms but do exist in metal clusters. The first shell closures (up to 40) can be simply understood by assuming an harmonic oscillator as a confining potential. It turns out that it is actually very close to a more realistic confining potential (computed numerically) created by ions for such low masses. The interesting aspect, though, is the fact that the enhanced stability is of electronic origin. We hence find that alkali clusters of small to moderate sizes follow a simple electron filling rule in a rather simple potential (a jellium potential as discussed in Section 5.1.5 would basically deliver the same result).

Shell closure manifests itself in many observables characterizing clusters. An interesting property of clusters is their ionization potential (IP), namely the energy needed to remove one electron from the cluster. Figure 5.8 shows the case of neutral K_N clusters with one valence electron per atom. N thus represents the number of atoms but also the total number of valence electrons. We observe again the scalability discussed in Section 5.3.1, that is a general trend of decrease of the IP towards the (still far below in that case) bulk values. But on top of this general behavior, there are clear jumps for well defined values of the electron numbers: the latter correspond to electronic shell

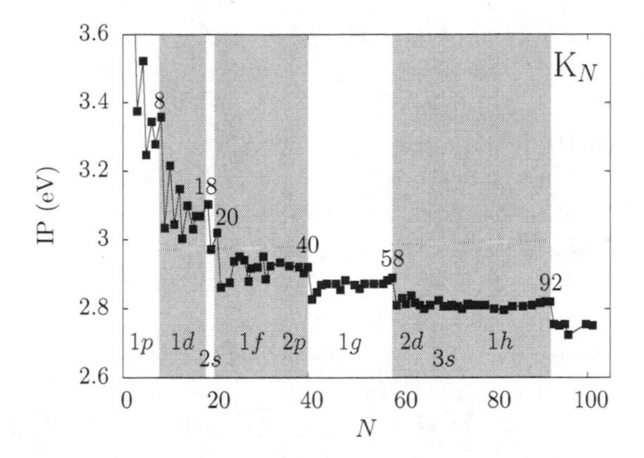

Figure 5.8: Ionization potential (IP) of neutral K_N clusters as a function of cluster size N, with shell closures highlighted by the successive rectangles and indicated by numbers. Adapted with permission from W. A. de Heer, Rev. Mod. Phys. 65, 611 (1993). Copyrighted (1993) by the American Physical Society.

closures. Indeed, at such a closure, there is a gain in the IP which abruptly drops when adding one extra atom/electron. These jumps reflect the fact that the extra electron occupies a new open shell which is much less bound than the previously completed one. The indicated nomenclature of the shells ($1p$, $1d$, $2s$, ...) corresponds to the shell sequence in a deformed harmonic oscillator shell model, called the Clemenger-Nilsson model. Not surprisingly, the overall effect decreases jumps in ionization potential with increasing cluster size. We observe the same shell closures in all alkaline clusters.

A specificity of clusters, and especially of metal clusters, lies in the fact that one can vary at will their size, at variance with atoms or nuclei where size variations are limited by nature itself. Indeed, the heaviest chemical element has atomic number $Z = 118$ (called 'oganesson'). At the side of nuclei, accounting for the various neutron/proton combinations in nuclei leads to about 3000 stable nuclei (see Section 7.1.1). On the other hand, metal cluster sizes vary between 2 and hundreds of thousands. Regularity (or irregularity) for example in abundance spectra can then be explored on a very large scale. Such studies reveal interesting features. First, strong electronic shell effects appear for almost any size range. Second, the initial trend, for moderate sizes, is a decrease of the amplitude of shell effects (in particular the shell separation), which is very similar to both the cases of atoms and nuclei. But when further increasing the cluster's size (which is impossible in atoms or nuclei), one observes a revival of shell effects. For even larger sizes one again observes a decrease of the effect and one speaks of "supershells".

One should also note that ions play an increasing role with increasing size with the appearance of ionic shell effects favoring the appearance of some finite size ionic structures. Ions then have a tendency to arrange in a more or less regular way (for instance, to form an icosahedron, see atomic structure of Ar_{13} in Figure 5.7). They somewhat build a precursor of a piece of solid following some crystal symmetries. And when a particularly compact and regular arrangement is attained, one observes an enhanced stability, which is then attributed to an ionic shell effect. There are therefore two kinds of shell closures, electronic and atomic ones. Note however that, while electronic shells can show up in metal clusters at large temperatures (around the melting point), the observation of atomic shells needs very low temperatures to minimize thermal shape fluctuations. Moreover, magic electron numbers can exhibit a dense sequence. On the contrary, closing an ionic shell is more demanding, for instance with the completion of a full icosahedral shape or at least the formation of clean flat faces. The spacing between two successive ionic shells is therefore larger than between two successive electronic shells.

As a final word, shell effects are known in basically all materials, metals or non metals. But while ionic shell effects (geometrical effects) are observed whatever the material, electronic quantum effects are only possible in metals because of the highly delocalized nature of the electron cloud.

5.3.4 Optical properties

Optical response

The term optical response labels the response of clusters and molecules to light. The term may be a bit misleading as the word 'optical' implies visible light. Indeed, many molecules and clusters do have a color and thus respond to light in the visible part of the optical spectrum. But in general the response covers a wider range of frequencies, typically from IR to UV or even above. The 'color' is an appreciable property of a system as it provides a fingerprint, characteristic of the system under study.

The case of metal clusters, that we shall specifically discuss here because of the quantum liquid character of their electronic cloud, introduces an important additional aspect called Mie plasmon. This collective mode directly stems from the metallic nature of binding and the associated quasi free motion of valence electrons. When applying an electric field to a metal cluster, the system tends to polarize by displacing the whole electron cloud with respect to the ionic background. When released, the electron cloud will then start to oscillate with respect to the ionic background. The associated frequency, in the range of visible light for alkaline metal clusters, is called the Mie plasmon frequency $\omega_{\rm Mie}$. Mie's work goes back to the early 20th century, well before metal cluster studies. In fact, the idea was to study the behavior of small metallic particles (recall metal inclusions in glass) in response to an external electric field or to light and to see how small metal particles can be analyzed that way.

Mie plasmon

The Mie plasmon corresponds to a collective excitation/response of the system, at variance with usual molecules. It can be classically estimated as oscillations of a spherical electronic cloud as a whole, against a spherical ionic background. Let us consider a model cluster represented by a jellium sphere (see Section 5.2.1) for the ionic background and an electron sphere representing the electronic cloud. Both spheres are assumed rigid and of radius R (therefore with volume $V = 4\pi R^3/3$). They contain N particles (we assume one valence electron per atom/ion) and have the same density $\varrho_I = \varrho_e = \varrho = N/V$. We first compute the electric field created by ions in a given point of space, using Gauss theorem. Inside a sphere of radius r is contained a charge $Q(r)$ and the (radial) electric field $E(r)$ then satisfies $4\pi r^2 E(r) = Q(r)/\varepsilon_0$ which gives the field $\mathbf{E}(\mathbf{r})$ all over space as

$$\mathbf{E}(\mathbf{r}) = \frac{\varrho e}{3\varepsilon_0}\mathbf{r}\,(\text{for}\, r < R) \quad \text{and} \quad \mathbf{E}(\mathbf{r}) = \frac{\varrho e}{3\varepsilon_0}\left(\frac{R}{r}\right)^3 \mathbf{r}\,(\text{for}\, r > R)$$

where e is the elementary electric charge.

We now imagine that the electron sphere is displaced by a vector $\mathbf{a}$, say along an arbitrary z-axis, and assume that $a \ll R$. The mismatch between electronic and ionic centers of mass creates a restoring force. In the center of mass frame of the ions, the restoring force on electrons can be written as an integral over electrons

$$\mathbf{F} = -\int \mathbf{E}\,\varrho\, e\, \mathrm{d}^3\mathbf{r} \simeq -\int_{r<R} \mathbf{E}(\mathbf{r}+\mathbf{a})\varrho(\mathbf{r})e\,\mathrm{d}^3\mathbf{r} \quad .$$

The fact that $a \ll R$ allows us to neglect surface effects so that integration becomes trivial and only the inner field ($r < R$) needs to be used, reducing the force to $\mathbf{F} = -N\varrho e^2/(3\varepsilon_0)\,\mathbf{a}$. The force is then simply proportional to the displacement a.

Let us now imagine that we have initially polarized the electron cloud, (for example by the action of an external electric field) and that, once we released this initial constraint, we let the system evolve freely in time. The equation of motion of the electron cloud is obtained by using the force $\mathbf{F}$ computed above. For a given electronic center of mass position $\mathbf{r}$, we can write:

$$\frac{\mathrm{d}^2\mathbf{r}}{\mathrm{d}t^2} + \frac{\varrho e^2}{3\varepsilon_0 m_e}\mathbf{r} = 0 \quad ,$$

where we have introduced the electron mass m_e. The system's size disappears from this equation as the total mass of the cloud is Nm_e and the total force is also proportional to N. We obtain a perfect sinusoidal oscillatory motion of pulsation

$$\omega_{\mathrm{Mie}}^2 = \frac{\varrho\, e^2}{3\varepsilon_0 m_e} \quad , \qquad (5.17)$$

known as Mie frequency. Notice that it is related to the volume plasmon ω_{p} given in Eq. (5.7) as $\omega_{\mathrm{Mie}} = \omega_{\mathrm{p}}/\sqrt{3}$. In sodium, for example, one finds numerically $\omega_{\mathrm{Mie}} \simeq 5.2 \times 10^{15}\ \mathrm{s}^{-1}$, corresponding to an energy $\hbar\omega_{\mathrm{Mie}} \simeq 3.2$ eV and associated to a period of oscillation $T = 2\pi/\omega_{\mathrm{Mie}} \simeq 1.2$ fs. Note that this crude model gives a very good estimate of the Mie plasmon range. Indeed as can be seen in Figure 5.9 the plasmon peak in Na_{21}^+ lies only about 0.3 eV below our simple estimate (there is a slight additional effect due to charge which does not alter the conclusion qualitatively).

We conclude with three additional remarks. First, it is interesting to note that the Mie frequency scales with the electronic density $\omega_{\text{Mie}} \propto \sqrt{\varrho}$ which is rather well verified experimentally when changing bulk material. Differences between Mie estimate (originally a macroscopic metallic sphere following Mie's idea) and realistic metal clusters responses stem from small quantum effects. This is not so surprising as the driving mechanism (attraction between electrons and ions) is purely classical. Second, the Mie period lies in the fs range which is a typical time scale for electronic motion. It will clearly interfere with any other electronic motion, which again points out the importance of optical effects in metal clusters. One should finally mention that the Mie plasmon resonance is very similar to the nuclear dipole giant resonance in which neutrons oscillate against protons (see Section 7.2.4).

An example of optical response

The Mie plasmon response is illustrated in Figure 5.9 for three small cationic Na clusters. Let us start with Na_{21}^+ (black full curve). One very clearly sees a

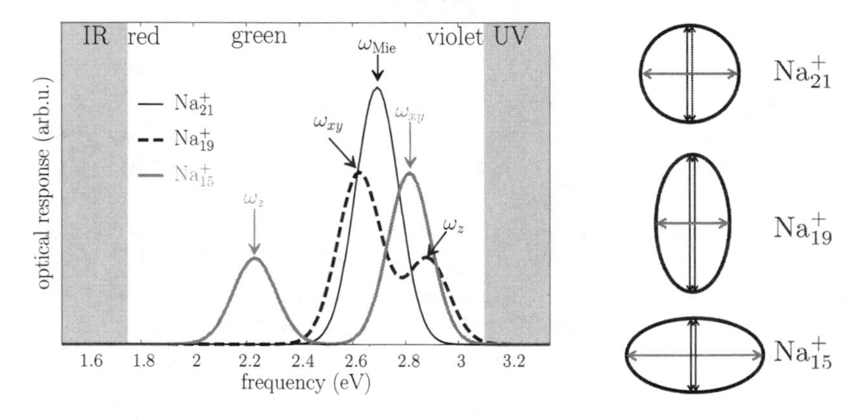

Figure 5.9: Optical response of three metal clusters, the spherical Na_{21}^+ (black curve), the oblate Na_{19}^+ (dashes) and the prolate Na_{15}^+ (gray curve). At the top of the figure, the frequency scale is associated to colors from the visible spectrum (such as red, green and violet), and the infra-red (IR) and the ultra-violet (UV) region. The shape of the three clusters are schematically depicted on the right.

Mie plasmon around 2.7 eV, lying in the visible part of light spectrum, which ranges from 1.75 eV (red) to 3.1 eV (violet). This collective response, although dominating the entrance channel of an optical excitation, nevertheless tends to degrade with time: the resonance has a finite lifetime due to the fact that electrons can progressively couple to individual excitations and break the initial collectivity of the motion. This is directly reflected in the width of the Mie plasmon. In the case of Na_{21}^+, this width is about 0.4 eV. The actual width of the resonance somewhat depends on cluster size. For very small clusters (< 10 atoms), it is usually very sharp because electronic level energies are typically well separated. As the electron level separation shrinks with increasing size (recall Figure 5.4 and the schematic construction of band structure in a solid), the lifetime of the Mie resonance tends to reduce, and correlatively, the width of the peak tends to increase, up to a maximum, because of the coupling of electrons with non-collective motion.

The size of the cluster also impacts another characteristic of the Mie plasmon, namely its position. Indeed, as for a sphere of radius R, we have $\omega_{\mathrm{Mie}} \propto \sqrt{\varrho} \propto R^{-3/2}$; the smaller the sphere, the higher ω_{Mie}. We can easily imagine that for a small radius, the electron cloud will oscillate back and forth more rapidly than for a larger radius. By playing with the size of the cluster, one can at will modify the value of ω_{Mie} and therefore change its color. This is exactly the property used by Roman glass makers to obtain colored glass with varying reflections. For instance, spherical gold nanoparticles with a diameter of 100 nm possess a rather broad Mie plasmon, at 570 ± 50 nm. This means that such a nanoparticle absorbs efficiently colors between green and orange, and appears blue. For a smaller diameter, say 30 nm, the Mie plasmon is located around 520 ± 30 nm. The gold nanoparticle then absorbs blues and greens, appearing red. Interestingly, we realize that the usual color we associate to a metal is bulk color, here yellow for gold. We here point out that at a nanometer scale, optical properties can be rather different from those in bulk. And the measurement of the Mie plasmon position is one of the main means to access cluster sizes.

Shape analysis

Up to here, we have discussed the case of spherical clusters only. One can ask what the pattern of the optical response is for a non-spherical cluster. To answer this question, let us consider once again the relation $\omega_{\mathrm{Mie}} \propto R^{-3/2}$. The power 3 in this behavior reflects the fact that there is no preferred spatial direction for the electron cloud oscillations with respect to the ionic background if the electron cloud is spherical. We have seen in Figure 5.9 that Na_{21}^{+} exhibits a unique plasmon peak. We also note that this cluster contains 20 electrons and thus corresponds to the electronic shell closure at 20 (see Section 5.3.3). An elementary quantum mechanical calculation shows that a shell closure leads to a spherical density. Therefore, a single peaked resonance tells us that the electron cloud (up to possible second order effects from the actual ionic structure) has indeed a spherical shape, as expected from simple quantum mechanics. This demonstrates the strength of electronic effects in such systems. This also holds true for the other magic electron numbers (8, 40, 58, 92, etc.).

Let us now consider the cases of Na_{19}^{+} and Na_{15}^{+} which are two clusters with open shells and henceforth no spherical electron cloud. Still, they exhibit two equivalent spatial directions. In Na_{19}^{+}, there are two 'long' equivalent axes (denoted by x and y) and a shorter one (denoted by z), providing an oblate or pancake-like shape. On the contrary, in Na_{15}^{+}, the electron cloud exhibits two 'short' equivalent axes and a longer one. This corresponds to a prolate or cigare-like shape. These shapes directly reflect in the optical responses. Indeed, let us start with the oblate Na_{19}^+. Oscillations of the electron cloud along the z direction will proceed at a higher frequency compared to the frequency along the x or y direction. In the corresponding optical response (dashes in Figure 5.9), we therefore observe a splitting into two peaks, one at low frequency (ω_{xy}) and another one at high frequency (ω_z). Moreover, because there are two modes orthogonal to the symmetry (z) axis, there is twice as much signal in the lower energy plasmon peak as in the upper one. For the prolate Na_{15}^+ (see gray curve), we rather have $\omega_z < \omega_{xy}$ and the plasmon peak at high frequency is twice higher than that at low frequency.

This direct relation between spectral splitting of Mie plasmon and cluster deformation has been exploited to deduce systematics of deformation in small to moderate size metal clusters. With size increase, the growing complexity of the electronic energy spectrum tends to widen optical spectra, as already noted above and thus possibly blurs the picture by filling peak separations. But the overall mechanism remains active whatever the size, as it is basically governed by geometrical effects. One might finally question where the actual deformation comes from. It actually matches the fact that electronic shells are not filled. But the actual cluster shape results from a subtle balance between electronic and ionic effects. Optical properties nevertheless point out the key importance of the quasi free electron fluid in metal clusters.

Chromophore effect and cancer therapy

Optical properties of metal clusters, and more generally speaking the very strong coupling of metal clusters with light, are important not only from a fundamental point of view. In order to illustrate this point we consider a recent application in an apparently far away domain, namely cancer therapy. The effect used for this purpose is linked to the plasmon response, which is especially important in coupling to light because of its resonant character. The actual effect in the present context is called chromophore effect. It covers the typical situation in which one inserts a metal nanoparticle into an otherwise optically inert compound, such as an insulator. The presence of the metal particle will strongly enhance the coupling of the compound to light, because of the strong response of the metal particle. In the present example this enhancement is used for improving cancer therapy methods by combining anti-cancer drugs to metal nanoparticles.

We take a specific example here in which one combines the anti-cancer drug cisplatin (a molecule containing one Pt atom) to 5 nm-diameter gold nanoparticles (GNP) before irradiation. In order to measure the impact of the GNP, one compares DNA films combined either with cisplatin alone, with GNP alone, or with both cisplatin and GNP. Irradiation proceeds in that case via collision by energetic (60 keV) electrons. The measured quantity is the number of double strand breaks of DNA. Indeed such double strand breaks are known to be hardly repairable, thus expected to improve efficiency in cell killing, which is precisely the goal of cancer therapy. The reference measurement is that without any addition (cisplatin nor GNP). As a next step, one feeds the system with GNP's (1 GNP for 1 DNA) and measures the enhancement factor as compared to the pure DNA case. The result is about a factor 2 of enhancement. One obtains about the same enhancement factor when feeding the system with cisplatin (2 cisplatin for 1 DNA). The interesting case is when one inserts both cisplatin and GNP (in the ratio of 2 cisplatin for 1 GNP and 1 DNA). The enhancement factor then raises up to about 7.5 as compared to the case of irradiation of pure DNA. The effect is obviously more than a mere addition of separate cisplatin and GNP ones. It demonstrates a remarkable synergy between GNP and cisplatin that concerns double stand break damages, and opens the door to important medical applications in an oncological context.

5.3.5 Wetting properties of metal clusters

We have already mentioned the importance of interactions in a (metal) cluster in Section 5.3.1, especially when it is in contact with an environment, as for instance when deposited on a surface. Let us be a bit more specific here and discuss the case of Na_6 deposited on an insulating surface as MgO. Recall that MgO is a typical ionic crystal which one could assimilate to a $Mg^{++}O^{--}$ system whence its insulating character. Here we we will compare the optical response of the Na_6 cluster without any environment (it is then said to be free) with the case when it is deposited. Figure 5.10 depicts these optical responses, as well as the corresponding ionic configurations. The free cluster

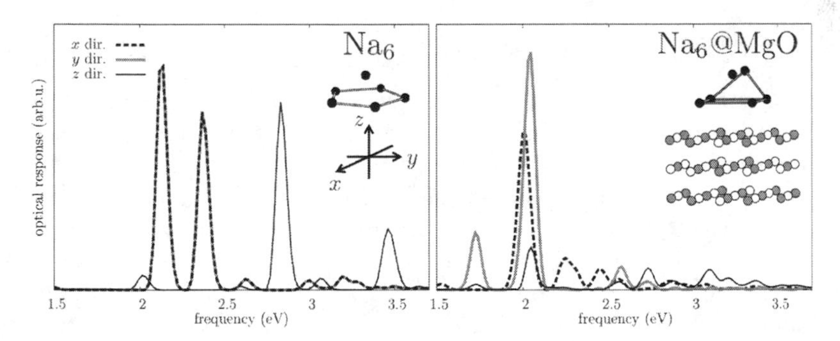

Figure 5.10: Left: optical response of free Na_6 in the three spatial directions x (black dashes), y (light gray curve), and z (dark gray curve) specified with respect to the ionic configuration as an inset. Right: the same for Na_6 deposited on a MgO surface, schematically shown as layers composed of white and gray balls.

consists in a pentagon (in the xy plane) and a top atom on the symmetry (z) axis of the pentagon. Its optical response (left panel) is quite fragmented but still exhibits some prominent plasmon peaks. Note that the spectrum in the x direction is identical to that in the y direction. As expected, the resonance peaks in the z direction lie at higher frequencies than those in the xy plane.

We now turn to the deposited Na_6 on MgO (right panel). From the flat geometry of the free Na_6, one might naively think that it would stick to the surface with almost no change of configuration. Looking at the actual ionic configuration, we observe that this is not the case: Na_6 is strongly deformed at the contact with the MgO surface. This is a posteriori not a surprise because the pentagonal basis of Na_6 does not accommodate well the square arrangement of Mg and O atoms at the surface.

This change of configuration between free and deposited cluster immediately shows up in the optical spectra which are very different. In particular, the x and y directions are not strictly equivalent any longer. Interestingly, we observe that the optical response in the z direction, that is normal to the surface, is very much suppressed. This is the result of strong interactions of the valence electrons of Na_6 with the Mg and O atoms of the surface. There are at the same time attractive polarization effects and strong repulsions which distort the electronic cloud and push it away from the surface in the z direction. The electronic cloud cannot therefore oscillate 'freely' in that direction, while there is more space in the xy plane. In addition, the spectrum in z is quite spread and fragmented. This reflects a strong modification of the electronic energy spectrum which is actually compressed. This induces more coupling of electrons with individual modes and this almost completely destroys the collective modes.

To conclude, we have seen that the optical spectrum constitutes a powerful tool to study the size, the geometry and, to some extent, the electronic density of states of a cluster.

5.4 CONCLUDING REMARKS

Electrons are everywhere and first show up as the key to bind atoms together. In any compound system, one is dealing with a set of interacting electrons whose arrangement provides the particular binding mechanism at work. We have quickly explored this binding mechanism in molecules and bulk matter. Metals constitute a particular class of materials in which electrons ensuring the binding of the system can be seen as nearly independent particles. Interactions are nevertheless again playing a key role. Although the net classical Coulomb effects are to a large extent compensated (electron-electron repulsion against ion-ion and ion-electron interactions), binding definitely stems from interaction, more precisely from the dominating, fully quantal, exchange contribution in this case.

The specific properties of the electron cloud in metals make them especially responsive to electromagnetic couplings, and therefore to light, leading the important optical properties of such systems. Conductivity is another key property of metals, which make them so useful in numerous applications based on electron transport. We have seen in particular that interactions tend to oppose natural electron flow and thus restrict transport properties. We shall see in Chapter 9 how metals, under specific conditions, may overcome this key limitation.

We have also considered in this chapter finite pieces of metals in the form of metal clusters. These systems offer a natural bridge between simple molecules and bulk metal and provide an invaluable source for understanding of how matter build up progressively when piling up particles. Metal clusters exhibit properties archetypal of finite quantum systems, as electronic shell closures which lead to a higher stability at specific cluster sizes. The case almost perfectly illustrates the properties of a finite droplet of interacting electrons, a finite droplet of quantum liquid. Still, one should not overlook the key role played by ions in delivering a potential well confining electrons. As such, metal clusters are not self-bound electron droplets.

CHAPTER 6

Atoms and Molecules

To dedicate a chapter to atoms and molecules in a book on quantum fluids may seem surprising at first glance. Indeed, atoms and molecules are both composite systems, possibly massive, and the first intuition is to follow the analysis of Chapter 1, where we have classically hierarchized phases of matter according to pressure, temperature and volume. However, when looking in more detail at a standard phase diagram (see for example the case of CO_2 or water in Figure 1.3), one soon realizes that it exhibits many other possible interesting paths. In particular, the low pressure/low temperature limit provides an interesting region to explore and actually provides a different scenario. Indeed, in this limit, the single possible phase transition is between gas and solid. But mind that it corresponds to a very low pressure/temperature domain. There are in particular no thermal effects possibly blurring the behavior of the system. We have already seen that boson and fermion gases in their degenerate limits at zero temperature exhibit remarkable features. this particular domain of vanishing temperatures. Both atoms and molecules, depending on their total spin, can be either fermions or bosons.

The chapter is organized following a somewhat historical perspective. Indeed the first excursions in that fascinating domain of low temperatures were performed long ago aiming at liquefying gases. This led to discovering the specific properties of helium which never becomes solid at low pressure. Properties of helium were in turn much studied in particular in terms of the remarkable properties of liquid helium in this low temperature regime. We thus start our discussion with the case of helium, especially ^{4}He which is a boson, and complement the results by the case of the fermionic ^{3}He. We then address the possibility to form and use helium nanodroplets as an ideal matrix. We end with the rapidly evolving field of ultracold gases and the experimental realization of BEC, involving both bosons and fermions.

6.1 HELIUM, THE ONLY LIQUID AT ABSOLUTE ZERO

A natural idea to liquefy gases is to compress them. This was the strategy adopted during early research along that line, in the mid 19th century. Several gases, however, resisted to be liquefied that way, including molecular oxygen, nitrogen and hydrogen, and were dubbed at the time as 'permanent' gases. We can understand where the problem lies by having a glance at Figure 1.3. When the critical temperature is much lower than the room temperature, it is necessary to first cool the gas down to cross the liquid-gas coexistence line and then compress the gas to liquefy it. By proceeding this way, the 'permanent' gases were finally liquefied leading to the exploration of rather low temperatures: oxygen with critical temperature $T_c = 154.6$ K and boiling temperature $T_b = 90.2$ K, nitrogen with $T_c = 126.2$ K and $T_b = 77.4$ K, and hydrogen with $T_c = 33.2$ K and $T_b = 20.3$ K. The last 'resisting' case was helium, discovered on Earth in 1895, and liquefied in 1908 by Kamerlingh Onnes by using a method similar to that currently employed to lower the temperature in a refrigerator: the evaporation of a liquid absorbs heat from the environment, then reducing the temperature. The experience led to the lowest temperature ever reached in a laboratory at the time: the normal boiling temperature of helium is indeed $T_b = 4.2$ K, while its critical one is $T_c = 5.2$ K. These extremely low values are related to the inert character of helium atoms.

But before going further, let us give or recall some general properties of these atoms. At the scale of the Universe, visible matter is mostly formed by hydrogen ($\approx 75\%$) and helium ($\approx 25\%$). These proportions result from the primordial nucleosynthesis and provide a strong argument in favor of the Big Bang theory, which precisely predicts these abundances (with some traces of lithium; see Section 8.1.2). However, these proportions are very different on Earth. In particular, helium represents only 5×10^{-4} % in the atmosphere and $\simeq 10^{-6}$ % in the crust. Natural helium exists in the form of two stable isotopes, ^{4}He and ^{3}He, the former being 99.99986 % of the total population. The atomic structure of the helium atom consists of two electrons around a nucleus, with two protons and one neutron in the case of ^{3}He, and one more neutron in the case of ^{4}He. Despite having identical chemical properties, ^{4}He and ^{3}He have completely different physical properties at very low temperatures. This is not due to their different nuclear masses but arises because they follow different quantum statistics. Indeed, the total spin of a composite system can be computed according to some quantum rules (that can be found in any quantum mechanics textbook). Let us recall here that a ^{4}He atom is a boson of spin zero (adding six spin-1/2 components, that is 2 neutrons, 2 protons and 2 electrons) and the ^{3}He atom is a fermion of spin 1/2 (adding five spin 1/2 components, that is 1 neutron, 2 protons and 2 electrons). The different statistical character of the two helium isotopes produces dramatic differences in their physical properties.

The ^{4}He isotope can be obtained from α-decay of minerals containing uranium, thorium and other heavy radioactive elements, which emit α particles, that is atomic nuclei of ^{4}He. However, some of the helium produced in the α-decay of heavy nuclei deep in the Earth are concentrated in deposits of natural gas. This is the source of the ^{4}He used in laboratories, industry or children's balloons. The ^{3}He isotope is basically obtained as a byproduct of nuclear reactions, mainly the disintegration of ^{3}H or tritium (one proton and two neutrons). Being an artificial product, related to thermonuclear bombs, ^{3}He is thus a much rarer isotope.

In general, the interaction between two atoms is due to the polarization of their electronic clouds, favored by the presence of weakly bound electrons. This was referred to as van der Waals binding in Section 5.1.1. Atoms of inert gases have closed electronic shells and are therefore characterized by a remarkable stability as compared to their neighboring atoms in the Mendeleev periodic table (see Appendix C). Inert atoms cannot easily react chemically, not only among themselves but also with other types of atoms. We have also seen that the weakness of the interaction between rare gas atoms has a direct consequence on the melting and boiling temperatures (particularly low for rare gas atoms); see Figure 2.6. The two electrons of a helium atom complete the first electronic $1s$ shell, which results in a very stable configuration: helium is actually the most stable of the elements. A measure of this stability is given by the value of the ionization potential energy, which is the minimum energy required to extract an electron from an atom. For helium, this value amounts to 24.5 eV, which is almost twice that of its left neighbor (hydrogen, 13.6 eV) and 4.5 times that of its right one (lithium, 5.4 eV). This is also related to the fact that the interaction between helium atoms is particularly weak. It is actually the weakest one among the rare gas: about four times smaller than for neon and about thirteen times smaller than for argon. The weakest force between helium atoms is another essential ingredient to understand their peculiarities.

6.1.1 Phase diagram of ^{4}He

As seen in Section 1.1.2, in the pressure-temperature diagram for a given substance, one can distinguish the existence regions of the solid, liquid and gas phases. The phase diagram of a typical pure substance presents a certain number of generic characteristics, as has been shown in Figure 1.3. We recall that, in a (P, T) plane, the triple point corresponds to the temperature and pressure values for which the three phases coexist. The liquid-gas coexistence line goes from the triple point to the critical point, beyond which liquid and gas phases cannot be distinguished.

We here consider the most abundant ^{4}He isotope, leaving until Section 6.2 the discussion of the ^{3}He phase diagram. The phase diagram of ^{4}He is displayed in Figure 6.1. Several remarkable facts can be noticed, as compared with any other substance: i) there is no triple point, ii) helium can remain liquid down to the absolute zero temperature, and iii) the liquid undergoes the so-called superfluid transition. For historical reasons, the normal and superfluid phases of ^{4}He have been coined helium-I and helium-II, respectively, separated in the phase diagram by the so-called λ-line. At saturated vapor pressure, the transition temperature value is 2.17 K.

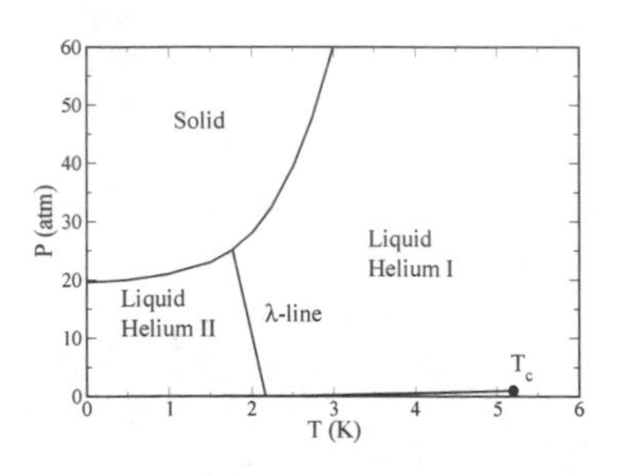

Figure 6.1: Phase diagram in the pressure-temperature map of ^{4}He.

The absence of a triple point means that, starting from the helium gas at not too high pressure and by gradually lowering down the temperature at constant pressure, the only phase transition that one can observe is from the gas to the liquid. Thus, even at absolute zero temperature, helium remains a liquid. This is against the classical idea about the absolute zero, and it is actually a purely quantum phenomenon.

We have seen in Section 1.1.3 that for a given substance, the sequential transitions from the gas to the liquid and then to the solid, reflect an increasing ordering. The existence of boiling and melting points therefore results from a balance between thermal motion and attractive forces between atoms or molecules of the substance. This attraction tends to freeze the constituents in a more or less regular way, while thermal energy continuously tends to break spatial order. Given the very weak interaction between helium atoms, it seems natural that very low temperatures are required to bring order in the gas phase and to produce liquid helium. We will see in the next section that another purely quantal effect hinders liquid helium to solidify at $T = 0$. We will also see in Section 6.1.3 that below the lambda point, He-II, although in the liquid state, is very different from a standard liquid.

6.1.2 Zero point motion

We have shown in Section 2.5.2 that a particle of mass m localized in a cube of volume L^3 possesses a zero point energy, or minimum kinetic energy, of order $h^2/(8mL^2)$, which is independent of the temperature. Henceforth, even at absolute zero, atoms are in motion because of the kinetic energy of the zero point motion. The attractive interactions between atoms tend however to stiffen the system. There is thus a competition between interactions and agitation, and at sufficiently low temperatures, interactions outweigh thermal effects and also the zero point energy for all substances except helium.

The helium-helium interaction, of van der Waals type, exhibits a strong short-range repulsion and a relatively weaker and longer-range attraction (see Figure 1.5). The strong repulsion is due to the fact that helium is a closed shell atom and thus electronic energy states of empty shells could be excited when neighboring atoms begin to overlap. Consequently, the repulsive energy scale is roughly ~ 1 eV when the distance is about 3 times the atom radius. For helium, this corresponds to ~ 1 Å. Beyond this distance, the helium-helium attraction is due to the dipole-induced-dipole interaction. The magnitude of the attraction between helium atoms is about 0.86 meV, which occurs outside the repulsive core at about 2.5 Å separation. As this interaction is only due to the electrons in the atom, it is the same for both ^{4}He and ^{3}He isotopes. The atom mass, nevertheless, sets the scale of zero-point motion in helium, with the kinetic energy being comparable to the potential energy even at $T = 0$. This is strictly a quantum effect, due to the low mass of the helium atoms and to the small space in which a typical helium atom is confined by strong repulsion from the neighboring atoms. But it is this zero-point energy which keeps the helium liquids from solidifying all the way down to absolute zero unless a significant external pressure is applied, more than 20 (30) atm for ^{4}He (^{3}He).

Table 6.1: Some characteristic properties of liquid heliums at their respective saturated vapor pressure near $T = 0$.

	^{4}He	^{3}He
Mass $\hbar^2/m$ (K Å^2)	12.12	16.08
Energy per particle (K)	-7.15	-2.49
Particle density (Å^{-3})	0.0218	0.0163
Volume per atom (Å^3)	45.9	61.3
Incompressibility (K)	27.3	12.1
Speed of sound (m/s)	238	183

Several characteristic properties of liquid ^{4}He and ^{3}He, extrapolated to $T = 0$, are reported in Table 6.1. Notice the units employed: lengths are given in angstrom, and energies in kelvin. Some of these properties can be qualitatively understood by taking into account that both isotopes feel the same interaction but have different masses and quantum statistics. The lighter mass of ^{3}He atoms implies a larger mobility and hence a smaller saturation density as well as a smaller binding energy per particle. The volume per particle occupied by liquid ^{4}He is smaller than that of ^{3}He. Therefore, it is easier to compress liquid ^{3}He (smaller incompressibility). As a consequence, the speed of sound in liquid ^{3}He is smaller than in liquid ^{4}He. Indeed, the propagation of sound in a medium proceeds from successive compressions and dilatations. In a crystal, this can be viewed as the propagation of displacements of atoms with respect to their equilibrium position. Schematically, the interaction between atoms can be modeled by springs. The less compressible the medium, the stiffer the springs, and the quicker the sound wave propagates. This is why the sound propagates faster in a solid compared to the liquid, and in the liquid compared to the gas. It is also worth noting that the ratio of the volume occupied by each atom ($V_{\text{atom}} = 4\pi R^3/3$, with $R \sim 1.3$ Å) to the available volume per atom is approximately 0.2 for ^{4}He and 0.15 for ^{3}He. This thus indicates that liquid heliums are highly packed systems and that therefore the influence of correlations is strong. As a comparison, this ratio is about 0.05 for nuclear matter.

The average distance between two atoms in liquid helium is about 2-3 Å. The zero-point energy, associated to a helium atom confined in a region of this size, is of the same order as the intensity of the attractive forces between two atoms. However, to solidify helium, each atom should be located in a region of an even smaller size. On the other hand, the zero-point kinetic energy would become even greater than in the liquid case (that is, greater than the depth of the attracting well), so that the solid cannot form, unless a (strong) external constraint is exerted. Helium was finally solidified in 1926 at a temperature of 4.2 K but by applying a very strong pressure of 150 atm to outweigh and compensate the zero-point motion and allow the formation of a solid phase. The presence of such a solid phase (or liquid anyway) then results from the competition between the energy gain associated to a more ordered state and the agitation of the zero-point motion. The absence of triple point for helium is explained by the fact that for pressures below about 25 atm, the solid phase is energetically less favorable at any temperature. Therefore, to explain the properties of helium at temperatures below a few K, quantum physics reveals itself essential.

In contrast, the atoms of other inert gases have on the one hand a higher mass, and accordingly, a negligible zero-point energy; and on the other hand, interact between them more strongly than in the case of helium, typically 3, 12, 16, 22 times in Ne, Ar, Kr, Xe respectively. As a consequence, neon, argon, etc. can be solidified by simply lowering down the temperature. The same reasoning holds for all other chemical elements, where the interatomic attraction is even stronger. Molecular hydrogen is a particular case because of its lighter mass and stronger interaction as compared with helium. It will be discussed in Section 6.1.4. In the following, we shall first concentrate on ^{4}He, and defer the discussion of the fermionic ^{3}He liquid to Section 6.2.

6.1.3 The $\lambda-$transition

At first glance, liquid helium hardly differs from other liquids apart from its lower density, and the fact that it does not solidify at very low temperature. Actually, near its boiling temperature (above 4.2 K), helium rather behaves as a dense gas of classical particles, and the ideal gas model qualitatively explains some of its properties. But at lower temperatures, a surprising behavior very contrary to our intuition appears. Firstly, we notice in Figure 6.1 that the solid/He-II coexistence line is almost horizontal at very low temperatures. This means that the solidifying pressure of He-II remains nearly constant at very low temperature. However, thermodynamics tells us that the temperature dependence of this pressure is related to the difference of entropy between the solid and He-II. Since entropy is related to the disorder of a system (see Section 3.2.2), the fact that this dependence is flat means that the solid is not more ordered than He-II. Even more surprising, experiments with X-rays do not detect any spatial order in He-II. This means that it is actually a different type of order, more subtle, which is at play here. This order is actually related to the bosonic nature of the constituents, as we shall discuss later on. The lambda point therefore introduces a clear separation between two very different phases of helium.

The physical properties of liquid ^{4}He are essentially related to the fact that this atom is a boson, while we know that a system of identical bosons has a very particular behavior at low temperatures. These properties dramatically change when crossing the λ-line. The different behavior can be clearly visualized by observing a beaker filled with liquid ^{4}He boiling under reduced pressure. In the He-I region (higher temperatures), the liquid is greatly agitated by bubbles of vapor which form in the whole volume, in as much the same way as bubbles are formed when an ordinary liquid is heated. However, once the λ-line is crossed (lower temperatures), the liquid He-II becomes quite calm and still.

In addition, He-II becomes a perfect conductor of heat, as indicated by the behavior of its specific heat, displayed in Figure 6.2 as squared points. We re-

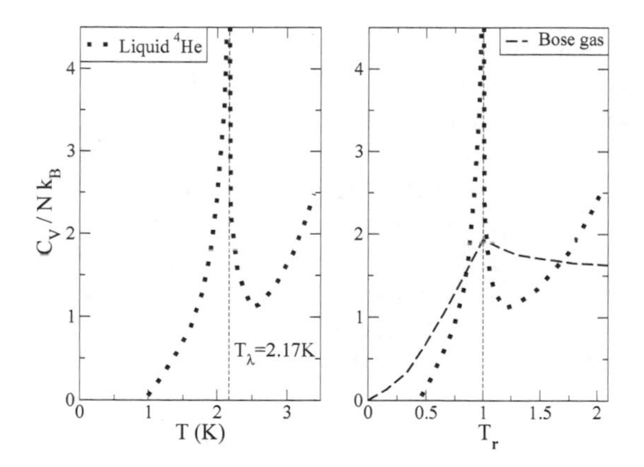

Figure 6.2: Specific heat C_V divided by the total number of particles N and the Boltzmann's constant k_B, as a function of a reduced temperature T_r, that is T/T_λ for ^{4}He atoms (squares, $T_\lambda = 2.17$ K), and T/T_{BEC} for an ideal Bose gas (dashes, $T_{BEC} = 3.13$ K).

call that this is a measure of the required energy to produce an increase of 1 K in 1 kg of matter. In a solid, the main contribution comes from phonons, that is specific vibrations of the lattice. More precisely, $C_V \propto T^3$ at low temperatures (see Section 5.2.3). And indeed, current experience shows that the specific heat of a substance decreases when the temperature decreases: the colder the substance, the easier to warm it up. But it turns out that at 2.17 K, the specific heat of ^{4}He increases abruptly by several orders of magnitude, and falls down to zero at lower temperature. As the curve representing the variations of the specific heat as a function of temperature has a shape similar to the greek letter λ, this was named the lambda transition, occurring at $T_\lambda = 2.17$ K. He-I is the phase above T_λ and is, in most respects, an ordinary liquid. The conduction of heat in the phase He-II is, on the contrary, almost instantaneous: a local increase of temperature is immediately transmitted throughout the liquid. No bubbles are formed in the volume and the evaporation takes place only at the surface of the liquid. Going from He-I to He-II, one observes a sudden stop of the boiling when the temperature becomes lower than T_λ.

He-II actually exhibits a whole range of unique phenomena, as the forementioned beaker experiment. They are related to the superfluidity of He-II, that is, the absence of viscosity. A first description of this 'superphase' is given in the next section in terms of Bose-Einstein condensation.

6.1.4 Superfluidity

Viscosity is a measure of the energy lost by a fluid in motion, due to friction with the walls of the container (see Section 1.4.2). The energy losses increase as the relative velocity between the fluid and the walls increases. On the contrary, He-II can freely flow through a capillary, with no apparent viscosity. More precisely, below the λ point, the viscosity of He-II flowing through a capillary of a micron diameter suddenly drops by a factor $\simeq 10^{11}$. This property is called superfluidity in analogy (a point to be further discussed, though) to the phenomenon of superconductivity presented in Section 9.1. In any case, He-II is far from behaving according to the usual hydrodynamic laws.

The identification of superfluidity goes back to the late 1930s and it was soon suggested that the effect could be related to the condensation of bosons discussed in Section 3.3. For an ideal gas of bosons of mass m, we have defined a critical temperature $T_{\rm BEC}$ for the Bose-Einstein condensate (BEC) by:

$$k_{\rm B} T_{\rm BEC} = 3.31 \, \frac{\hbar^2}{m} \, \varrho^{2/3} \quad , \tag{6.1}$$

below which the lowest energy state is macroscopically occupied. Inserting in this expression the ^{4}He mass and density values (see Table 6.1), one finds a value of 3.13 K for the BEC critical temperature, which is surprisingly close to T_λ. A striking similarity shows up in Figure 6.2 which compares the specific heat capacity C_V of liquid helium (squares) and that of a BEC of helium gas (dashes). To allow the comparison, C_V is plotted as a function of a reduced temperature, that is T/T_λ for liquid helium and $T/T_{\rm BEC}$ for a helium BEC. There are nevertheless two differences between these curves that are worth mentioning. First, at the critical temperature, there is a simple cusp in the BEC curve, while we have a much sharper peak in liquid helium, very close to being logarithmic. This peak is very well represented by a power law $|T - T_\lambda|^{-\alpha}$ with a critical exponent $\alpha \simeq 0.009$. Second, as $T \to 0$, the specific heat tends to zero as the power T^3 for He-II, in contrast to $T^{3/2}$ for the BEC (see Section 3.3).

These discrepancies lead us to ask the following question: can superfluid helium be completely identified with a BEC? The answer is no. Superfluidity of ^{4}He and BEC present undoubtedly strong similarities. But the two situations are not strictly identical. Indeed, the interaction between helium atoms does significantly modify the distribution of energy levels as compared to that of an ideal gas of bosons. This results in a condensed fraction smaller than that of the ideal Bose gas. Indeed, in the latter case, at $T = 0$, all bosons condense in the lowest energy state ε_0. In He-II instead, even if there is a macroscopic number of atoms which condense into a single quantum state at $T = 0$, the number of atoms having an energy ε_0 is only about 9%, although the liquid is 100% superfluid. We can say that the existence of a condensate is a necessary but not a sufficient condition for superfluidity. We postpone a more detailed description of He-II to Section 9.2 and now turn to the ^{3}He isotope.

To end with condensates of bosonic atoms, let us consider hydrogen, in connection with the previous arguments on zero-point motion and Eq. (6.1). Atomic hydrogen is a boson (made of two fermions, that is a proton and an electron) and has a smaller mass than helium atom. Therefore at low temperatures, one can expect a larger zero-point motion and enhanced quantum effects. Actually hydrogen appears in nature in the form of diatomic molecules, consisting of two electrons and two protons. The electrons are coupled to zero angular momentum. The nuclear spin may be 0 or 1, and these possibilities are known as parahydrogen (pH_2) and orthohydrogen (oH_2) respectively. The most stable form is parahydrogen, since the lowest energy level of oH_2 is higher than the lowest one of pH_2 by 170.50 K. However, at room temperature, equilibrium hydrogen is 75% ortho and 25% para, that is the ratio of statistical weights. At low temperatures, near the boiling point of 20.3 K, hydrogen is nearly all in the para state. We recall that, using the values of ^{4}He mass and density of liquid ^{4}He, Eq. (6.1) gives $T_{\mathrm{BEC}} \simeq 3$ K which is very close to the superfluid transition temperature T_λ. As $m(\mathrm{pH_2}) \approx m(\mathrm{He})/2$, Eq. (6.1) gives a rough estimate $T_{\mathrm{BEC}} \simeq 6$ K. This value suggests that superfluid molecular hydrogen would be more easily observed... provided it remains liquid at such a low temperature. However, the interaction between pH_2 molecules is so strong that the phase diagram of molecular hydrogen has a triple point at 13.8 K and 7.2 atm. The critical point is characterized by $T_c = 33$ K and $P_c = 13$ atm. It also has a very low density $\varrho = 0.0216$ molecules per Å^3 at the normal boiling point. At low temperature and normal pressure, hydrogen condenses into a solid phase.

Several ways have been proposed to observe superfluidity in pH_2. No evidence of superfluidity was found in experiments with the liquid supercooled at temperatures as low as 1.3 K. Another possibility is to study the behavior of pH_2 in restricted geometries to increase the surface, thus reducing the importance of the interaction. Experiments on thin pH_2 films or pH_2 in porous media have not been successful so far. At present, the only evidence shows up in experiments with small clusters of pH_2, as will be seen in Section 9.2.5.

6.2 HELIUM THREE

6.2.1 Phase diagram of ^{3}He

We here discuss some physical properties of ^{3}He with the help of its phase diagram. The latter is depicted in Figure 6.3, together with that of ^{4}He, to ease the comparison between both isotopes. As already mentioned, the electronic

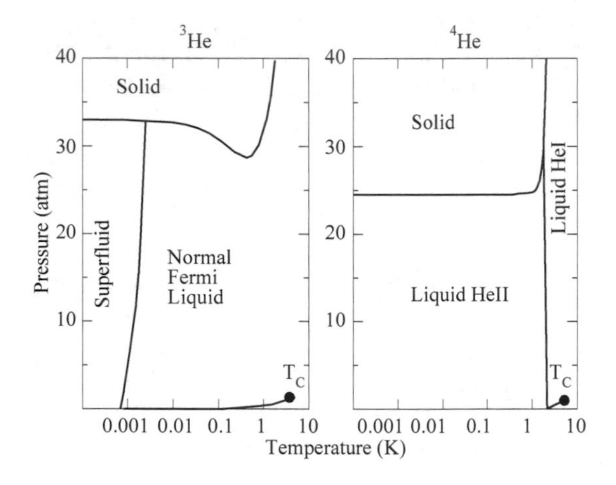

Figure 6.3: Phase diagram in the pressure-temperature map of ^{3}He (left) and ^{4}He (right). Note that the temperature scale is logarithmic.

shells in both ^{4}He and ^{3}He atoms are identical, the effect of the nuclear mass being negligible for practical purposes. Therefore, near their respective boiling points, both liquids behave like a dense classical gas. At low temperatures instead, the physical properties of both isotopes differ from those of a classical liquid. Similarly to ^{4}He, there is no triple point in ^{3}He: it remains liquid down to $T = 0$ and becomes superfluid, but at much lower temperature than ^{4}He. The critical temperature of liquid ^{3}He is $T_c = 3.32$ K, and its normal boiling temperature $T_b = 3.19$ K.

There are however qualitative differences between both isotopes. Notice first the minimum in the melting curve, at about 3 MPa and 0.32 K, with a rise in the melting pressure to about 3.5 MPa when $T \to 0$. This means that, below 0.3 K, the variation $\partial P/\partial T$ is negative. This unusual behavior is related to the fact that ^{3}He atoms are spin-1/2 fermions. Due to this spin, these atoms can be 'spin-polarized'. The atoms in the solid are more localized than in the liquid and the atomic wave functions less overlap. Therefore, spins of various atoms do not interact significantly and globally, the direction of the spin vectors is random: the solid is paramagnetic. This holds true down to $\simeq 0.01$ K. At lower temperatures, a transition to an antiferromagnetic solid (anti-aligned spins) takes place, and eventually the entropy goes to zero at absolute zero. In the liquid, the spin disorder is removed at higher temperatures, and the entropy drops off rapidly.

In the region between about 3 mK and 3 K, the behavior of liquid ^{3}He is different from liquid ^{4}He and, of course, different from the behavior expected for a classical liquid. It is a quantum liquid made of spin-1/2 fermions, which is called a 'normal' Fermi liquid. Below about 3 mK, ^{3}He becomes superfluid, but has a more complex structure than ^{4}He, as we shall see in Section 9.2.4.

6.2.2 The normal liquid ^{3}He

We now give some characteristics of normal liquid ^{3}He which refers to the phase for a temperature between 1 mK and 3 K, and a pressure between the saturated vapor pressure and about 3.4 MPa. The Fermi gas presented in Section 3.4 provides here some telling estimates. At zero temperature, all energy levels are filled up to the Fermi energy $\varepsilon_F = \hbar^2 k_F^2/2m$, where the Fermi momentum has the value $k_F = (3\pi^2 \varrho)^{1/3}$. For the density value $\varrho = 0.0163$ Å^{-3} (see Table 6.1), one obtains $k_F = 0.78$ Å^{-1}, and the Fermi temperature is $T_F = \varepsilon_F/k_{\mathrm{B}} = 4.9$ K. Therefore the temperatures where superfluidity occurs are at least three orders of magnitude lower than the energy scale set by the Fermi energy ε_F.

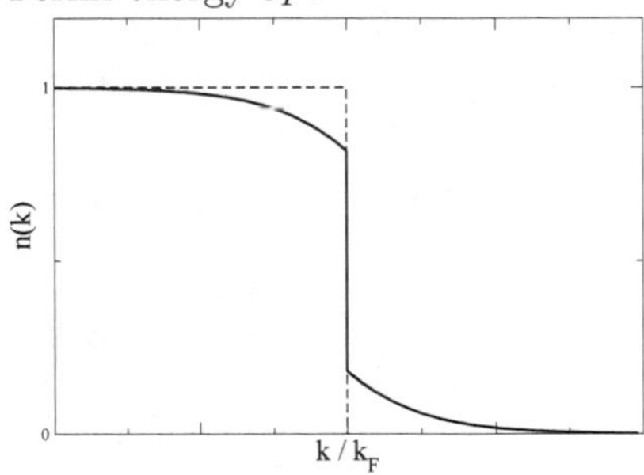

Figure 6.4: Momentum distributions of the ideal Fermi gas (dashes) and an interacting Fermi liquid (solid line) at $T = 0$.

The ground state of an ideal Fermi gas is characterized by a step function in the energy or the momentum distribution (see Figure 3.8 or dashed line in Figure 6.4). At zero temperature, all energy states below the Fermi surface are filled and there is no occupied state above k_F. The discontinuity at $k = k_F$ is equal to 1. This holds strictly true only in a pure Fermi gas with no interaction. In general, the discontinuity at the Fermi momentum persists in presence of interactions but it is reduced, as shown by the full curve in Figure 6.4. The deviation of the discontinuity from 1 is a measure of the strength of the interactions. Fermi liquids that have this property are called 'normal' Fermi liquids. The discontinuity disappears for a superfluid.

Both liquid ^{3}He and the ideal Fermi gas share some qualitative properties: the specific heat varies linearly with temperature, some transport coefficients such as viscosity and thermal conductivity have the same dependence with temperature... These behaviors indicate that the normal liquid ^{3}He could be approximated by a degenerate Fermi gas, with some corrections due to the interactions between atoms. The persistence of the discontinuity at k_F in the case of a Fermi liquid lies at the basis of the celebrated phenomenological theory of normal Fermi liquids, developed by Landau in 1956. Landau realized that at low T, this parallel between both systems can be justified within some simple assumptions. It is worth stressing that Landau's theory has also been extended to other Fermi liquids such as electrons in metals, nuclear matter as well as to finite systems such as metal clusters, helium droplets or atomic nuclei. It is thus a quite generic approach for quantum Fermi fluids and it is worth being discussed a bit on the example of ^{3}He.

Landau's theory does not attempt to describe the ground state, but only excitations of the system near Fermi surface. Low-energy excitations involve only particles close to the Fermi surface. Such excitations are described in terms of an equivalent gas of weakly interacting fermions, denoted as 'quasi-particles', whose equilibrium distribution is given by the Fermi distribution function. The validity of the theory is restricted to small momenta ($< k_F$), small excitation energies ($< \varepsilon_F$) and small temperatures ($< T_F$). The quasi-particles are not to be identified with single atoms, but represent the collective motion of several (or many) of them. As a consequence they are characterized by an effective mass m^* and an effective interaction between them, both being determined by fit to some experimental observables. For instance, the effective mass is directly accessible through measurements of the specific heat per unit volume. For a non-interacting gas (see Section 3.4.2), the expression is $C_V = \gamma T$ with $\gamma = k_{\mathrm{B}}^2/(\hbar k_F)^2 N \pi^2 m$. The same form holds for a normal Fermi liquid, by simply replacing the bare mass m with an effective mass m^*. We have also seen in Section 5.2.3 that, in a bulk metal, an effective mass appears in C_V, stemming from interactions of valence electrons with the lattice and between them. From the measurement of C_V, one finds an effective mass for ^{3}He which varies with the density. For instance, we have $m^*/m = 2.8$ at saturation vapor pressure, and $m^*/m \simeq 6$ at melting pressure. It is also worth noting that, as the effective interaction is defined on the Fermi surface at k_F, the parameters of this model depend on the density.

Landau's theory is not limited to a phenomenological description of a number of properties in terms of a few adjustable parameters. It also predicts new phenomena, such as the excitation known as 'zero sound'. The ordinary (or first) sound, as discussed in relation with the incompressibility in Section 6.1.2, propagates in a medium by successive compressions and dilatations in a regime where the pressure wave period is much larger than the typical collision time between particles. This corresponds to the so-called hydrodynamic regime. Zero sound instead is a collective excitation mode of the system, originated by the effective interaction, which appears at sufficiently low temperatures in the collisionless regime, that is when the collision time is longer than the pressure wave period. The mode propagates with no attenuation if the interaction among quasiparticles is repulsive, and is damped if the interaction is attractive. Two regimes are found, the low temperature one in which sound propagation takes place at velocity c_0 and the ordinary sound, which takes place at high temperature, with velocity c_1 slighlty smaller than c_0.

6.2.3 Isotopical mixtures of liquid helium

Isotopes ^{4}He and ^{3}He have different masses, interact through the same (weak) interaction, but follow different quantum statistics, which lead to different properties for their liquids. A liquid formed by a mixture of both isotopes is interesting in several respects. One can expect that the addition of a few ^{3}He atoms to liquid ^{4}He will produce changes similar to that observed when adding impurities into a liquid. For instance, the properties of the free surface of liquid ^{4}He change dramatically when a small amount of ^{3}He atoms are added. The ^{4}He surface tension is substantially lowered, even for a relatively small ^{3}He concentration. This effect is explained in terms of the different zero-point motion of each isotope. The excess in kinetic energy pushes the ^{3}He atoms to the surface, where they form a quasi two-dimensional system.

We define the ^{3}He concentration in the mixed liquid as $X_3 = N_3/(N_4+N_3)$, where $N_{4,3}$ refer to the total number of atoms of each species. Figure 6.5

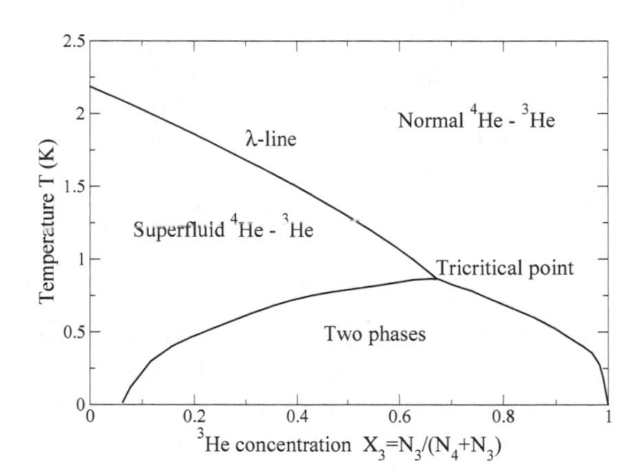

Figure 6.5: Schematic phase diagram of liquid ^{3}He-^{4}He mixtures containing N_3 ^{3}He atoms and N_4 ^{4}He atoms, in the (T, X_3) plane where $X_3 = N_3/(N_3 + N_4)$ is the ^{3}He concentration.

presents the phase diagram of the isotopically mixed liquid in the (T, X_3) plane. There are three regions made of mainly normal liquid ^{4}He-^{3}He, mainly superfluid ^{4}He with some dissolved ^{3}He, and a two-phase region in which the liquid separates into ^{4}He rich and ^{3}He rich solutions, with the lighter ^{3}He-rich phase floating on top of the more dense ^{4}He-rich phase. Note that the addition of ^{3}He to liquid ^{4}He lowers the lambda temperature down to 0.87 K which is called the tricritical point. Above that temperature, liquid ^{3}He and ^{4}He are miscible in all proportions. At zero temperature, the maximum solubility of ^{3}He in ^{4}He is about 6.6% at the saturating vapor pressure. The maximum solubility increases as pressure is increased until it reaches a maximum of about 9.5% at about 10 atm, lowering down afterwards. Near the $X_3 = 1$ concentration, the solubility of ^{4}He in ^{3}He approaches zero exponentially as temperature tends to zero, and consequently only a few experiments have been done in that region.

6.3 DROPLETS OF QUANTUM LIQUIDS

At first glance, superfluidity is a macroscopic phenomenon related to a macroscopic number of atoms. It was first observed in the flow of liquid helium through a capillary, and afterwards superfluidity has also been observed in aerogels, porous media, thin helium films on the surfaces of solids... All these systems contain a very large number of atoms, and one can wonder whether superfluidity could also appear in systems containing a small number, much smaller than Avogadro's number. This question raised the interest in studying helium nanodroplets, namely self-bound systems of helium atoms with size up to about 100 nm, containing less than a few million atoms. The knowledge and understanding of helium droplets has experienced a tremendous growth since the 1990s, and has attracted the attention of researchers from many-body theory, atomic and molecular spectroscopy, and quantum chemistry. The possibility of having a quantum liquid both as bulk and finite droplets is an important issue. Moreover, helium droplets have revealed to be very useful for studying properties of atoms and molecules embedded in them.

6.3.1 The difficult production of helium droplets

For a long time, the experimental research of superfluidity in finite helium systems was hindered by the difficulties in producing and controlling nanodroplets. Nowadays, beams of helium clusters are produced by adiabatically expanding the high purity gas at a high pressure ($10 - 200$ atm) and low temperature ($3 - 20$ K) into a vacuum. Owing to evaporation, the temperature decreases very rapidly, and in a few milliseconds the final temperatures of the clusters in the expanded gas drop down to 0.38 K in the case of ^{4}He, and 0.15 K in ^{3}He. According to the respective phase diagrams, ^{4}He clusters are in the superfluid liquid region, whereas ^{3}He are in the normal liquid region. The number of atoms in the droplet is measured by scattering experiments, crossing the beam of droplets with a secondary beam of a heavy rare gas. It has been observed that the different helium clusters in the cryogenic free expansion all have the same sharp velocity v, with a small relative dispersion $\Delta v/v \approx 2\%$. Since the velocity of the droplet beam is sharply peaked, the distribution of the deflection angles with respect to the incident direction can be directly related to the droplet mass. Varying the conditions in the source, it is possible to adjust the cluster size in the beam. As an example, pressures of 20 atm and temperatures of 12 – 15 K produce droplets with about $10^3 - 10^4$ atoms. Increasing temperature at fixed pressure yields smaller sizes, and vice versa.

The study of very light helium nanodroplets, containing up to a few dozen atoms, is an experimental challenge requiring a great ingenuity. The lightest nanodroplets have been mass selected by diffracting a molecular beam from a transmission grating. Nanodroplets are indeed quantal objects, which can produce interferences related to their de Broglie wavelength $\lambda = h/(Mv)$, where v is their velocity and $M = Nm$ the mass of a droplet containing N atoms of mass m. The de Broglie wavelength λ takes typical values about $0.1/N$ nm. After diffraction by a grating, the first order diffraction peak appears at an angle $\theta \simeq \lambda/d$, where d is the spatial period of the grating. This diffraction technique has in particular provided conclusive evidence of the existence of the dimer $(^4\text{He})_2$, and has also led to the determination of its bond length of 52 ± 4 Å, from which a binding energy of order 1 mK was estimated. The weak helium-helium interaction makes the dimer $(^4\text{He})_2$ an extremely fragile system, very close to the limit of stability, with the zero-point energy almost identical to the potential depth. Experiments show that from the dimer on, ^{4}He droplets are bound for any number of atoms.

Very small nanodroplets are also a challenge for theoretical methods, since their binding energy is very small. Calculations show that pure ^{4}He droplets are bound for any number of atoms. The binding energy of the dimer $(^4\text{He})_2$ lies between about 0.8 and 1.6 mK, depending on the particular potential employed, which is in agreement with the experimental estimate. In contrast, a minimum number of constituents is required to have a bound system made of only ^{3}He atoms: even if the interaction is the same, the smaller mass of ^{3}He atoms and its fermionic character cooperate against the attractive part of the interaction to form a stable system. An upper bound to that minimum has been established to 30 atoms.

One may wonder if this difference between ^{4}He and ^{3}He droplets is due to statistics or to zero-point motion. Very small systems formed by a mixture of both helium isotopes are particularly interesting, with a competition between the binding tendency due to ^{4}He atoms and the opposite due to ^{3}He atoms. Given that 30 is the minimum number of ^{3}He atoms to be self-bound, one expects that the additional ^{4}He atoms will lower the required number of fermions to produce a bound state. Theoretical calculations show that indeed four ^{4}He atoms can bound any number of ^{3}He fermions. All in all, both theory and experiment indicate that both statistics and mass difference have a comparable effect.

6.3.2 Helium nanodroplets as an ideal matrix

A fundamental property of helium droplets is their ability to pick up any species with which they collide. As a consequence, they can be doped with nearly any kind of atomic or molecular species, forming new complexes. Virtually, all closed shell atoms and molecules are heliophilic, that is, when captured, they are localized inside the drop. The important exception are alkali and alkaline earth atoms (respectively in the first and the second columns of the periodic table of elements) their interactions with helium atoms are so weak that they cannot support a bound state. They therefore remain on the surface of the droplet. Note that this explains why these are the only known surfaces which are not wetted by liquid helium. Indeed, we have seen in Section 1.4.1 that wettability of a surface by a liquid is all the more efficient when interactions between the atoms of the liquid and those of the solid surface are strong. The extreme weakness of helium-alkali and helium-alkali earth interactions hinders helium to wet such surfaces.

Superfluid ^{4}He droplets provide ideal inert, transparent and cold matrices to perform various types of studies on embedded atoms and molecules, due to their low temperatures and to the weak interaction of helium with all other substances. Dopants are captured by helium droplets during their formation in the supersonic beam expansion, when going through a pickup chamber in which a gas of atoms or molecules at very low pressure has been introduced. The number of dopants in a given droplet can be controlled by changing the pressure and temperature in the chamber. These possibilities have drawn the attention of researchers from different fields, so that the study of doped helium droplets has become an interdisciplinary field. We mention here a few examples.

As soon as the dopant enters the droplet, it is instantaneously cooled to 0.37 K. This low temperature opens the possibility to carry out high resolution spectroscopy studies. In spite of the weak interactions between the dopant and the helium atoms, the latters suffice to indeed absorb energy from the dopant and effectively cool it down, even when the dopant is artificially excited. The point is illustrated in Figure 6.6 where we show the fragmentation time of small potassium clusters attached to helium droplets of various sizes. Such small metal clusters, when excited by a similar laser in vacuum do fragment in about 1 ps. When attached to the helium droplets, this fragmentation time is strongly enhanced passing from about 6 ps for a droplet of about 7000 atoms to about 30 ps in a droplet containing 15000 atoms. The effect is attributed to the capability of the helium droplet to absorb the energy excess deposited by the laser in the metal cluster. The larger the system, the larger the possible energy absorption by the droplet and the longer the ensuing fragmentation of the metal cluster. The time scales also tell us that the energy transfer from metal cluster to droplet takes place quickly, within less than the vacuum fragmentation time of 1 ps. Nevertheless, the droplet cannot accommodate the excess energy given by the metal cluster and energy is then released by helium atom evaporation (typically 2000 per ps in that case). All in all, these experiments clearly show that the droplet plays the role of an energy absorber with respect to the metal cluster.

The low temperature of the droplet host allows to carry out high resolution

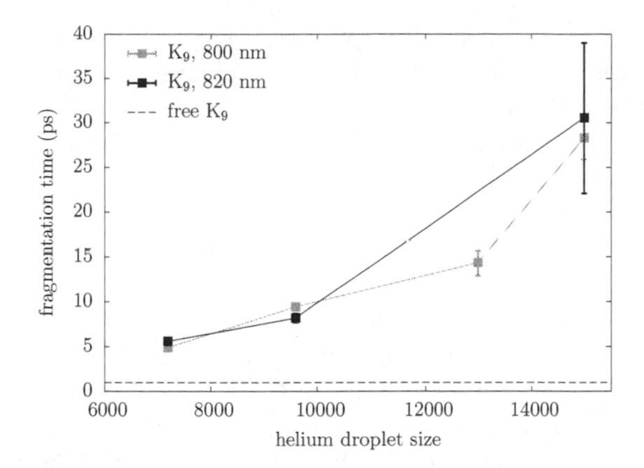

Figure 6.6: Fragmentation time of the K_9 cluster after irradiation by a laser pulse of wavelength 800 nm (circles) or 820 nm (squares) as a function of the mean size of the embedding helium droplet. The horizontal dashed line indicates the fragmentation time for the free K_9. Adapted with permission from F. Stienkemeier and K. K. Lehmann, J. Phys. B 39 (2006) R127.

spectroscopy studies of atoms or molecules embedded in the droplet. Since their interaction with helium is weak, and the host medium is liquid, one avoids a basic problem that arises when the dopant is isolated in the interior of crystals of rare gas or hydrogen. Indeed, in such situations, the interaction with the crystal can significantly affect the geometry of the foreign molecule. Since the number of dopants in a given droplet can be controlled by changing the pressure and temperature in the pickup chamber, one may also avoid another problem often appearing in low temperature spectroscopic studies, such as molecular aggregation. Indeed, at low enough pressures, it is possible to have a single impurity per doped droplet. Since the first reports in 1992 of infrared spectra of molecules embedded into helium nanodroplets, the impact of high resolution spectroscopy in this field has become crucial, making room for a series of studies where the superfluid character of ^{4}He has played a key role. Spectroscopic investigations on the electronic transitions of atoms, molecules and clusters embedded in helium droplets have also been carried out. Organic molecules can therefore be spectroscopically resolved with a precision never achieved with other methods. The results can be used as benchmarks to validate theoretical models, or to identify absorption lines not yet assigned in spectra of stellar species.

The possibility of controlling the number of captured impurities permits other interesting studies. The pickup chamber may contain several species which, when captured, form a complex which may react on contact and frequently adopt unique self-organized structures. For instance, hydrogen cyanide (HCN) is a polar molecule. Once inside the droplet, each additional molecule is aligned by the electric field of its predecessors. As a result, they combine to form linear chains, instead of the usual cyclic isomers. A polymer chain containing up to seven HCN molecules was observed in that way. Another example is with water molecules. The structure of ice-II is a rhombohedrical crystal, whose unit cell is a planar cyclic ring formed by six water molecules. The formation of isolated cyclic hexamer (the 'smallest piece of ice') was observed in ^{4}He nanodroplets and studied with infrared spectroscopy. Alternatively, a reaction can be initiated by laser excitation of one of the constituents of the complex to a specific vibrational and/or electronic state. The nature and quantum states of the newly formed products can then be probed via either laser induced fluorescence or by laser induced depletion of the droplet signal measured further downstream from the reaction region by a sensitive detector. One may also form metal clusters with electronic structure quite different from the respective free one, as in the case of alkali atom clusters, or study the Coulomb explosion of metal clusters grown inside He drops after they are excited by intense femtosecond laser pulses. The few examples mentioned here point to the great potential of helium droplets for the study of a wide range of dynamic processes at very low temperatures.

6.4 ULTRACOLD REAL GASES

We have seen that liquid ^{4}He cannot be considered as a true realization of Bose-Einstein (BE) condensation due to the presence of interatomic interactions, even weak, and its high density. Indeed, in Section 3.3.1, BE condensation was described for an ideal gas of bosons, therefore without any interaction. Leaving the case of liquid helium aside, we will focus the discussion in this section on the experimental realizations of a BEC. A priori, it seems easy to create a BEC, as it suffices to make a boson gas extremely cold until quantum effects show up. As shown by Eq. (6.1) relating ϱ and T_{BEC}, the latter temperature being in general very small, the gas should be extremely rarefied and at the same time, the temperature extremely low ($< T_{\mathrm{BEC}}$). But when lowering down temperature, the gas undergoes a phase transition to liquid and then to solid. This means that the system has to be maintained in a metastable gas phase during a time long enough to produce and observe a BEC. In the 1970s, it was suggested that spin-polarized hydrogen could be an ideal candidate to observe the BE condensation. Indeed it has no bound state and hence remains a gas down to zero temperature. However, the first experimental attempts failed for a series of practical reasons, one of them being that, at very low temperatures, hydrogen atoms stuck to the walls of the container and recombined. In the 1990s, an entirely different type of cold-atom physics and technology has been developed. The first physical realizations of BEC were finally obtained in very dilute gases of interacting alkali atoms confined in a trap, which we now discuss.

6.4.1 Alkali atoms and BEC

In 1995 three different groups reported the first observations of BEC in cooled and dilute vapors of ^{87}Rb, ^{23}Na and ^{7}Li atoms. These isotopes contain odd numbers of electrons and protons, and an even number of neutrons, so that the total number of fermions is even and these atoms are bosons (see below).

A BEC experiment requires typically three steps. First, the gas is precooled by means of an incident laser beam, tuned to specific frequencies to excite the atoms and reduce their speed. After de-excitation, the scattered photons carry away on average more energy than the amount absorbed by the atoms, resulting in a net cooling.

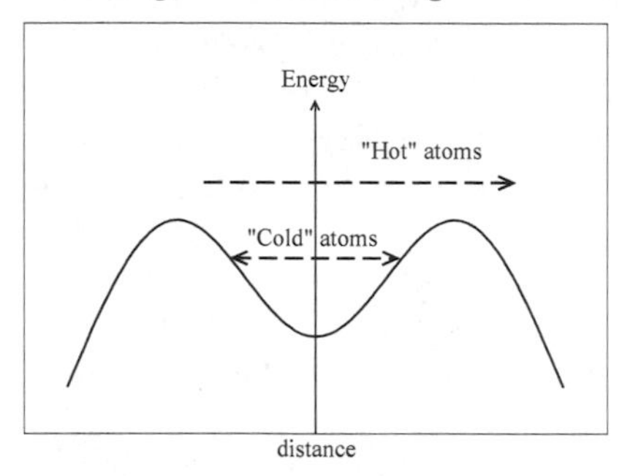

Figure 6.7: Schematic spatial representation of a magnetic trap in terms of energy.

As a second step, the atoms being now cold enough, they are confined in a magnetic trap and because there is no wall, the atom cannot stick to any surface. The magnetic field produces a splitting between atomic degenerate states and the trap produces a local minimum in potential energy. As schematically shown in Figure 6.7, atoms with high kinetic energy ('hot' atoms) are not bound by the trap and then escape, while atoms with lower kinetic energy ('cold' atoms) are trapped in the dip.

As a third step, the geometric characteristics of the trap are modified to allow the most energetic atoms to escape from the trap, resulting in an evaporative cooling. The remaining atoms thermalize and reach a lower temperature. The condensed cloud is then imaged by optical methods. Typical densities of atoms in the traps are of order $10^{11} - 10^{15}$ cm^{-3} (many orders of magnitude less than the helium density $\simeq 10^{22}$ cm^{-3} near $T = 0$; see Table 6.1), involving from a few hundred to a billion atoms. Most BEC experiments reach quantum degeneracy, that is the fact that a macroscopic number of atoms condense in the lowest energy state, between 500 nK and 1 μK.

These techniques were also successfully applied to hydrogen. But BECs of alkali atoms offer a great variety of interesting situations because of their internal energy-level structure. Alkali metal atoms are in the first column of the periodic table (see Appendix B and Section 5.1.1). Therefore, they possess an odd number of electrons with a single valence electron in their outermost s-orbital. According to quantum rules on the addition of spins, the total electron spin is here 1/2. As for the nucleus, ^{23}Na and ^{87}Rb atoms used for the first BEC realizations have a nuclear spin of 3/2. Combining the electron and the nuclear spins leads to states with a total spin of either 2 or 1. These states have slightly different energies due to the so-called hyperfine interaction, between the electron and atomic nucleus spins. Each of these states consists in fact in 5 or 3 degenerate states, with different values of the z-component of the spin.

This degeneracy is broken in a magnetic field (Zeeman effect), and varying the intensity of the field thus allows to control those states which can be located in the local minimum of the trap.

One should keep in mind that a system consisting of a number of atoms much smaller than Avogadro's number, here confined by a magnetic trap, cannot be directly assimilated to a gas. The thermodynamical limit is never reached exactly and, strictly speaking, no phase transition and no discontinuity in the thermodynamical functions are to be observed in such a finite system. In practice however, the occupation of the lowest state by a large number of atoms occurs suddenly as temperature is lowered and it then makes sense to refer to a critical temperature and a phase transition. Another important difference with respect to a gas is that the system is inhomogeneous. This is actually an interesting fact because it allows BEC to be observed in both momentum and configuration spaces. Indeed, the experimental detection of BEC has been produced in two different situations. In the first one, the trap is switched off and the condensate expands freely. The measure of its density with light absorption at several successive instants can be directly related to the initial momentum distribution. In the second method, the density of the atoms in the condensate is directly measured by means of dispersive light scattering.

Figure 6.8 displays a schematic representation of the velocity distribution of a cloud of trapped alkali atoms at three temperatures, similar to that obtained in the experiment at Boulder giving the first evidence of a BEC with ^{87}Rb atoms. The left panel corresponds to a thermal gas, just above the critical

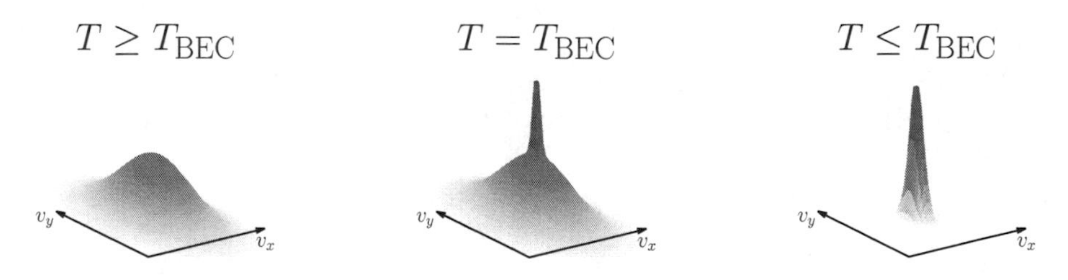

Figure 6.8: Schematic representation of the observed velocity distribution of an ensemble of trapped atoms. The darker the color, the higher the number of atoms. Left: just before the appearance of the Bose-Einstein condensate. Middle: just after the appearance of the condensate. Right: a further evaporation leaves a sample of nearly pure condensate.

temperature for condensation (about 100 nK in the conditions of the experiment). Cooling this cloud down a little bit (middle panel), a spike appears: that is the condensate appearing in the middle of yet non-condensed atoms. Cooling further (right panel), one is left with a pure condensate. The longest dimension of the cloud in the experiment is about 40-50 mm, much bigger than the light wavelength. The picture is thus an actual snapshot of a single macroscopic wave function of a few thousand atoms. Notice that the central peak of the condensate is not infinitely narrow, contrary to a BEC of an ideal gas of bosons. This is simply a consequence of the interactions in the real gas.

An important consequence of BEC is the occurrence of coherence effects in mesoscopic systems (with about 10^6 atoms), directly related to the phase of the condensate wave function. Coherence of matter waves has been directly observed through the interferences produced when two initially separated BEC are released from the trap, expand and eventually overlap. Sometimes, a source of such coherent matter waves is referred to as 'matter laser', using the word laser to point out an analogy between electromagnetic waves, consisting of photons, and matter waves, consisting of atoms. An obvious difference between both waves is that the number of photons in a laser is not fixed, contrary to the number of atoms in a BEC. The analogy is nevertheless very useful, suggesting experiments with BEC similar to those existing with coherent photon beams.

6.4.2 The Gross-Pitaevskii equation

To study the condensates actually observed, one has to obtain the solutions of the Schrödinger equation for a dilute system consisting of a relatively large number of interacting atoms, confined by an external potential (the magnetic trap), at finite but low temperature. Instead of directly solving that complicated theoretical problem, Gross and Pitaevskii independently developed in 1961 the basic theory for studying nonuniform dilute Bose gases at low temperature, which leads to the Gross-Pitaevskii equation. It is a time-dependent mean-field equation (see Section 4.4) whose solution provides a macroscopic wave function (also called order parameter) for the condensate, assuming that the atoms are subject to an effective potential $V_{\rm ext}(\mathbf{r})$ due to the magnetic trap, and to an effective potential $V_{\rm eff}(\mathbf{r})$ originating from atom-atom interactions. Many of the relevant properties of cold condensed gases, such as collective oscillations, density profiles, structure of vortices, etc. can be accounted for by means of this non-linear equation, by assuming relatively simple approximations for these effective potentials. We hence describe it a little bit here.

A good approximation for available traps is a harmonic potential (see Eq. (2.17) in Section 2.4.2) of frequency $\omega_{\rm HO}$. Neglecting the atom-atom interaction, almost all predictions are analytical and relatively simple. The lowest single-particle levels are Gaussians, and the size of the condensate can be characterized by the length scale (oscillator parameter)

$$a_{\rm HO} = \left(\frac{\hbar}{m\omega_{\rm HO}}\right)^{1/2}, \qquad (6.2)$$

which is typically of order a few microns. m is here the atomic mass. The shape of the trap may be modified to break the spherical symmetry. In the extreme cases of highly anisotropic harmonic potentials, one approaches the limits of quasi-two-dimensional or quasi-one-dimensional systems, thus allowing the investigation of the effects of reduced dimensionality.

Let us discuss now the atom-atom effective potential. Dilute gas means that the range r_0 of the interatomic interactions is much smaller than the average distance r_s between atoms, which is fixed by the density of the gas: $4\pi r_s^3/3 = 1/\varrho$. The condition of dilute gas can then be expressed as $\varrho r_0^3 \ll 1$. This condition can be also written in terms of an experimentally accessible length, namely the s-wave scattering length a, as $\varrho|a|^3 \ll 1$. Indeed, if the average distance is large, as in a dilute gas, the relevant properties of the atom-atom interaction can be derived from the asymptotic behavior of the relative motion of a pair of atoms. More precisely, low-energy scattering properties do not depend on the detailed form of the interaction, since any form of interaction giving the same value of a would lead to the same low-energy scattering description. To understand the meaning of the scattering length and why it provides such a simple indicator, let us take the simpler (but similar) case of a low energy particle interacting off a scatterer. Because the particle has a low energy, its de Broglie wavelength is very large and the particle cannot resolve the structure of the scatterer. Details of the potential from which the particle is scattered therefore become unimportant and what matters is only how the potential does behave at large distances.

The question now is to analyze how the wave function of the scattered particle behaves. One can expand it in components of various angular momenta of the outgoing wave, the same way as one analyzes the electrostatic field created by a charge distribution in terms of multipole components. As at very low energy the scattered particle does not see any structure, the scattered wave is dominated by a spherically symmetric wave (the same way as from very far away a charge distribution can be assimilated to a point charge), that is its s-wave scattering component (with angular momentum $l = 0$). The whole interaction process is then reduced, in this low energy limit, to characterize the scattered outgoing spherical wave, which is achieved by the value of the scattering length a. For the three mentioned alkali atoms, the value of parameter a is 5.77 nm for ^{87}Rb, 2.75 nm for ^{23}Na, and -1.45 nm for ^{7}Li. Positive (negative) values of a correspond to an effective repulsion (attraction) between atoms. Typical values of density range from 10^{11} to 10^{15} cm^{-3}, so that the condition $\varrho|a|^3 \ll 1$ of a dilute gas is satisfied.

As suggested above, low-energy scattering properties do not depend on the detailed form of the interaction. For the sake of simplicity, one can thus assume the atom-atom interaction to be a zero-range one with a strength g fixed to the s-wave scattering length as $g = 4\pi a\hbar^2/m$. One can consider the effective potential as resulting from the mean-field approximation, that is as the average interaction felt by an atom when it interacts with the others through the above defined zero-range potential. In that case, the effective potential becomes simply proportional to the density $V_{\rm eff}(\mathbf{r}) = g\varrho(\mathbf{r})$.

With these ingredients, the time-independent form of the Gross-Pitaevskii equation satisfied by atom i is then written as:

$$\left(-\frac{\hbar^2}{m}\Delta + \frac{1}{2}m\omega_{\rm HO}^2\mathbf{r}^2 + g\varrho(\mathbf{r})\right)\psi_i(\mathbf{r}) = \varepsilon_i\psi_i(\mathbf{r}) \quad , \tag{6.3}$$

where Δ is the Laplacian operator and the density is given by $\varrho(\mathbf{r}) = \sum_i n_i|\psi_i(\mathbf{r})|^2$, with occupation numbers $n_i = \left\{e^{\beta(\varepsilon_i-\mu)} - 1\right\}^{-1}$ given by the Bose-Einstein distribution functions. The Gross-Pitaevskii equation is thus a non-linear differential equation, which has to be solved consistently: the effective potential depends on the density, which in turn depends on the single-particle wave functions.

Before closing this section, we briefly address the question of the relation between the dilute nature of a gas and the role of atom-atom interactions. The existence of a dilute gas does not imply that the atom-atom interactions are negligible. Actually, it has observable consequences on most quantities, due to the combined effect of BEC and of the nonuniform nature of the system. To properly estimate the importance of the interactions, one should compare the associated potential energy with the kinetic energy of the atoms in the trap. A rough estimate can be made as follows. On the one hand, $E_{\rm pot} \simeq Ng\varrho$, with $\varrho \simeq N/a_{\rm HO}^3$. On the other hand, $E_{\rm kin} \simeq N\hbar\omega_{\rm HO}$. After some manipulations we obtain

$$\frac{|E_{\rm pot}|}{E_{\rm kin}} \simeq 4\pi N\frac{|a|}{a_{\rm HO}} \quad ,$$

which provides an indicator on the importance of the atom-atom interaction compared to the kinetic energy. It can be easily larger than 1 even if $\varrho|a|^3 \ll 1$, so that very dilute gases can also exhibit an important nonideal behavior. For instance, in the first experiments on BEC of alkali atoms (see Section 6.4.1), the value of the ratio was larger than 1 in ^{87}Rb, of the order $10^3 - 10^4$ in ^{23}Na, and much smaller than 1 in ^{7}Li.

6.4.3 Trapping ultracold atoms in optical lattices

We have mentioned the use of magnetic traps for the confinement of BEC which avoid the use of a container and interaction with its walls. The manipulation of BEC has been largely enriched with the experimental possibility to trap ultracold atoms in optical lattices. Such a lattice is a periodic array of micro-traps generated by counter-propagating optical laser beams. Optical lattices provide a versatile tool for studying fundamental quantum phenomena and find applications in a variety of fields such as condensed matter, many-body theories, atomic and molecular physics, quantum optics or quantum information.

To understand how they proceed, we start considering the interaction of atoms with a laser field. The oscillating electric field $\mathbf{E}$ from a laser induces a dipole moment $\mathbf{p} = \alpha\mathbf{E}$ in an atom, which oscillates at the frequency of the electric field. The constant α is the so-called dipole polarizability of the atom. In turn, the induced dipole moment interacts with the field, producing a change in the energy of the system. This change can be regarded as an effective potential, which is proportional to the intensity of the laser field and the dipole polarizability of the atom. The laser may be tuned to avoid spontaneous atomic emission effects occurring at specific atomic resonance frequencies. The dipole polarizability is positive for laser frequencies larger than the atomic resonance frequency (blue detuning), and negative for the opposite case (red detuning). The effective potential felt by the atom could be tuned to be either repulsive (blue detuning) or attractive (red detuning), forcing the atom to move towards regions of lower or higher fields, respectively.

Suppose that two counter-propagating laser beams are applied to a condensate. The interference between the laser fields forms a standing wave, resulting in a periodic potential. Controlling the frequency and the intensity of the laser one can obtain an array of trapped condensates. With more interfering laser beams one can obtain periodic trapping potentials in 1D, 2D and 3D, arranged in different geometrical configurations. These so-called artificial crystals of light are highly tunable and easily controllable, thus providing an ideal laboratory for studying a large variety of correlated systems.

A BEC placed in an optical lattice exhibits many features similar to electrons moving in the periodic potential of a crystal lattice. There are several differences, however. In the BEC there is a coherent wave function, which extends over the entire lattice, with essentially a single momentum. Another difference is the typical length of the lattice, about 100 nm for BEC versus a few angstroms (tenths of nm) for electrons.

6.4.4 Atomic Fermi gases

The mechanisms previously described for cooling and trapping a gas are in fact independent of the quantum statistics of the atoms. The first realization of a BEC with bosons launched the possibility of studying fermion gases, with the main objective of observing superfluidity at low temperatures. A system of fermions has a characteristic temperature scale, fixed by $T_F = \varepsilon_F/k_B$ (see Section 3.4). In terms of the density of atoms $k_B T_F \propto \hbar^2 \varrho^{2/3}/m$ has the same expression as given in Eq. (6.1) for $k_B T_{BEC}$, apart from a factor of order unity. Actually, one could have deduced these expressions from dimensional grounds. The presence of the reduced Planck's constant $\hbar$ reflects quantum effects in both cases, but the physical meaning is different. While T_{BEC} is associated with a phase transition to a condensate of bosons, T_F does not describe any phase transition but rather corresponds to a smooth change of point of view, that is from a classical to a quantum behavior of a system of fermions. One therefore expects that superfluidity in a fermion gas would take place at a temperature much lower than T_F, similarly to what happens in the liquid ^{3}He.

Cooled Fermi gases can thus reach quantum degeneracy once their temperature drops below T_F, as is illustrated in a temperature in the range of the

Fermi energy ε_F. This is illustrated in Figure 6.9 where the average energy per atom (scaled to the classical value) as a function of the temperature T of the system is plotted. At sufficiently low temperatures ($T < T_F/2$), there

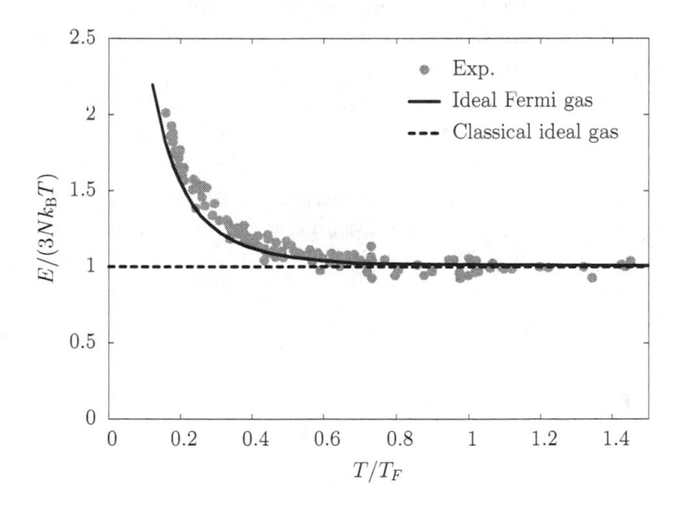

Figure 6.9: Appearance of quantum degeneracy in trapped Fermi gases of ^{40}K atoms. The average energy per particle (filled and open circles, energy scaled to the classical estimate) is plotted as a function of the temperature, in unit is the Fermi temperature $T_F = \varepsilon_F/k_B$. Quantum degeneracy occurs for temperatures $T < T_F$, data agree well with the ideal Fermi gas prediction (solid line). The horizontal dashed line corresponds to the classical gas. Adapted with permission from B. DeMarco, S. B. Papp, and D. S. Jin, Phys. Rev. Lett. 86, 5409 (2001). Copyrighted (2001) by the American Physical Society.

is a clear deviation from the classical estimate, which reflects the expected transition towards the degenerate fermion gas.

6.5 CONCLUDING REMARKS

Atoms and molecules may exhibit surprising behaviors when put together at low temperature. The usual, classical picture of a system is that, at sufficiently low temperature, it freezes into a solid phase. This is what occurs for all materials but for helium the reason being that interactions between helium atoms are so weak that zero point motion is enough to prevent freezing, at least at normal pressure. We have studied in some detail this very particular material which, again because of the the faintness of interactions, not only remains fluid at low temperature but even becomes a superfluid, with no associated viscosity, at sufficiently low temperature. The effect is well known in bosonic ^{4}He and bears some relation to the Bose-Einstein condensation, although, again, interactions play here a crucial role.

Helium is also an interesting material as it can be formed in droplets at very low temperatures. The latters (in superfluid state or not) provide a unique environment for studying molecular properties of attached species. Such setups allow to keep the molecule at very low temperature in contact with a (close to) non-interacting medium so that one can have access under, ideal conditions, to the properties of the molecule itself.

We have concluded this chapter by discussing new directions of research for studying atomic and molecular assemblies under well defined environmental conditions. Such setups can now be attained in ultracold atomic gases where environment conditions can be monitored almost at will. Such setups also allow us to tune the level of interaction experienced by the members of the gas. This provides fascinating new directions of investigation for the understudying of assemblies of quantum particles.

CHAPTER 7

Nuclei

In Chapters 5 and 6, we have discussed two types of quantum liquids, namely the electron fluids and liquid helium. One difference between the two comes from the fact the electron fluid cannot exist without a positively charged background, while liquid helium systems are self-bound systems. Another difference stands in the temperatures at which these two liquids are observable: electron fluid exists at ambient temperature while liquid helium is only attainable at low enough temperatures. There is however a common feature: while both exist as bulk, in metals for electrons, in infinite helium phases for atoms, we have also considered droplets of these quantum liquids.

In this chapter, we consider a quantum liquid very different from these two. This holds true not only because of the different constituents and the different interactions between them but also, because, in the laboratory, it exists only in the form of droplets, namely atomic nuclei. These nuclear droplets are self-bound systems. Theoretically, it is sometimes useful to consider bulk nuclear liquid which is rather an idealized system. We will however show that it is relevant to infer the equation of state of nuclear matter.

This chapter is organized as follows: we first give some basics on nuclei and on their liquid nature, especially through the liquid drop model. We then move to individual and collective properties, both stemming from strong quantum effects. We finally end with some words on the nuclear equation of state.

7.1 NUCLEAR DROPLETS

Nuclei are constituted of neutrons and protons which have different electrical charge (0 for neutrons, $+e$ for protons) but very similar masses of order 10^{-27} kg, differing in about 0.1%. Strong interactions among neutrons, between protons and neutrons and among protons are very similar. Still, because of the charge difference and the small mass difference, one nevertheless cannot define a true symmetry between protons and neutrons. It is replaced by an approximate one called isospin symmetry. This symmetry introduces the abstract notion of nucleon as being an object with two possible states, either a neutron or a proton, analogous to a spin-1/2 particle. In practice, one then will often use the term nucleon as a generic one to describe any (neutron or proton) constituent of the nucleus. The isospin-up and -down states are conventionally assigned to proton and neutron, respectively. The total number of nucleons is usually denoted by $A = N + Z$ with N the number of neutrons and Z that of protons. A is also called mass number or simply mass of the nucleus.

7.1.1 Basics of nuclei

Nuclei and nucleons

The different electrical charge of neutrons and protons has important consequences for the stability of nuclei. The strong interaction alone would prefer symmetric nuclei with equal number of neutrons N and protons Z. But Coulomb repulsion between protons renders the increase of charge Z unfavorable and eventually destabilizes the nucleus. Neutrons can then help: increasing N at frozen Z stabilizes the nucleus. Therefore neutrons somewhat act as the analog of a buffer solution in chemistry. We recall that a given chemical element is fully characterized by Z and that atoms possessing a same Z but different N are isotopes. But neutron number cannot be enhanced indefinitely and a very large value of N cannot fully compensate Coulomb repulsion. Hence, there exists an upper limit to observable values of Z and N. Coulomb pressure indeed renders super-heavy nuclei unstable against spontaneous fission.

Figure 7.1 provides the so-called nuclear chart which gathers the known (either natural or artificially synthetized) nuclei in a (N, Z) diagram. There are 254 stable nuclei, plus 34 with a half-life larger than 4.5×10^9 years, the solar system age. About 60% of stable nuclei are doubly-even nuclei (i.e. even values of both Z and N), and there are only five doubly-odd stable nuclei. These nuclei appear as black squares in Figure 7.1 and form what is called the stability valley. The stability valley roughly corresponds to symmetric nuclei ($N \simeq Z$) for light masses. Typical examples are oxygen ^{16}O ($N = Z = 8$) or calcium ^{40}Ca ($N = Z = 20$). For heavier nuclei, Coulomb repulsion can only be counterbalanced by increasing the number of neutrons ($N > Z$). Heavy nuclei in the stability valley therefore typically have about 1.5 more neutrons than protons. An example is the lead isotope ^{208}Pb ($N = 126$ and $Z = 82$) which is the heaviest stable nucleus. We shall come back to this chart in Section 7.2.1, especially to discuss the so-called 'magic numbers' highlighted by horizontal or vertical double-dashed lines in the nuclear chart.

'Independent' nucleons

The nuclear strong interaction, which ensures the binding of nucleons inside nuclei, is attractive at medium range (about 1 fm $= 10^{-15}$m). It is strongly repulsive at much shorter distances because of the quark structure of nucleons (Pauli principle between quarks). The attractive component is nevertheless strong enough to ensure binding, but within favoring the range of 1 fm, as

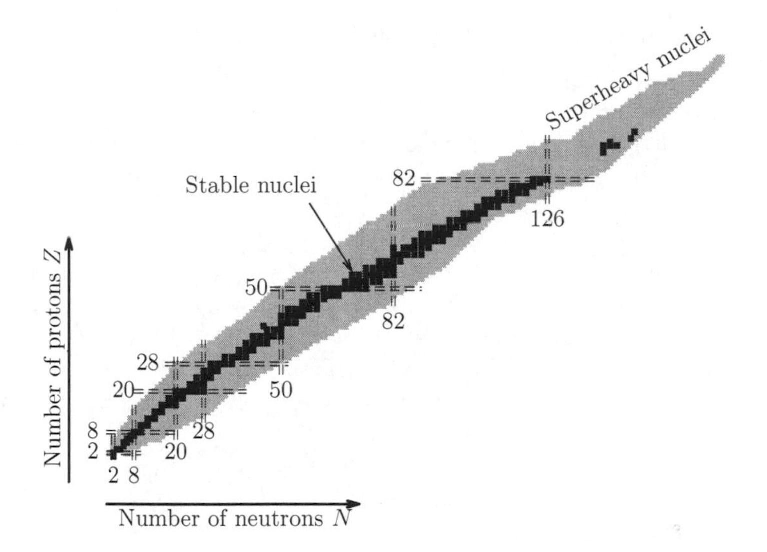

Figure 7.1: Nuclear chart presenting as a gray area the existing nuclei in the $N - Z$ plane with N the number of neutrons and Z the number of protons in the nucleus. The stable nuclei and very long-lived nuclei are depicted by black squares. Horizontal and vertical dashes highlight magic numbers of N or Z (see Section 7.2.1).

we shall see in more detail below. In fact, the short range repulsion, characterizing the interaction between two nucleons in vacuum, is strongly softened inside a nucleus. Indeed, because of Pauli principle which tends to repel nucleons to each other, the latters effectively feel a softened interaction with a much reduced short-range singularity. The net result is that the many-nucleon problem can then be reformulated in terms of a softer (no strong, nor singular, short range repulsion) 'effective' nuclear interaction in the nuclear medium. The situation is a bit analogous to the molecular case (see Sections 1.3.2 and 5.1.2) in which only valence electrons do enter actual binding and thus only explore medium range processes where the singularity of Coulombic effects is washed out, shielded by the (core) electrons remaining bound to their parent ion.

The resulting 'softness' of the effective nuclear interaction makes the mean free path of nucleons inside nuclei very large, typically of the order magnitude of the nuclear radius. This, as we shall see, provides a great simplification for our understanding of the nuclear many-body problem. Indeed, the picture which stems from such a situation is the one of nucleons moving more or less independently from each other. They are of course confined in a potential, but this potential is itself a simple average over all interactions between nucleons: this is the well known mean-field picture of nuclei. This simple picture turns out to be a good approximation and at the same time extremely rich and fruitful. Hence, we shall keep it in mind for our forthcoming discussions. Still, before proceeding, let us characterize nuclei a bit more globally, especially in terms of size.

Saturation, sizes and density profiles
Although the attractive part of the nuclear interaction provides the binding mechanism, this binding is constrained by the repulsive core of the interaction. Combined with the Pauli principle, that repulsion defines a closest packing which applies to nuclei of any size. In other words, the average distance between nucleons is about the same in all nuclei, and this then also holds true for the central density of nuclei. This property is called *saturation*. And it has the immediate consequence that one can view nuclei in the simple picture of finite drops of a (hardly compressible) nuclear fluid, very similar to a classical liquid such as water. Saturation manifests itself immediately in terms of the scaling of nuclear size with mass number A.

Electron scattering experiments provide information about sizes and spatial distribution of nuclei by analyzing the electron beam deflected by the nuclear charge density. At low beam energy, one explores mostly the outer layers of the nucleus and access its radius this way. It turns out that nuclear radii exhibit a scaling law with A. Remarkably enough, this leads to an almost linear dependence with $A^{1/3}$ (up to an additional constant of small value), $R \simeq r_0 A^{1/3}$, where $r_0 \simeq 1.14$ fm is a constant almost identical for all (not too small) nuclei. This almost perfect scaling is directly related to saturation. It means that each nucleon can be seen as occupying the same elementary 'volume' $V_0 = 4\pi r_0^3/3$. When adding one nucleon to a nucleus, the nuclear volume is increased by V_0, independently of possible internal rearrangements. An immediate consequence is that the average density inside nuclei is about the same whatever the nucleus $\varrho_0 = 3/(4\pi r_0^3) \approx 0.16$ fm^{-3}. This value is known as nuclear matter equilibrium (or saturation) density. It corresponds to the density of the ground state of a hypothetical infinite phase of proton-neutron symmetric nuclear matter (without Coulomb effects), as could be observed in an astrophysical context (see Section 7.3). It is worth noting that, in macroscopic units, the mass density of nuclear matter is 2×10^{18} kg/m^3, that is 15 orders of magnitude greater than water.

The saturation property also has consequences for the spatial distribution of the density inside a nucleus which basically becomes constant inside the nucleus. Figure 7.2 displays the resulting density distributions for a few nuclei. Very clearly, the density approaches a nearly constant value in the nuclear interior, especially in the case of the larger nuclei, which allows a smooth transition zone at the surface. The case of smaller nuclei is a bit different, to the extent that surface represents most of the system, leaving little to the interior region. Remember by the way that the nuclear radius systematics is not fulfilled, correspondingly, in such small nuclei.

The nuclear radius and nuclear density are interesting quantities, as they fix the gross characteristics of the nuclear potential in which nucleons evolve. Indeed, infinite nuclear matter, a piece thereof somewhat constituting nuclear interior, allows one to estimate average depth of the potential. As it is fixed

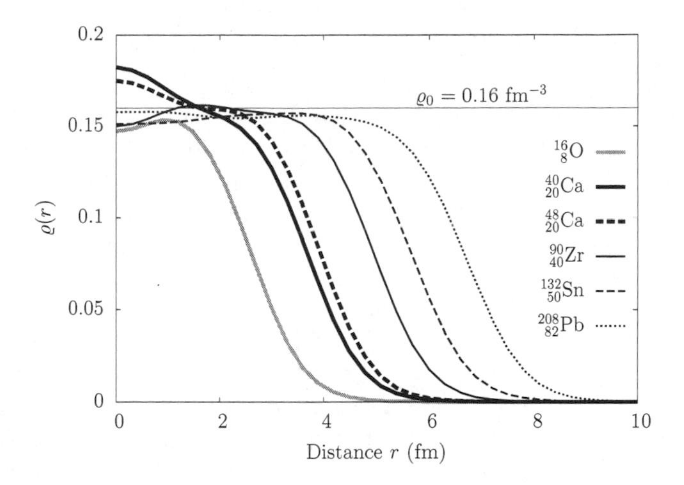

Figure 7.2: Radial number density distribution $\varrho(r)$ for a variety of magical nuclei (see Section 7.2.1) as indicated. The horizontal line highlights the value of the saturation density ϱ_0.

by the nuclear density, it is thus, up to details, about the same for all nuclei. The spatial extension is directly related to the nuclear radius, which matches the short range of nucleon-nucleon interaction. Altogether, we see a picture emerging in which appears a simple potential well with about fixed depth and extension propotional to $A^{1/3}$. It should be noted that such a reasoning is generic for all saturating fermionic systems as electrons in metal clusters or ^{3}He droplets.

Analogously to electron fluids and ^{3}He liquid, one can use the Fermi gas results to get some estimates. . These results have been given in Section 3.4 and are directly used. Taking into account that as nucleons are spin-1/2 and isospin-1/2 states, the degeneracy factor is $g = 2{\times}2 = 4$. Using this degeneracy number in Eq. (3.23) to express the total number of particles in terms of the Fermi momentum, one obtains the following relation between number density and Fermi momentum for a nuclear Fermi gas:

$$\varrho_0 = \frac{2k_F^3}{3\pi^2} \tag{7.1}$$

Using the saturation density value $\varrho_0 = 0.16\ \text{fm}^{-3}$, one obtains $k_F = 1.33\ \text{fm}^{-1}$. To calculate other quantities involving the nucleon mass m, it is customary to take $m = (m_n + m_p)/2$, so that $mc^2 = 938.92$ MeV and $\hbar^2/m = 41.47$ MeV fm^2. One then obtains the saturation values of the Fermi energy $\varepsilon_F = 36.7$ MeV, and the associated Fermi temperature $T_F = \varepsilon_F/k_B = 4.3 \times 10^{11}$ K.

7.1.2 The Liquid Drop Model

Nuclear masses and the Liquid Drop Model

The above discussions and gross properties of nuclei allow us to better characterize and justify what could be a Liquid Drop Model (LDM) of nuclei. The LDM was historically the first model proposed to explain nuclear properties. The original idea was suggested in the late 1920s by G. Gamow, even before a proper knowledge of nuclear interactions and constituents was achieved. It turns out that subsequent experiments and explorations indeed confirmed this original intuition. We shall therefore assimilate the nucleus to a spherical liquid droplet of mass Am (where m is the nucleon mass) and constant density with however some corrections. This again leads to a radius proportional to $A^{1/3}$ (and thus a volume proportional to A), in agreement with systematics of nuclear radii. We are here just making the simplest picture out of what we have discussed previously. The LDM model soon turned out to be surprisingly rich, in spite of its apparent simplicity. It proved to be particularly successful in estimating nuclear binding energies by means of the so-called semiempirical mass formula. The binding energy of a system is the energy required to break it into its separate constituents. For a bound system, it is obviously a positive quantity. For an atomic nucleus (A, Z), it is defined as $B(A, Z)/c^2 = Mm_n + Zm_p - M(A, Z)$, where $M(A, Z)$ is the nuclear mass. In turn, the nuclear mass is deduced from the atomic mass as $M_{\text{atom}}(A, Z) = M(A, Z) + Zm_e - B_e(Z)/c^2$. The electron binding energy $B_e(Z)$ gives here a relatively small correction, which can be parametrized by a power law $B_e(Z) = 14.33Z^{2.39}$ eV.

The empirical nuclear binding energies per nucleon are displayed in Figure 7.3. Leaving aside the lightest nuclei, one can see that the value of B/A is roughly constant, about 8 MeV for all nuclei. The initial LDM semiempirical mass formula appeared in the 1930s and has been constantly refined over the years to attain a remarkable degree of accuracy. The point is to express the total energy of a nucleus $E(A, Z) = -B(A, Z)$ as a function of its mass A and charge Z and a few general parameters fixed for all nuclei. Let us discuss which terms should be included in the total (negative) energy of a nucleus.

Mass formula

In general, one expects that in an ensemble of A particles interacting with a long-range interaction, the potential energy is proportional to the number of pairs of interacting particles, namely $A(A - 1)/2$ which is about A^2 for large masses A. Because the nuclear interaction is short-range, a given nucleon only interacts with its closest neighbors. This energy is then only proportional to A, and not to the total number of pairs $A(A - 1)/2$. We must include a 'volume' term proportional to the number of particles A, which we will write as $-a_{\text{v}}A$, with a_{v} positive. This is again, expressed in a different form, the saturation property of nuclear matter that we are facing here. This volume term of the mass formula hence simply represents the energy of what would be a piece of infinite nuclear matter of mass A.

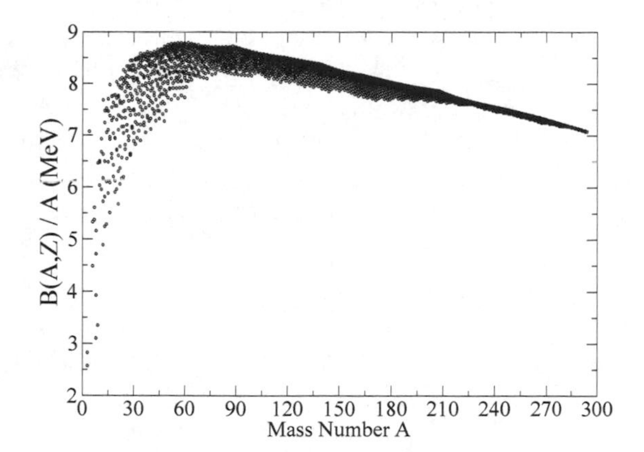

Figure 7.3: Experimental binding energy per nucleon as a function of the mass number A, for all nuclei collected in the AME2016 evaluation (M. Wang *et al.* Chinese Phys. C **41** (2017) 03003).

But nuclei are finite objects with a well defined surface. As in any drop of liquid, the situation is different on the surface as compared to the bulk. Indeed, a nucleon at the surface of the droplet interacts with fewer particles than a nucleon inside the nucleus. The surface then induces a correction to the volume contribution which has an opposite sign. The simplest approximation is to introduce a surface term proportional to the surface, itself scaling as R^2 (square of nuclear radius R, itself proportional to $A^{1/3}$). The net result is a (positive) surface correction term of the form $a_{\rm sf}A^{2/3}$.

We should also include the (repulsive) Coulomb contribution due to the repulsion between protons. It is estimated from a simple electrostatic picture by considering a uniformly charged sphere of radius R and charge Ze. In such a simple geometry, the calculation is analytical and leads to the expression $3/5\, e^2 Z^2/(4\pi\epsilon_0 R)$. With the standard scaling $R \sim A^{1/3}$, one immediately obtains the (positive) correction term $a_{\rm c} Z^2\, A^{-1/3}$.

These three terms—volume, surface and Coulomb—suffice for gross estimates of the binding energy. However, to reproduce accurately the measured energies, one has to consider further correcting terms. To reflect the fact that the nuclear interaction favors symmetric nuclei $N \sim Z$, one thus includes a so-called symmetry energy term, depending on the isospin asymmetry parameter $I = (N-Z)/A$. As that correction cannot depend on the sign of $N-Z$, it is written as $a_{\rm sy}AI^2$, further complemented by a surface-symmetry correction term $-a_{\rm ss}A^{2/3}I^2$ with $a_{\rm sy}$ and $a_{\rm ss}$ positive constants.

One should also consider the fact that about 60% of stable nuclei are doubly-even (even N and Z), while there are only five doubly-odd stable nuclei. Hence, there is an energy gain when two neutrons or protons can form a pair of spin-up/spin-down particles. This is analogous to the pairing encountered in the context of superconducting and superfluid transitions (see Section 9.3). In the present case, the slight energy gain is included as a pairing correction term, which can be simply taken as $\Delta(A,Z) = \Delta_n + \Delta_p$, with $\Delta_{n,p} = \pm\delta$ depending whether N, Z are even or odd. Other parameterizations are sometimes considered.

A typical mass formula can finally be written as

$$E_{\rm LDM}(A,Z) = -a_{\rm v}A + a_{\rm sf}A^{2/3} + a_{\rm c}\frac{Z^2}{A^{1/3}} + \left(a_{\rm sy} - a_{\rm ss}A^{-1/3}\right)AI^2 + \Delta(A,Z) \; . \quad (7.2)$$

This expression can be compared to the common expansion of a given observable in a metal cluster, as terms of powers of $N^{-1/3}$ (see Section 5.3.1). This reflects again a property of the Fermi gas in which a fermion typically occupies a sphere of radius proportional to $\varrho^{-1/3}$ (see Sections 3.4.1 and 7.1.1).

The parameters $(a_{\rm v}, a_{\rm sf}, a_{\rm c}, a_{\rm sy}, a_{\rm ss}, \delta)$ are fitted on measured binding energies of nuclei. Typical values (in MeV) of these parameters are: $a_{\rm v} \simeq 16$, $a_{\rm sf} \simeq 18$, $a_{\rm c} \simeq 0.7$, $a_{\rm sy} \simeq 26$, $a_{\rm ss} \simeq 17$, $\delta = -1.25$, the precise values depending on the selected nuclei for the fit. This fit turns out to be remarkably good because experimental binding energies and the LDM estimates agree very well over the whole range of masses. In particular, the LDM nicely reproduces the observed maximum of binding per nucleon in the region of iron-nickel (see Figure 7.3). The fact that iron possesses the highest binding energy has strong consequences in the stellar evolution of massive stars (see Section 8.1.2).

Limitations of the Liquid Drop Model

There are however some interesting differences between the LDM and the experimental data, at well identified values of N and Z, in which the experimental binding energy is larger than that obtained from the LDM. This is exemplified in Figure 7.4 which shows the difference between the experimental and the LDM binding energy as a function of N (left) or Z (right). The highest differences are observed for N and Z equal to 2, 8, 20, 28, 50, 82 and 126. These numbers were already indicated in the nuclear chart in Figure 7.1. For historical reasons, they are referred to as magic numbers. They are associated to a purely quantal effect analogous to shell closure of electrons in atoms, as we will discuss in Section 7.2.1. Up to this effect (which is small relative to the total energies), one should keep in mind that the performance of the LDM is remarkably good.

One could wonder whether similar LDM modelings are conceivable for other quantum liquids such as He droplets or the electron cloud of metal clusters. In spite of the huge differences between the interactions inside He and nuclei, it indeed turns out that a LDM can be applied in He droplets with, of course, a proper adaptation to specificities of He (no Coulomb term

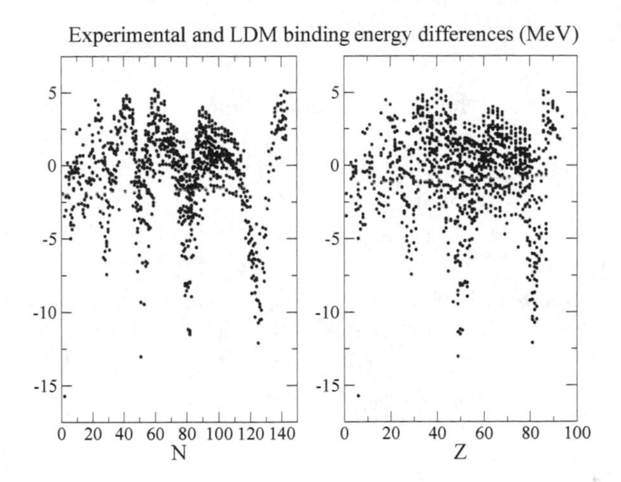

Figure 7.4: Differences between experimental and liquid drop model energies of known nuclei as a function of N (left) and Z (right).

in particular). Nevertheless the scaling laws observed in nuclei, which allow to link together nuclei of very different masses, do not apply the same way in He droplets. Due to the extreme faintness of the interaction between helium atoms, one needs at least several hundreds or even a thousand of atoms to allow the formation of a helium droplet. This is at variance to the case of nuclei which are well described by the LDM already for sizes as small as $A \simeq 16$. The situation is different in metal clusters in which the governing interaction is the long-range Coulomb interaction (with different scaling properties than those obtained with the short-range nuclear interaction). In addition, metal clusters are compound of electrons and ions, the latters playing a key role in determining cluster shapes. The situation in that case is therefore more involved than in nuclei.

7.1.3 Fission: an illustrative example

The original LDM was refined by N. Bohr in the mid 1930s to describe nuclear collisions induced by neutrons which led to the model of compound nucleus. Indeed, it is after the absorption of a neutron by a uranium nucleus and its subsequent disintegration that barium ($Z = 56$) nuclei were observed in 1938. Soon afterwards, the LDM was used to understand what was going on, namely nuclear fission: a heavy nucleus as uranium is split in two smaller ones plus 2-3 neutrons. The energy released is typically 6 orders of magnitude larger than that released in chemical reactions involving atoms. The released neutrons can in turn induce the fission of more nuclei, leading to a chain reaction. This is the principle of A-bombs and nuclear power plants.

Energetics

It may seem surprising that an atomic nucleus can be broken and release energy. This means that some heavy nuclei are 'so to speak' ready to be broken, or in other words easily unstable with respect to some excitation. The energy release is understood by looking at Figure 7.3. The binding energy per particle of heavy nuclei is smaller than the one of lighter nuclei. It is instructive, using the LDM energy Eq. (7.2), to roughly estimate this energy difference. For simplicity, consider a heavy nucleus (A, Z) undergoing a symmetric fission into two nuclei $(A/2, Z/2)$. In that case, there is no contribution coming from the symmetry terms in Eq. (7.2). We can also drop the small pairing term for this estimate. The energy difference is then

$$E(A,Z) - 2E\left(\frac{A}{2},\frac{Z}{2}\right) = \left(1 - 2^{1/3}\right) a_{\mathrm{sf}} A^{2/3} + \left(1 - 2^{-2/3}\right) a_{\mathrm{c}} \frac{Z^2}{A^{1/3}} \ . \quad (7.3)$$

That difference is positive (thus favoring fission) when

$$\frac{Z^2}{A} > \frac{2^{1/3} - 1}{1 - 2^{-2/3}} \frac{a_{\mathrm{sf}}}{a_{\mathrm{c}}} \simeq 17.5 \quad . \quad (7.4)$$

Assuming $Z = A/2$ or $Z = A/3$, we see that, for any nucleus with $A \geq 70$ or $A \geq 158$, it is energetically favorable to split into two lighter ones. The interest of this estimate is to point out the basic mechanism at play in a fission: it results from a competition between Coulomb energy, which decreases as the two fission fragments are set apart, and the surface energy, which opposes to the surface increase. This is the same mechanism at play when one increases the charge of a liquid droplet by electrostatic means, until the surface tension is no longer sufficient to compensate the Coulomb repulsion of charges inside the droplet. This technique is used experimentally to produce droplets containing some molecule under study, starting from an initial drop of water. Mind however that the surface tension in water, equal to 73 dyn/cm at room temperature (see Section 1.4), has nothing in common with the surface tension in nuclei, estimated at $a_s/(4\pi r_0^2) \simeq 1$ MeV/fm$^2 \simeq 10^{22}$ dyn/cm.

Fission barrier

Let us come back to nuclei. Spontaneous fission does not occur in nature, except for a few long-lived unstable nuclei. What is the mechanism that prevents fission to spontaneously occur? From an energetic point of view, this means that the two systems, one single nucleus on the one hand and two fission fragments on the other hand, do have the same energy, and that they are separated by a potential barrier (just as in a chemical reaction), called the fission barrier. A heavy nucleus undergoes a deformation before splitting into two lighter nuclei. The fission barrier provides a picture of the potential energy surface in which the nucleus lies. An example is shown in Figure 7.5,

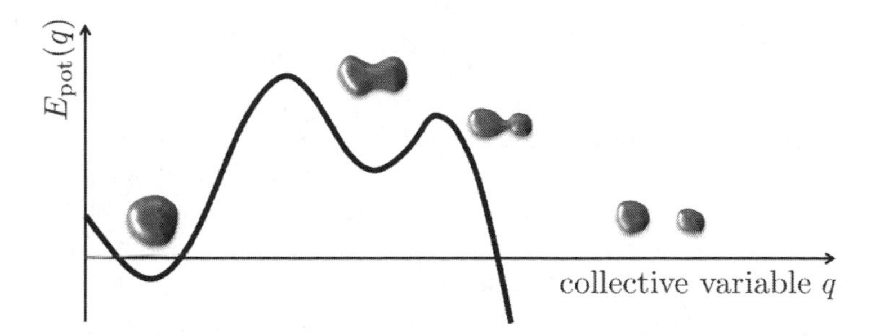

Figure 7.5: Typical potential energy E_{pot} of a nucleus which can fission, as a function of a collective variable q (see text for more details).

also illustrating the deformation of the fissioning nucleus along the fission path by a few snapshots. Let us define what is plotted and how. The curve depicts the potential energy of a nucleus as a function of a variable q measuring the deformation of the nucleus. There is no universal definition for such a variable. It can be for instance the quadrupole moment of the system (defined in Section 7.2.2). The actual barrier shown on Figure 7.5 is interesting in many aspects. One in particular observes indeed a barrier but: i) the ground state configuration in realistic systems corresponds to a finite deformation; and ii) the barrier exhibits an isomeric state (local minimum at large deformation) similar to a chemical situation. Both effects are purely quantal, as we shall see below.

The existence of the fission barrier explains the stability of the original nucleus. Without it, the nucleus would disappear instantaneously. In heavy nuclei, this fission barrier is usually not very high, typically less than the energy needed to extract one nucleon from the nucleus. One then needs only a little amount of energy to provoke fission by overcoming the fission barrier. For example, in ${}^{238}_{92}\mathrm{U}$, the fission barrier is about 5 MeV, to be compared to that of ${}^{96}_{40}\mathrm{Zr}$ which is several tens of MeV. This situation looks similar to that encountered in chemistry, with a thermally activated breaking of molecules, in particular molecules of biological interest. This is for example a basic mechanism used in cooking! Nuclear fission can also be initiated by thermal activation. But the energy may also be deposited by projectiles hitting the nucleus. This is exactly what occurs in the core of nuclear power plants where fissions of uranium nuclei are activated by a flux of neutrons. We finally mention that fission in some heavy nuclei can also occur almost 'spontaneously' via quantum tunneling (see Section 2.5.1), especially when the fission barrier height is not very high.

Fission mechanism

Nuclear fission is a remarkable example where the quantum liquid nature of nuclei shows up. Indeed, fission associates a global behavior, typical of a macroscopic classical droplet, to important quantum effects. It is worth noting that fission also occurs in charged metal clusters, which correspond to droplets of an electronic liquid (see Section 5.3.2). Indeed, an external electric field can extract some electrons from the system, leading to a net charge in the cluster. The cluster then rearranges itself by deformation to accommodate this net charge while remaining in a single piece, at least for some time (the time to overcome the fission barrier). As in a classical droplet, fission in a metal cluster stems from Coulomb repulsion between charges which overcomes internal binding, the latter being a mixture of attraction to ionic background and electron exchange. Increasing the net charge provokes the decrease of the fission barrier. Ultimately, if this charge is too high, there is no fission barrier anymore and the metal cluster immediately fissions.

If the nuclear case looks similar, there is a noticeable difference with metal clusters or classical droplets: charges (from protons) are already present in the system, even in its ground state configuration, and the repulsion between protons in nuclei is counterbalanced by the attraction between nucleons (proton-proton, proton-neutron and neutron-neutron). The existence of a fission barrier is then purely stemming from deformation effects, the latter ones stemming themselves from quantal effects, and does not need the presence of any external agent. In addition, the presence of a dip in the middle of the fission barrier (see Figure 7.5) is also due to quantum effects. It actually corresponds to fission isomers, that is strongly deformed (metastable) shapes of the nucleus. Nuclear fission thus mixes macroscopic features associated to a liquid droplet to microscopic ones stemming from quantum mechanics. In other words, a fissioning heavy nucleus can well be regarded as a droplet, but a droplet of quantum liquid. This interplay between classical and quantal features manifests itself in other situations in nuclei. Some of them will be discussed in the following sections.

7.2 INDIVIDUAL AND COLLECTIVE

As compared to the Coulomb interaction between electrons, the nucleon-nucleon interaction is much more complex because nucleons are composite objects. We have mentioned in Section 1.2.3 that it results as a residue of the strong interaction between underlying quarks and gluons, and it is described solely in terms of nucleon degrees of freedom. As a function of the distance between nucleons, the nucleon-nucleon interaction is repulsive below about 0.7 fm, attractive behind and vanishing beyond 3 fm. However, solving the Schrödinger equation for nucleons in a nucleus with such a 'realistic' nucleon-nucleon interaction is a highly complicated problem. Fortunately enough, as discussed in Section 4.4, sound approximations can be used for many purposes. Therefore, one can consider that a nucleon in a nucleus is moving in a mean field generated by the average interaction with the other nucleons. This mean field is modeled at several levels of complexity, to capture the essential facts beyond the specific problem one is interested in. We will see that even a rather schematic interaction suffices to understand the values of magic numbers. This is related to a description of the nucleus in terms of individual degrees of freedom. Other nuclear properties are best reproduced in terms of collective degrees of freedom, involving many nucleons and considering the nucleus as a quantum liquid drop.

7.2.1 Shell effects

We have seen that, although the liquid drop model provides a fair average description of the nuclear binding energies, it misses what we have interpreted as shell effects (see Figure 7.3). These effects simply stem from two facts. First, nucleons are quantal objects immersed in a potential (the mean field resulting from their mutual interactions), and henceforth their energies are quantized. Second, nucleons are fermions and therefore subject to Pauli exclusion principle, which limits the maximum number of fermions in a given energy level or 'shell'. A closed shell is related to a larger binding energy and enhanced stability. For historical reasons, the numbers of neutrons and/or protons closing a shell are referred to as 'magic numbers' and are given by the sequence 2, 8, 20, 28, 50, 82 and 126. They were highlighted by vertical and horizontal lines in the nuclear chart in Figure 7.1. We have also met this sequence when we plotted the differences between experimental and LDM binding energies (see Figure 7.4). Among other observables, they manifest themselves in the separation energy, which is the energy required to remove one nucleon from the nucleus. This energy increases for magic numbers, similarly as the ionization potential in closed shell atoms or metal clusters, implying an enhanced stability (see Figure 5.8). The doubly-magic nuclei (having neutron and proton shell closures) ${}^{4}_{2}\mathrm{He}_{2}$, ${}^{16}_{8}\mathrm{O}_{8}$, ${}^{40}_{20}\mathrm{Ca}_{20}$, ${}^{48}_{20}\mathrm{Ca}_{28}$, or ${}^{208}_{82}\mathrm{Pb}_{126}$ are particularly stable nuclei relative to their neighbors. This enhanced stability shows up as larger abundances of these nuclei in nature. The doubly magic ${}^{208}\mathrm{Pb}$ is for example much more abundant than its (non magic-Z) neighbor ${}^{197}_{79}\mathrm{Au}_{118}$. And accordingly, the less abundant gold is much more precious than the more abundant lead. The enhanced nuclear stability shows up also in abundances of elements formed in stars. Note also that a nucleus with a magic Z (resp. N) number possesses correlatively a large number of isotopes (resp. isotones).

The situation is very similar to what is observed in atoms (see Section 5.1) or metal clusters (see Section 5.3.3). Because the dominant interaction is different, the shape of the confining potential is different and the sequence of energy levels and shell closures are different as well. The number of electrons at which there is shell closure in atoms is 2, 10, 18, 36, 54, 86, ..., and in simple metal clusters 2, 8, 20, 40, 58, 92, 138, ... In these two systems, the electrons are moving in an external potential due to the Coulomb interaction with either the atomic nucleus or the lattice of positive ions, plus the electron-electron repulsion. However, as no external potential exists in a nucleus, it took very long to understand the mechanism behind the observed sequence of nuclear shell closure, hence the name of 'magic numbers'. As discussed in Sections 4.4 and 7.1.1, a nucleon in a nucleus feels a mean field created by the interaction with the remaining nucleons. This mean field provides the equivalent of the confining external potential in atoms or metal clusters, and it can be modeled by a simple ansatz such as a spherical harmonic oscillator. However, the resulting sequence of magic numbers associated to such a potential does not fit the observed one. To recover the right sequence, a spin-orbit coupling has to be included in the nuclear Hamiltonian.

Actually, in atoms, the spin-orbit coupling also plays an important role, especially in the case of heavy atoms. Indeed, a proper description of the sequence of atomic levels is accounted for by adding a spin-orbit term to the Hamiltonian, in the form of the scalar product $\mathbf{L} \cdot \mathbf{S}$ where $\mathbf{L}$ and $\mathbf{S}$ are respectively the angular momentum and the spin vector of the electron. The spin of an electron is fixed but its angular momentum around the nucleus (which reflects both its 'distance' to the nucleus and its velocity) depends on the energy level occupied by the electron: the higher the energy level, the higher the angular momentum quantum number. It is then natural that spin-orbit effects play an increasing role with increasing atomic numbers since more and more electrons and levels are to be taken into account. However, in the case of atoms (and the same holds true for metal clusters), the actual sequence of shell closures stems from the nature of the confining potential (shape of the Coulomb potential in both cases). Spin orbit does alter the detail of the sequence of levels within a shell but does not change the gross structure of shells and therefore shell closures.

In the late 1940s, M. Goeppert-Mayer and J. H. D. Jensen developed the basis of the nuclear shell model which provides the simplest realistic way for understanding nuclear shells. The observed sequence of shell closure can be reproduced assuming a confining potential of the form $V_C(r) + \kappa \mathbf{L} \cdot \mathbf{S}$ where $V_C(r)$ is a central potential. A number of schematic shapes may be used for the confining potential $V_C(r)$ but it turns out that a harmonic oscillator (see Eq. (2.17)) with frequency $\hbar\omega \simeq 40A^{-1/3}$ MeV works well for this purpose. This illustrated in Figure 7.6 which displays the sequence of nucleon levels in a spherical harmonic oscillator potential, with and without the spin-orbit term. The left part of the figure indicates the level sequence notation and the corresponding magic numbers (that is the total number of nucleons) given by the harmonic oscillator without the spin-orbit coupling, as already encountered in atoms and metal clusters. We recall that in the spectroscopic notation, the integer stands for the principal quantum number n while the letter s, p, d, etc. corresponds to $l = 0$, $l = 1$, $l = 2$, etc. For a given n and l, the energy level degeneracy is equal to $2l + 1$. The inclusion of spin-orbit coupling partially breaks this degeneracy and rearranges the levels such that new magic numbers appear (see the numbers on the right part of the figure). They now agree with the empirical observations. We have also introduced here the spectroscopic nomenclature nl_j where j is the quantum number related to the total angular momentum vector $\mathbf{J} = \mathbf{L} + \mathbf{S}$. One can show that for spin-1/2 particles, j can take values equal to $(2k + 1)/2$, k being a non-negative integer.

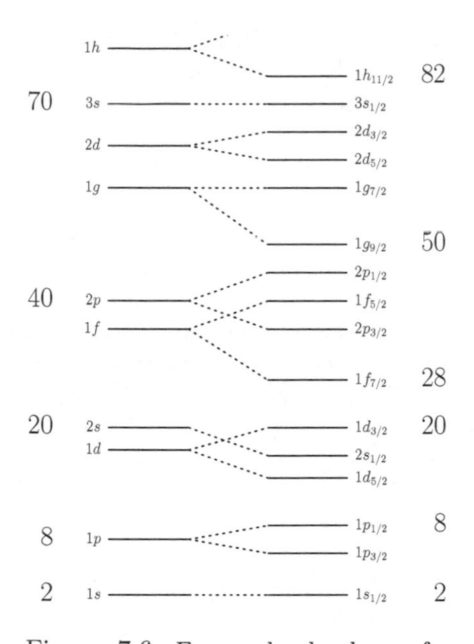

Figure 7.6: Energy level scheme for a harmonic oscillator without (left) or with (right) spin-orbit coupling. A number on the left or on the right indicates a magic number obtained by adding up the numbers of nucleons from the ground state level up to that level.

The key role of spin-orbit in the nuclear case may come as a surprise. In fact, it points to an interesting difference between nuclei and atoms or simple metal clusters (where spin-orbit effects are negligible), associated to differences in the type of binding. Indeed, in atoms, electrons are primarily bound to the nucleus. Although they do interact with each other (which produces a repulsive component), the final binding, and the electronic spectrum thereof, mostly stems from the Coulomb field delivered by the nucleus. The electron-electron repulsion then comes as a correction and the spin-orbit coupling as, so to speak, a higher order effect. Therefore, it is not surprising that the spin-orbit is indeed a correction only acting in massive atoms. The nuclear case is very different to the extent that there is no external confining agent. The nucleus is said to be self-bound: the confining mean field potential stems from the nucleon-nucleon interactions themselves. The spin-orbit term hence naturally gains importance as coming as a first order (and not a second order) correction.

7.2.2 Rotating nuclei and deformation

We have focused our previous discussions on the structural properties of nuclei. But nuclear dynamics also provides remarkable behavioral examples reflecting the quantum liquid nature of nuclei. We have already illustrated such effects in the case of fission in Section 7.1.3. We here complement the point by a few other examples, starting with rotations and deformations of nuclei.

There exist only very few magic nuclei and even less doubly magic nuclei. Actually there are only five doubly magic nuclei: ^{4}He, ^{16}O, ^{40}Ca, ^{48}Ca and ^{208}Pb. They possess a remarkable property that we have not mentioned yet: they are perfectly spherical. This may look surprising, especially for light systems, but we should keep in mind that nuclei are quantal objects with widely spread wave functions. This allows doubly magic nuclei, even for a small number of nucleons, to lead to a perfectly spherical distribution of matter. Actually, the wave functions of the various nucleons basically add up in a non-trivial way to finally form a spherical distribution. This is of course directly linked to Pauli exclusion principle. The surprise, though, comes from the fact that the wave functions, which are not all spherical (see Section 2.4.4) are nevertheless 'soft enough' to 'fill holes' between themselves and finally lead to the global spherical shape.

But this remarkable compensation works only in doubly magic nuclei. Indeed, imagine that we add a nucleon to a doubly magic nucleus. Where will the new nucleon take place? Pauli principle forbids it to penetrate the sphere formed by the other nucleons, so that it can only remain outside. Even if its wave function can somewhat spread over the inner sphere, this cannot suffice to reconstitute a new spherical distribution, a simple reason being spin polarization. Indeed, all magic numbers are even numbers because Pauli principle allows two nucleons to coexist with the same energy (and the same wave function) but opposite spins. The resulting nucleus (doubly magic plus one nucleon) is then bound to be deformed. Furthermore, one will not get a spherical distribution by adding more and more nucleons, until one reaches the next magic number. As a consequence, the immense majority of nuclei, with the exception of doubly magic nuclei, are not spherical and exhibit a deformed shape instead. This short discussion on nuclear shapes might look academic in view of the small size of these systems. In fact, it turns out that measuring nuclear shapes leads to important insight into pairing effects, already present in the phenomenological LDM mass formula Eq. (7.2). This pairing is quite similar to superconductivity and helium superfluidity (see Section 9.3).

Before addressing this issue, let us explain how shapes can be measured. A basic way to observe shapes is to look at rotations, as seeing a rotation is by itself a signal that the system is deformed. The rotation of a perfectly spherical nucleus cannot be observed. Indeed, for a perfect sphere (even a macroscopic one) which would rotate on itself about an arbitrary axis going through its center, there is no way to see it rotate... except if there is an imperfection on the surface. One can then identify the rotation by looking at the motion of the imperfection. The same holds for nuclei. And contrary to macroscopic objects where one could imagine mechanisms to 'mark' the surface, this is not feasible in nuclei, just because this would imply an interaction with the system. And this, in turn, would very much affect its properties, as usual in the microscopic world. Therefore, rotation of doubly magic nuclei cannot be observed. And this consequently demonstrates that they are spherical. On the contrary, other nuclei exhibit rotational motion and one can very precisely measure the properties of this motion and thus characterize nuclear shapes.

The basic measured quantity characterizing rotational motion is energy. In classical mechanics, the rotational energy E_{rot} is proportional to the square of the angular momentum l divided by the moment of inertia $\mathcal{I}$ of the system with respect to the rotation axis: $E_{\mathrm{rot}} = l^2/(2\mathcal{I})$. The shape of the system shows up in the moment of inertia which characterizes the mass distribution around the rotation axis. A deformed system exhibits different moments of inertia, depending on the axis around which rotation occurs. For example, an elongated cylinder has a small moment of inertia $\mathcal{I}_{//}$ when considering a rotation along its symmetry axis, and a much larger one $\mathcal{I}_\perp$ when considering a rotation along an axis perpendicular to this symmetry axis. The same holds in quantum mechanics but now with a quantized angular momentum vector $\mathbf{L}$. We have seen in Section 2.4.3 that the square of $\mathbf{L}$ takes values $\hbar^2 l(l+1)$, with l the associated quantum number. Therefore, the quantal rotational energy is $E_{\rm rot} = \hbar^2 l(l+1)/(2\mathcal{I})$. This delivers a very characteristic spectrum with level separation proportional to $l/\mathcal{I}$. The proportionality constant provides a direct measure of the moment of inertia and hence of the shape of the system.

Note that, contrary to a solid body, in nuclei, the situation is complicated by the fact that the system is liquid. This makes the evaluation of moment of inertia delicate, involving in particular viscosity aspects. Indeed, a very viscous system more or less looks like a solid object, so that rotation is hampered by mass. On the contrary, a very fluid-like object (and even more so, a superfluid one) will rotate more easily. Such differences on the nature of the nuclear fluid directly show up at the side of rotational energies through moments of inertia which are larger for solid-like systems. Rotational energies thus provide a rather direct hint on the fluidic nature of nuclei.

The above discussions have pointed out the quantum origin of nuclear deformation. These quantum effects are sizable (regarding the variety of nuclear shapes) but also to some extent fragile. By this, we mean that nuclear deformation is sensitive in particular to temperature. Of course, we have in mind here nuclear temperatures, of the order of several MeV, as have been observed for example in nuclear collisions leading to a heating up of a compound system created by partially aggregating projectile and target partners of a collision. We do not want to discuss these dynamical processes here. It is sufficient to note for our purpose that actual temperatures can be delivered this way to nuclei (we shall also find 'hot nuclei' in massive stars; see Section 8.3). And one speaks of temperature when it becomes comparable to the Fermi temperature, typically a few MeV. Not surprisingly, such temperatures are expected to affect the structure of nuclei as they become comparable to typical energy scale. We have seen in Section 7.2.1 that the typical nuclear shell spacing behaves as $40A^{-1/3}$, taking values from about 12-13 MeV in small nuclei down to about 6 MeV in large ones. Temperatures of order 3-4 MeV then mean a washing out of shell effects in large nuclei and we expect correlatively a disappearance of deformations at such temperatures in large nuclei.

This point is illustrated in Figure 7.7 in the case of the deformed ^{168}Yb nucleus whose shape is evaluated as a function of temperature. This shape is quantified by the quadrupolar moment Q. It is expressed in barn which is a common unit in nuclear physics (1 barn= 10^{-24} cm^2). At zero temperature, the minimum of free energy is around $Q = 19$ barns and the nucleus is prolate (cigar-like shape with one long axis and two equal short ones). But there is also a local minimum at $Q \simeq -14$ barns for which the nucleus is oblate (pancake-like shape with two long axes and a short one), at a higher energy though. As the temperature increases, the energy difference between the oblate and the prolate configurations fades away and we finally end with a spherical

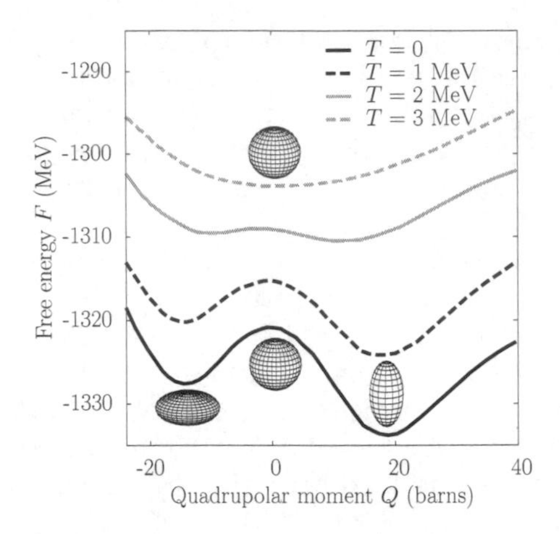

Figure 7.7: Free energy $F = U - TS$ of the nucleus ^{168}Yb ($Z = 70$) as a function of the quadrupolar moment Q at different temperatures T. The curves for $T = 1, 2, 3$ are upshifted by 20, 72 and 145 MeV respectively to allow a better visibility. Adapted from M. Brack and P. Quentin, Phys. Lett. B 52, 159, Copyright (1974), with permission from Elsevier.

shape ($Q = 0$) with the lowest energy at high temperature. This provides an instructive example of the competition between quantal and thermal effects in nuclei.

7.2.3 Superdeformed nuclei

We have seen that most nuclei are deformed. But it turns out that there exist specific nuclei whose deformation is especially large. They are called super deformed nuclei and exhibit remarkably elongated or planar shapes. Interestingly enough such exceptional shapes are again associated to shell effects and shell closures, the same way as spherical shapes are associated to shell closure in a spherical potential. But in order to understand such effects one now needs to consider highly deformed potentials. Again, and in spite of its limitations, the harmonic oscillator provides a simple way to understand such mechanisms. The deformation is included by choosing different oscillator frequencies $\omega_x, \omega_y, \omega_z$ along the three spatial directions. For the sake of simplicity, we assume axial symmetry, with two nuclear shapes either prolate (long axis z, short equivalent axes x, y) or oblate (long equivalent axes x, y, short axis z).

The potential can therefore be written as

$$V_{\text{HO}} = \frac{1}{2} m\omega_x^2 (x^2 + y^2) + \frac{1}{2} m\omega_z^2 z^2$$

and following the same reasoning as for the spherical harmonic oscillator the associated single particle energies can be written as

$$\varepsilon_{n_x,n_y,n_z} = \varepsilon_{n_x} + \varepsilon_{n_y} + \varepsilon_{n_z} = \hbar\omega_x(n_x + n_y + 1) + \hbar\omega_z(n_z + 1/2) \quad ,$$

each exhibiting now the three quantum numbers n_x, n_y and n_z. We could again follow the same reasoning as in the spherical case and see how levels will be progressively populated. But it is immediately clear that as soon as one will consider a deformed potential, namely with $\omega_z \neq \omega_x$ we shall lose the level degeneracy and the associated sequence of shell closures. On the other hand, and on purely mathematical basis, one can imagine that new degeneracies could occur for proper choices of the frequencies ω_x and ω_z. And it turns out that it is exactly what occurs. The deformed potential well is associated to a deformed nucleon cloud which we can characterize by its major axes $a_x = a_y$ and a_z, the analogs of the radius for a spherical shape. And actual values of these axes are directly linked to the oscillator frequencies. One can then show, for example, that for the simplest case of $a_x = a_y = a_z/2$, which corresponds to a strongly elongated (cigar like) system, one recovers a new set of shell closures (in this particular case corresponding to a new sequence of 'deformed' magic numbers): 2, 4, 10, 16, 28, 40, 60, 80, 110 and 140. This (as in a spherical case) ensures an enhanced stability of the corresponding nucleus (again as compared to neighbor nuclei).

The above discussion has been led in the schematic case of the deformed harmonic oscillator which lacks some important features of a realistic nuclear potential. Still, it is quite illustrative and it turns out that the essence of what we have observed, namely the appearance of new shell closures for specific (large) deformations is absolutely correct and observed in more realistic computations with realistic potentials and spin orbit corrections. And it provides a nice explanation to the observation of the super deformed nuclei, which typically occur for the new shell closures attainable in deformed systems.

7.2.4 Resonances

Rotation is clearly a collective motion. It is hard to associate an amplitude to it but the latter, if any, could take whatever value. Nuclei exhibit many other collective motions, which again provide very interesting insights into nuclear properties. We have already discussed the case of fission which is associated to a collective large amplitude deformation of the nucleus. Low amplitude motions, for example of vibrational nature, also constitute important properties of nuclei. They are known as giant resonances, and we shall explain the terminology later on.

Collective modes

In nuclei, nucleons constantly move. This is mostly due to Pauli principle and this is usually called Fermi motion (see Section 3.4). This is what makes nucleons inside a nucleus a quantum fluid. Now Fermi motion averages out to zero. This means that there is no net momentum resulting from the superposition of the motion of all nucleons. This is characteristic of an equilibrium (lowest energy) situation. One can however imagine excitations around this ground state configuration. And precisely because we are considering motion around an equilibrium, we expect small amplitude vibrations around it resulting from the superposition of incoherent motion of nucleons. The situation considered here can be modeled by a set of coupled harmonic oscillators, each one representing the binding of one nucleon to the nucleus. Elementary mechanics tells us that, in such a system of coupled strings, emerges a set of vibration eigenfrequencies which represent, in a simpler form, a vibrational motion of the ensemble of particles. Such eigenmodes may be more or less close to each other which can be interpreted in terms of collectivity. In the extreme case where all eigenfrequencies are the same, the motion is fully collective and there is no mechanism to damp it. In general, eigenmodes are distributed over a more or less wide energy range, reflecting the existence of a collective motion (more or less corresponding to the dominant eigenfrequencies) but with a finite lifetime. Indeed the presence of side peaks around the major one induces a degradation of collectivity on the long term. In other words, two particles vibrating collectively may still 'speak to each other' and progressively lose their coherence (identical vibration frequency) to the benefit of a nearby frequency. This picture of a distribution of eigenvibrations is also at the origin of the term 'Giant Resonances' (GR), which reflects the peaked nature of the distribution, the amplitude of the effect as compared to the background of vibration frequencies, and also the fact that in nuclei the energy of such collective modes lies above the separation energies of nucleons. The three main GR are schematically depicted in Figure 7.8.

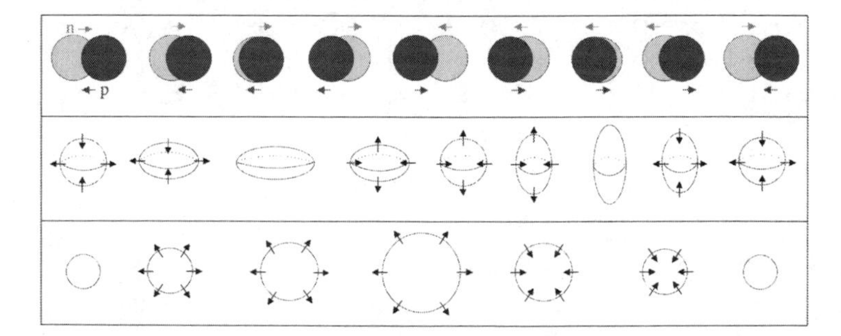

Figure 7.8: Nuclear giant resonances. Top: giant dipole resonance. Middle: giant quadrupole resonance. Bottom: giant monopole resonance. See text for more details.

The giant dipole resonance

The first identified GR, in the late 1940s, is the Giant Dipole Resonance (GDR) in which neutrons oscillate against protons, as sketched in the top panel of Figure 7.8. This is a small amplitude motion with displacement of the center of mass of protons against the one of neutrons within a fraction of fm, while the global center of mass (neutrons plus protons) remains fixed. It should be noted here that, as in other GRs, the collective motion superimposes on the Fermi motion. One could thus view the GDR as two Fermi spheres (one for neutrons and another one for protons) oscillating against each other.

The GDR tests the nuclear interaction, especially the attraction between neutrons and protons. The associated resonance frequency can indeed be linked to the symmetry term a_{sym} of the LDM formula Eq. (7.2). The GDR bears some similarities with the optical response observed in simple metal clusters (see Section 5.3.4). In both cases, the vibration is of dipolar nature: separation of neutron and proton centers of mass in nuclei, separation of electron and ionic centers of mass in metal clusters. There is indeed only a small deformation of the vibrating clouds as compared to their ground state shapes so that only the positions of their center of mass are affected.

The giant quadrupole resonance

The Giant Quadrupole Resonance (GQR) represents the simplest GR associated to a shape oscillation of a nucleus. Let us take the case of a spherical nucleus. The GQR corresponds to a successive modification of the nuclear shape from spherical to pancake shape, then back to spherical and finally to cigar-like shape and back (see middle panel of Figure 7.8). And the process continues that way again and again. Along the whole process, it is the surface tension which constantly brings back the system to its original shape, as in a classical liquid drop (see Section 1.4.1). The GQR vibration energy is hence linked to the surface term a_{sf} of the LDM formula Eq. (7.2). The quadrupolar deformation (sphere to pancake or cigar) in fact reflects a very specific property of nuclei, directly linked to the quantal nature of the nuclear fluid. Let us be more precise here.

In a normal fluid such as water, the collective motion remains, to a large extent, independent from the individual motion of constituents. Mind that in a water droplet, temperature is responsible for a constant agitation of water molecules. Both motions (intrinsic thermal and collective) basically superimpose on each other without interfering. Think for instance about a water droplet deposited on a vibrating plate. More precisely, at each instant during the motion, any small portion of the water droplet remains at equilibrium *around* the collective velocity of the droplet. The velocity distribution of the water molecules is then just the superposition of the thermal velocity distribution and the translation velocity. Such a situation reflects the fact that the characteristic time of return to *local* equilibrium is much shorter than the macroscopic time scale associated to the global vibration.

This is not what occurs in the GQR. In this case, local equilibrium (mediated by Fermi motion) and global motion time scales match and one cannot separate them anymore. As a consequence, one cannot consider that locally nucleons have relaxed (at each instant during the GQR motion) towards a local Fermi sphere. Indeed one can show that the instantaneous local momentum distribution itself possesses a quadrupolar deformation. But the latter is out of phase with the spatial distribution. In other words, the momentum distribution is cigar-like when the spatial distribution is pancake-like and vice versa. This is a very subtle quantal effect which strongly differentiates the nuclear fluid from a classical one.

The giant monopole resonance

The last resonance we will discuss is the Giant Monopole Resonance (GMR), which does not correspond to a shape vibration but to a size vibration (see bottom panel of Figure 7.8). While both the GDR and GQR occur at basically constant nucleon density, the GMR explores the then left degree of freedom, namely the nucleon density (or vibrations of the nuclear radius). It consists in a succession of compressions and dilatations, and is often called a breathing mode. In turn, the GMR does not alter the shape of the system, only its size. The resonance therefore provides insights on the compressibility of the nuclear fluid constituting nuclei. The situation is here very analogous to the one in a gas where the compression modulus reflects the variation of pressure associated to a variation of volume. More precisely, if K_A denotes the compression modulus (also called the incompressibility) of a nucleus of mass A, one can show that it is related to the energy of the GMR E_{GMR} by

$$K_A = \frac{m}{\hbar^2} \langle r^2 \rangle E_{\text{GMR}}^2 \quad , \tag{7.5}$$

where m is the nucleon mass and $\langle r^2 \rangle = \int \mathrm{d}^3\mathbf{r}\varrho(\mathbf{r})r^2$ is the radius mean square (rms) of the nucleus. We will see in Section 7.3.1 how one can exploit such data to explore the equation of state of nuclear matter.

Nuclei are not very compressible, which is basically the case of most dense liquids. And nuclei do represent the densest finite objects known (see Section 7.1.1). Only in the core of some massive stars, somewhat larger densities (a few times the nuclear density) can be temporarily reached (see Section 8.3). Correlatively, nuclei are only little compressible. This is connected to the fact that the average distance between two nucleons in a nucleus is not really larger than twice the size of each nucleon, thus leaving little space for possible compressional effects. To some extent, overlooking the ever present Fermi motion and the quantal nature of nucleons with their necessarily extended wave functions, nucleons look as somewhat packed on each other. And Pauli repulsion has obviously also its share in this low compressibility.

Energies

We would like to conclude this mostly qualitative discussion with a few more quantitative estimates. Standard giant resonances all possess vibration frequency in the same energy range, and basically scale the same way with size. Not surprisingly, collectivity reflects itself in this scaling. Experiments show that, to a large extent, all GR frequencies can be written in the same form $\hbar\omega_{\text{GR}} \sim \kappa_{ls} A^{-1/3}$ where we recognize the nuclear radius ($\propto A^{-1/3}$) and where the parameter κ_{ls} characterizes a given mode. In this quantity, l corresponds to the multipolarity (0 for GMR, 1 for GDR, 2 for GQR, ...) and $s = 0$ for modes where neutrons and protons move in phase and $s = 1$ when motion is in opposite phase. We have discussed here the three cases $l = 0, s = 0,$

$l = 1, s = 1$ and $l = 2, s = 0$ for which the value of κ_{ls} is in all cases of the order of 80 MeV. This puts such modes in the range between about 12 MeV for heavy nuclei up to about 25 MeV for light ones. The high energy of these collective modes partly explains their finite lifetime and/or the associated energy width which makes them painful to identify experimentally. Indeed a typical damping mechanism of a GR is the emission of nucleons (immediately possible because the resonance lies above emission threshold) which breaks the collectivity of the motion. Another difficulty is also due to the usual energetic vicinity of the various modes which makes their actual identification complicated.

7.2.5 Superheavy and exotic nuclei

Let us end this short review of nuclei by coming back to the nuclear chart plotted in Figure 7.1. Stable nuclei and very long-lived nuclei are indicated with black squares. The remaining nuclei (more than 3000) have been produced artificially, and decay to stable nuclei by emission of alpha particle (^{4}He nucleus), or beta particles (electron or positrons). As already mentioned, magic numbers are associated to particularly stable nuclei. The magic numbers 2, 8, 20, 28, 50, and 82 are the same for neutrons and protons, and the value 126 occurs only for neutrons. The ending point of the so-called 'stability valley' corresponds to the doubly-magic nucleus ^{208}Pb, with $N = 126$, $Z = 82$. Beyond that, there are long-lived nuclei up to uranium, as well as artificially produced heavy nuclei, with a very short lifetime. Indeed, their mass is too high for the strong interaction alone to bind them, as the electrostatic repulsion between protons becomes intense enough to counteract the nuclear attraction. Such a heavy nucleus emits spontaneously protons, neutrons or alpha particles in order to reduce its mass, until reaching the one of a stable nucleus.

Are there more magic numbers? This could provide an additional stability and possibly imply the existence of long-lived superheavy nuclei. Theoretically one would expect additional magic numbers: near 184 for neutrons, and around 114, 120, or 126 for protons (the precise value depends on the theoretical inputs). They have not been confirmed by experiments, although there are hints of an increase of lifetimes in the heavy elements around the value $Z = 114$, which corresponds to flerovium, of order of seconds. In 2016 four new elements were included into the periodic table. These are nihonium ($Z = 113$, symbol Nh), moscovium ($Z = 115$, Mc), tennessine ($Z = 117$, Ts), and oganesson ($Z = 118$, Og), with shorter half-lives of order millisecond.

Another group of artificial nuclei displayed in Figure 7.1 refers to nuclei having a neutron/proton ratio which deviates much from the value found in nature, which is close to 1. These nuclei are referred to as 'exotic nuclei' because of their surprising properties as compared to more ordinary nuclei, and include the formation of neutron or proton halos, changes of magic numbers, new types of radioactivity, soft-dipole excitations... Their lifetime is short, from milliseconds to a few years, because they decay by emiting electrons or positrons to reduce the excess of neutrons or protons, respectively, in a process governed by the nuclear weak interaction (β radioactivity). As N/Z is increased or decreased from the stable ratio, a point is reached where no more neutrons or protons can be added because that nucleus is not bound with that number of nucleons. The upper limit of the grey region in Figure 7.1 is the proton drip-line, which is believed to be (almost) completed. On the contrary, it is still an open question to find the neutron drip-line, which will be the lower bound in Figure 7.1.

Exotic nuclei are currently produced by breaking heavy nuclei either by bombarding them with high-energy protons or making them to collide with a target of light nuclei, usually beryllium. Research on exotic nuclei is important at least for two reasons. On the one side, it can reveal novel properties and increase the knowledge of nuclear properties. Theoretical models can be tested and extended, thus acquiring new insights on both the strong and weak nuclear interactions. On the other side, it will help to understand the formation of chemical elements in stellar nucleosynthesis.

7.3 THE EQUATION OF STATE OF NUCLEAR MATTER

7.3.1 The notion of nuclear matter

In the previous sections, we have considered atomic nuclei as droplets of a nuclear liquid. We discuss now some properties of that same liquid, which is usually referred to as nuclear matter. Obviously, a homogeneous medium made of interacting nucleons is not a system that can be experimentally studied in the laboratory. However, this is a very useful concept because it provides information about the inner part of finite nuclei and, owing to its relative simplicity, it can be used to model some parts of neutron stars. Nuclear matter is thus an idealized system made of an infinite number of protons and neutrons, but having a well-defined number density ϱ, and a fixed asymmetry $I = (N - Z)/A$. The idealization is completed by neglecting the Coulomb repulsion among protons, so that nucleons interact only via the strong nuclear force. Clearly, such an idealized system can be considered only from a theoretical point of view, with however two main restrictions regarding the theoretical description. First, it must be consistent with that of finite nuclei, so that the nucleon-nucleon interaction is the same. Second, nuclear matter results must be consistent with the empirical LDM results. Indeed, the LDM with its remarkable reproduction of nuclear binding energies provides a natural link between actual nuclei and nuclear matter.

The equation of state of a system (see Section 1.1.1) is a functional relationship between thermodynamical parameters describing the system in thermal equilibrium. For gases and liquids the equation of state usually means the pressure P as a function of density and temperature, all quantities being experimentally accessible. However, the starting point in nuclear matter is the energy per particle $\mathcal{E} = E/A$ as a function of the density and temperature, from which the relevant nuclear matter properties are obtained. In particular the pressure is obtained as $P = \varrho^2 \partial\mathcal{E}/\partial\varrho$ at fixed T. This is the reason why in this field the expression equation of state applies indistinctly to both $\mathcal{E}$ and P as a function of density and temperature. Actually $\mathcal{E}$ depends on the nucleon density ϱ, the asymmetry parameter I and the temperature T.

The theoretical study of nuclear matter is performed following two main roads: either a microscopic treatment (which may include relativistic effects), based on a realistic nucleon-nucleon interaction, or a mean field approach, based on an effective interaction. This effective interaction depends on the relative nucleon-nucleon distance as well as the spin, isospin, angular momentum, etc. of the interacting nucleons. It is constructed in a semi-phenomenological way, and contains a certain number of free parameters which are fixed to reproduce experimental properties of finite nuclei and some known properties of nuclear matter, related to the LDM formula (see Section 7.1.2). That way, there is a direct connection between nuclear drops and bulk liquid.

The major constraints imposed by LDM to nuclear matter concern the saturation point, especially energy per particle at saturation density, for which we shall take the well established values $\varrho_0 \simeq 0.16$ fm^{-3} and $\mathcal{E}_0 \simeq -16$ MeV. Mind that these values correspond to symmetric nuclear matter, namely for $I = 0$. Other values of equilibrium density and energy would be associated to other asymmetry parameter (see Figure 7.9). There is another quantity characterizing the saturation point which has motivated numerous studies and that we shall briefly discuss. It is the incompressibility modulus K_0 of nuclear matter, which practically provides the curvature of the equation of state at saturation point. It is measured in nuclei through the Giant Monopole Resonance (see Section 7.2.4) which provides values of the compressibility of finite nuclei K_A via Eq. (7.5). Following the mass formula Eq. (7.2) of the LDM, one can then consider a phenomenological expansion of K_A as

$$K_A = K_{\mathrm{v}} + K_{\mathrm{sf}} A^{-1/3} + K_{\mathrm{c}} Z^2 A^{-4/3} + K_{A,\tau} I^2 \quad , \tag{7.6}$$

with successively the contribution to K_A from the volume, the surface, the Coulomb interaction and the isospin asymmetry. One usually identifies K_{v} and K_0, thus corresponding to symmetric nuclear matter. The values of parameters entering Eq. (7.6) can be determined through Eq. (7.5), to get K_A from the experimental measurements of the GMR energies. There are limitations due to data accuracy, shell effects, etc., which reflect in relatively large error bars. There is a general consensus on a value $K_0 = 240 \pm 20$ MeV, even if some fits of experimental measurements lead to larger error bars and possibly slightly larger average values. The uncertainties on the other parameters entering Eq. (7.6) are larger, even very large, and we shall not discuss them further. It is enough here to realize the difficulties encountered in extracting trustable values of such parameters from GMR energies. Isospin effects, in particular turn out to lead to wide variations of the associated corrective term $K_{A,\tau}$ in Eq. (7.6) which again points out the difficulties associated to research associated to 'exotic' nuclei (discussed in Section 7.2.5). To make this discussion on the nuclear equation of state more quantitative, we present and discuss in the following some results obtained within a mean-field approximation (see Section 4.4).

7.3.2 A phenomenological equation of state

We consider now the mean field description of nuclear matter at zero temperature. In this context one considers nuclear matter as a fully degenerate Fermi gas, with nucleon wave functions described as plane waves. The kinetic energy has been given in Section 3.4.1, and the potential energy is obtained by taking the expectation value of the interaction. Let us explicitly show the dependence of the kinetic energy on the asymmetry parameter. To this end, we consider two liquids, one made of neutrons and one made of protons. We associate a Fermi momentum for each of them as $k_{F,n}^3 = 3\pi^2 \varrho_n$ and similarly for protons. Equation (3.26), valid for an ideal gas of spin-1/2 fermions, then provides an energy proportional to $k_{F,n}^5$ or $k_{F,p}^5$ respectively. From the definition of the asymmetry parameter $I = (N - Z)/A$, we write the neutron and proton densities as $\varrho_n = \varrho(1 + I)/2$ and $\varrho_p = \varrho(1 - I)/2$, where ϱ is the nucleon density. Finally, the kinetic energy per nucleon is written as

$$E_{\text{kin}}/A = \frac{3}{5}\frac{\hbar^2}{2m}\left(\frac{3\pi^2}{2}\right)^{2/3} \varrho^{2/3}\,\frac{1}{2}\left[(1+I)^{5/3} + (1-I)^{5/3}\right] \tag{7.7}$$

In the same way, the potential energy is also given in terms of ϱ and I. There are several families of effective interactions, according to the dependence on the relative coordinate. It can be a Gaussian (e^{-r^2/a^2}), a Yukawa-like ($\mathrm{e}^{-r/a}/r$) or a zero-range interaction with a momentum dependence. The latter is the popular Skyrme family, largely used because of its simplicity. In the case of symmetric nuclear matter, the simplest Skyrme interaction potential can be reduced to an oversimplified two-term parameterization:

$$V = t_0 \varrho + t_3 \varrho^{1+\sigma} \quad . \tag{7.8}$$

In this expression, t_0 takes a (large) negative value providing the attractive part of the nuclear interaction, and t_3 a (larger) positive value, mimicking the short range repulsion. The parameter σ directly impacts the incompressibility modulus K_0 and takes values of order 1/3. The calculation of a more realistic potential energy is a bit more involved and we will not enter into details. In the next figures, we present typical results as obtained with a Skyrme interaction. This suffices for the qualitative discussion we are interested in.

Figure 7.9 displays the energy per particle $\mathcal{E}$ (left panel) and the pressure P (right panel) as a function of ϱ for different values of I. One can see that whatever the density, the minimum energy is always obtained for $I = 0$, that is for symmetric nuclear matter. This result was to be expected by inspecting the LDM formula Eq. (7.2). When there are more neutrons in the system,

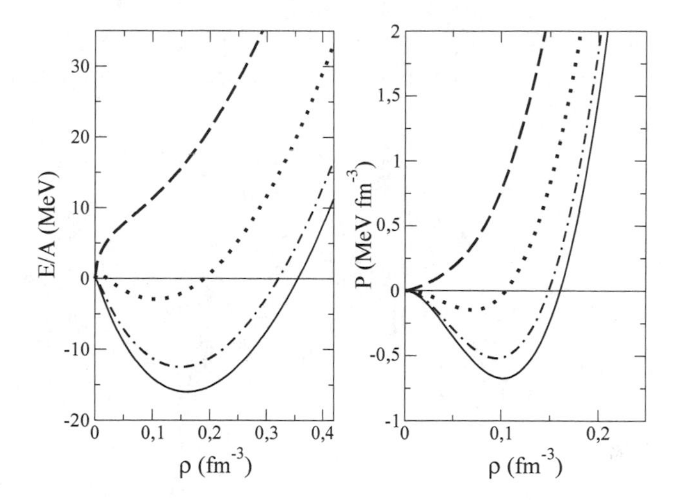

Figure 7.9: Energy per particle E/A (left) and pressure P (right) as a function of the number density ϱ for several asymmetry parameters $I = (N - Z)/A$. Solid lines: $I = 0$ (symmetric nuclear matter). Dot-dashed lines: $I = 1/3$. Dotted lines: $I = 2/3$. Dashed lines: $I = 1$ (pure neutron matter).

it becomes less bound. In the extreme case of pure neutron matter ($I = 1$), there is no binding at all (the energy is always positive). We here anticipate that an external field (namely, a gravitational attraction) is necessary to bind a neutron system (see Section 8.3). In the case of symmetric nuclear matter, the minimum corresponds to the saturation values: $\varrho_0 \simeq 0.16$ fm^{-3} and $\mathcal{E}_0 \simeq -16$ MeV. The phenomenological interactions should reproduce these values, within some acceptable error bars.

The pressure P is plotted in the right panel of Figure 7.9. The equilibrium of the system (minimum energy) corresponds to a vanishing pressure. For neutron matter, this value is only obtained at zero density. Except for that case, the pressure has a minimum. On the right side of the minimum the slope of the curves is positive: $\partial P/\partial\varrho > 0$, which corresponds to a positive compressibility. This condition implies the stability of nuclear matter against density oscillations, in particular around the saturation density ϱ_0. However, on the left side of the minimum, the slope is negative, thus implying a negative compressibility: an increase of pressure implies a decrease of density, which is a non-physical result. Actually, this is the indication of a phase transition, which we will consider in the next section.

7.3.3 The liquid-gas transition

We discuss now the possible liquid-gas transition of symmetric nuclear matter. Typical isotherms $P(\varrho, T)$ are displayed on the left panel of Figure 7.10 as obtained in the mean-field approximation with Skyrme-like interactions. The

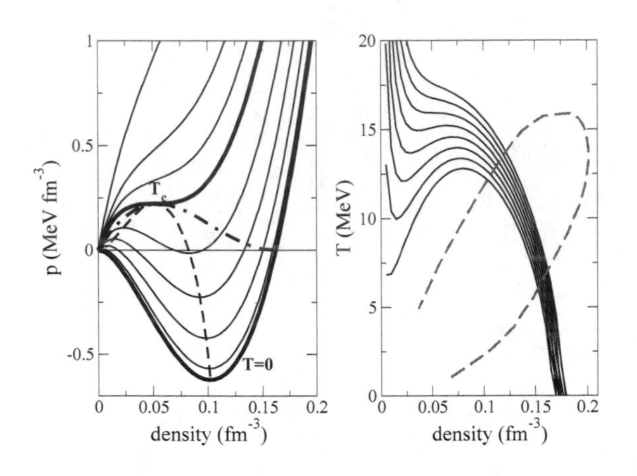

Figure 7.10: Typical nuclear matter equation of state. Left: isotherms with thick curves corresponding to critical and zero temperatures. Dashed and dot-dashed curves delimit the spinodal and the gas-liquid coexistence regions. Right: isobars with a possible trajectory (dashed curve) explored in a heavy-ion collision experiment.

similarities with the van der Waals isotherms, discussed in Section 1.3.4 are striking. We recall that the van der Waals interaction binds some types of atoms or molecules together (see Sections 5.1 and 6.1). In spite of the fact that the nuclear interaction is at least five orders of magnitude stronger than the typical interaction between atoms or molecules, the qualitative behavior of the isotherms is the same, so that we can draw similar conclusions about nuclear matter by analogy with the behavior of any other liquid. This is why in the mid-1970s, a liquid-gas phase transition was predicted in nuclei. More precisely, a 'vaporization' line was inferred in the phase diagram of nuclear matter, and experiments on heavy nuclei collisions were launched to detect it.

The two isotherms in thick lines correspond to critical and zero temperatures. Below T_c, the isotherms exhibit a minimal pressure for some non-vanishing value of density. From that minimum down to the maximum located at the left, the slope of the pressure is negative. This is a non-physical situation, as mentioned in Section 7.3.2. The dashed line in the figure delimits the 'spinodal' instability region. As with the van der Waals equation of state, it is thermodynamically more favorable to have the coexistence of a liquid and a gas phase. The dotted-dashed line separates the liquid phase, appearing as finite nuclear droplets, namely nuclei, from the gas phase, consisting of 'evaporated' nucleons and thus also called the hadron gas.

Depending on the specific microscopic model and the choice of the effective interaction, the critical point lies at $\varrho_c = 0.03 - 0.04$ nucleon/fm^3 and $T_c = 16 - 18$ MeV. Above the vaporization curve, nucleons should not be bound anymore. These results can be applied to real nuclei but not directly. Specific statistical physics tools are required to deal with finite systems, including also surface and Coulomb effects. A finite size results in a reduction of T_c of about 2-6 MeV, while the Coulomb interaction further reduces it by 1-3 MeV. All contributions taken into account, this leads to a 'critical' temperature around 10 MeV in nuclei of hot pieces of nuclear matter produced in heavy ion collisions.

Experimentally, a way to explore the liquid-gas phase transition is achieved through collision between heavy nuclei or, more precisely, heavy ions. Indeed by a partial ionization of atoms, ions can be accelerated by means of electromagnetic fields and projected against targets to produce nuclear collisions. Under certain conditions, nuclei are 'heated' to temperatures of a few MeV, and some of the nuclear liquid evaporates. As in liquid water, or any other liquid, there is coexistence between liquid and gas phases below the critical temperature. Therefore, heavy ions collisions allow us to explore the phase diagram of nuclear matter.

In collisions at intermediate energies of some MeV$/A$, many fragments (with $Z > 2$) are experimentally observed and the term multifragmentation has been introduced to describe such observations. Multifragmentation has been predicted already by Bohr in 1936 and has been studied experimentally and theoretically since the 1980s. It has been observed that it sets in above 3 MeV$/A$ and is maximal around 9 MeV$/A$, precisely close to the binding energy in nuclei. The question of multifragmentation as a signature of a vaporization of nuclear matter is quite hard to address. Heavy fragments resulting from fission or fusion are usually thought as signatures of the liquid phase, while isolated nucleons and light nuclei would be a trace of the hadron gas. The advent of the so-called 4π detectors, allowing the complete reconstruction of the collision event and the detection of all produced fragments, constitutes a great progress in this field. At the side of the theory, models face an involved many-body problem. Some of them use techniques of statistical mechanics at equilibrium to describe the freeze-out of the hot nuclear matter after the collision, while other models aim at describing the dynamics of the collision itself. The latter ones however are bound to only describe the first stages of the collision and become unable to account for multifragmentation which results from fluctuations about the most probable trajectory. Stochastic mean-field approaches have been developed to account for these density fluctuations.

7.4 CONCLUDING REMARKS

Nuclei are typical examples of self bound droplets of quantum fluid, in that case made of neutrons and protons. This makes them very peculiar, in particular because they naturally exist as finite droplets, as opposed to other systems we have been investigating, in particular electron fluids, and, to a large extent, helium systems. This droplet nature is especially visible from the liquid drop model, which mixes classical liquid to purely quantal features and which turns out to be surprisingly rich, even to study complex processes such as fission.

Not surprisingly finite quantum systems such as nuclei exhibit marked shell closure effects, visible in numerous observables. Maybe more surprising are the consequences of the quantum properties of these assemblies of nucleons, in particular in terms of shapes. The latter aspect directly brings us back to the droplet picture with its a priori variable shape. A last interesting aspect we have discussed in some detail concern collective modes, in particular giant resonates, which bear some similarities with the collective modes discussed at the side of the electron cloud in metal clusters.

We have concluded this chapter by discussing the (partly) abstract concept of the nuclear equation of state which corresponds to an idealized infinite phase of nuclear matter and which has played a key role in our understanding of nuclear systems. The nuclear matter equation of state in particular exhibits an interesting phase transition of liquid-gas nature, in which the composed nucleus is broken into its constituents. Such phenomena are observed in collisions between finite nuclei (with all the precautions to be taken to perform such analysis) and have focused numerous investigations over the past decades.

CHAPTER 8

High densities

We have discussed in Chapter 7 some properties of nuclei which can be described as very special quantal liquid droplets. However, up to now, we have restricted ourselves to nuclear matter close to the saturation density $\varrho_0 = 0.16$ fm^{-3}. And, as we did at many places in this book, we may wonder what occurs if we change the nucleon density and/or the temperature of the nuclear matter in large amounts. This question concerns, stated in other words, the phase diagram of nuclear matter, similar to what we have already encountered in other systems. Still, there is an important difference here, stemming from the fact that, while around nuclear saturation density and at moderate temperatures (see Section 7.3.1), nuclear matter is indeed constituted of nucleons, it may not be always the case when considering large densities and/or temperatures. This reflects the fact that nucleons are made of quarks whose degrees of freedom may become fully visible for sufficiently high energy deposits, at variance with other systems where we consider only one kind of degrees of freedom all over the phase diagram.

Studying the phase diagram of nuclear matter far away from the saturation point then naturally implies the appearance of new phases not constituted of nucleons. This is not the case, for example of electrons, because they are elementary particles with no underlying substructures. Hence, one can study electrons in whatever temperature/density regime without need of referring to other particles. It turns out that electrons as well can experience exotic thermodynamical situations. While on Earth they may exist as more or less free systems, for example in metals and/or plasmas, they naturally exist as a very dense phase in white dwarfs. These stars correspond to the late stage of evolution of non massive stars (with masses typically around or below the solar mass). In this case, the large density makes the Fermi energy of electrons very large and thus fully disconnected from their parent ions.

This chapter will address these surprising cases of high density matter, and show again the importance of quantum effects in these systems. Because of the surprising features of dense nuclear matter, as compared to the more usual electron system, we start with a discussion on the phase diagram of nuclear matter, schematically depicted in Figure 8.1 in the temperature-density plane. Nuclei lie basically at $T = 0$ and about the saturation density ϱ_0. The

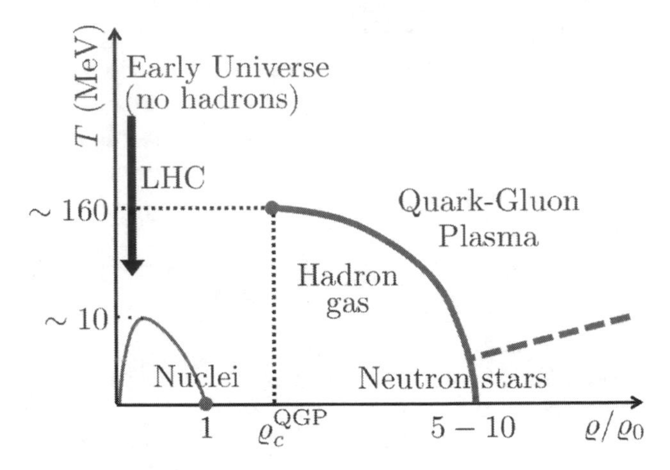

Figure 8.1: Phase diagram of nuclear matter, in the plane temperature T vs. reduced density ϱ/ϱ_0 with $\varrho_0 = 0.16$ fm^{-3}. See text for details.

thick line stands for the liquid-gas phase transition in nuclei, discussed in Section 7.3.3. We recall that this concerns a transition between nuclei, seen as a condensed (liquid) phase of nucleons, and the hadron gas in which nucleons are not confined anymore in nuclei. As visible in this figure, there are many other phases of nuclear matter that might exist under certain conditions in the Universe. We will start in Section 8.1 with matter in the Early Universe. We will also give some clues to understand the stellar evolution. We will then focus on two particular stars, namely white dwarfs in Section 8.2, which represent the latest stage of the lifetime of low mass stars, and neutron stars in Section 8.3, which are remnants of a supernova explosion. They both constitute an infinite fermionic liquid phase, of electrons in the former and of neutrons in the latter. We will see that the stability of such objects strongly relies on the fermionic character of electrons and neutrons. We will end this chapter by discussing in Section 8.4 some aspects of the quark-gluon plasma which is a phase of matter predicted at very high temperatures and/or densities.

8.1 THE ORIGINS OF MATTER IN THE UNIVERSE

8.1.1 A brief history of the Early Universe

We briefly discuss here the early stages of the Universe which involves a succession of various phase transitions leading to the composition of today's Universe. We here focus on the various phase transitions which took place at very high temperatures and at very small densities (see gray arrow in Figure 8.1). The fact that this region of the phase diagram corresponds to the

Early Universe might look counterintuitive since it is believed that the Early Universe was the place of tremendously high densities. A word of caution is in place here: right after the Big Bang, the *energy* density was indeed gigantic but matter in the common sense did not exist yet. Mind that in Figure 8.1, the horizontal axis corresponds to the number density ϱ of baryons (see Appendix C), which include protons and neutrons. It is thus sound to position the Early Universe in the region $T \to \infty$ and $\varrho \to 0$.

At the latest estimates, the Universe is 13.7×10^9 years old and possesses billions of galaxies. On the other hand, we have seen in Chapter 7 that there are about 250 stable nuclei, going from the lightest ^{1}H to the heaviest ^{208}Pb. One might wonder what the origin of the large-scaled galaxies and that of the stable nuclei at the microscopic level are. Here and in the following section, we will give some elements of answer to these two questions.

Let us start with the very first instants of the Universe and discuss how the four fundamental interactions operated and how the elementary particles (see Appendix C) appeared. A timeline is presented in Figure 8.2. Before the

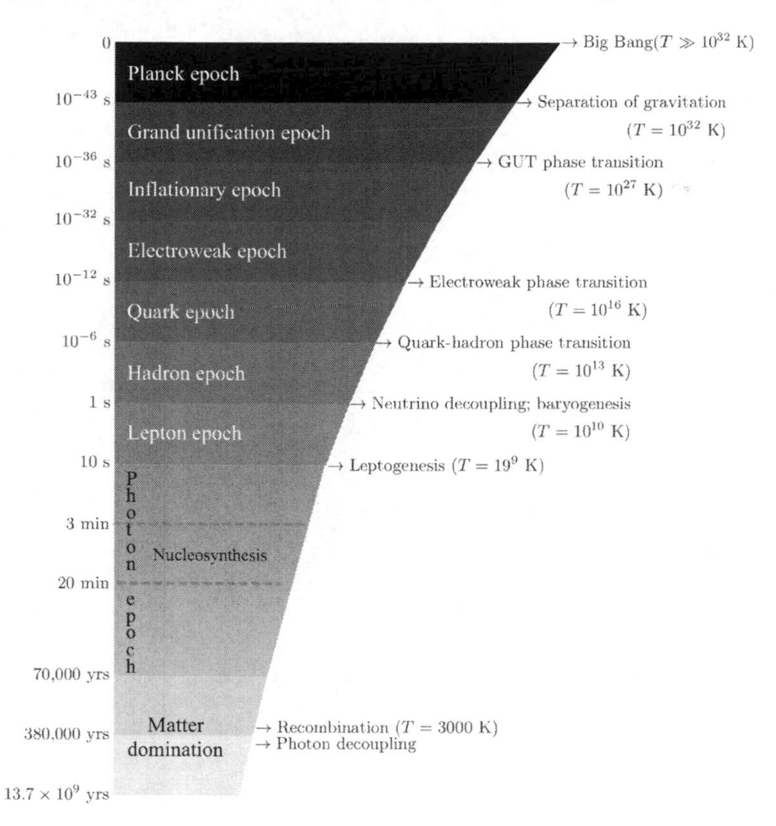

Figure 8.2: Timeline of the Early Universe, from the Big Bang to the Recombination.

Planck time $t_P = \sqrt{G\hbar/c} = 5.4 \times 10^{-44}$ s, the Universe was smaller than the Planck length $l_P = \sqrt{G\hbar/c^3} = 1.6 \times 10^{-35}$ m and the temperature was higher than the Planck temperature $T_P = \sqrt{\hbar c^5/(Gk_B^2)} = 1.4 \times 10^{32}$ K. This corresponds to the Planck epoch where gravity is quantal and physics is described by the Theory of Everything in which the four fundamental interactions are unified. We basically know nothing from this epoch and corresponding theories are very speculative. At $t = t_P$, a first phase transition occurs and gravity separates from the three other fundamental interactions, the latters forming the so-called Grand Unified Theory (GUT). We enter the grand unification epoch and the Universe is distributed in a uniform space-time. At $t = 10^{-36}$ s, the temperature and the size of the Universe are $T = 10^{27}$ K and $l = 6 \times 10^{-28}$ m respectively, and the strong interaction separates from the electroweak interaction: this is the GUT phase transition where matter and space-time separate and particles are born. The so-called Standard Model applies from this time on. Gluons and pairs of quarks and antiquarks are deconfined to form a Quark-Gluon Plasma (QGP), and coexist with pairs of leptons and antileptons (a lepton is a particle which does not contain any quark and which is sensitive to the electromagnetic and/or the weak interactions only). Between 10^{-36} s and 10^{-32} s, the Universe undergoes a tremendously rapid exponential expansion in which its volume is multiplied at least by a factor 10^{26} (some theories give a factor as high as 10^{50}), reaching eventually about the size of a proton (0.1 fm). This stage is followed by the electroweak epoch, where the Universe is composed of the QGP, Higgs bosons, and the $W^\pm$ and Z^0 particles which mediate the electroweak interaction. The electroweak interaction is a unified picture of the electromagnetic and the weak interactions. At $t = 10^{-12}$ s, the electroweak phase transition occurs: all $W^\pm$ and Z^0 decay, the weak interaction becomes short-ranged and decouples from the electromagnetic interaction. The Universe has a size of about 10 cm and its temperature is $T = 10^{16}$ K. We then enter the quark epoch where the QGP, leptons and photons fulfill the Universe. At 10^{-6} s, the Universe has sufficiently cooled down to $T = 10^{13}$ K to confine quarks, antiquarks and gluons together into hadrons: this is the quark-hadron transition. There exist two types of hadrons: baryons which consist in three quarks, and mesons which are bound states with a quark and an antiquark. The most common baryons are the proton and the neutron.

At $t = 1$ s, the Universe temperature has decreased down to $T = 10^{10}$ K. Meanwhile, baryons and antibaryons have annihilated to produce photons. However we all know that the Universe is not empty of matter. This means that the quantity of matter was not strictly equal to that of antimatter. Some estimations give 1 extra baryon for every billion of baryon/antibaryon pairs. This tiny imbalance was sufficient to create matter that will dominate the Universe later on. Discussing the origin of such a baryon asymmetry would lead us too far away from the topic of this book and therefore we shall not enter these details here. The electroweak theory in the Standard Model is able to provide the ingredients for the production of a baryon excess. Unfortunately, it is not efficient enough to produce the baryon asymmetry that one can deduce from the cosmic background radiation. There exist variants of the electroweak theory but their validations from ultra-high-energy experiments are still awaited.

At the time of the baryogenesis ($t = 1$ s), neutrinos decouple and do not interact anymore with matter. We enter the lepton epoch in which the Universe is dominated by leptons and antileptons. This epoch ends at $t = 10$ s: there is then no more creation of lepton/antilepton pairs. As in the baryogenesis, the annihilation of leptons and antileptons leaves a tiny excess of leptons, in the same amount as the baryon excess. The mechanism of leptogenesis is probably strongly connected to that of the baryogenesis. The end of the lepton epoch at $t = 10$ s is the starting point of the photon epoch which lasts until 70 000 years: the Universe is dominated by photon energies. During the photon epoch, from $t = 3$ min to $t = 20$ min, the temperature of the Universe, of about 10^9 K, becomes low enough to allow the formation of light nuclei: this is the primordial nucleosynthesis. Nuclei of ^{1}H, ^{2}H, ^{3}H, ^{3}He, ^{4}He, ^{6}Li, ^{7}Li, ^{7}Be and ^{8}Be are produced. The primordial nucleosynthesis was probably widespread, leaving mass abundances of 76% of ^{1}H, 24% of ^{4}He, 0.01% of ^{2}H and traces of Li and Be.

At $t = 70\,000$ years, the energy densities of matter and photon equalize and the Universe enters an epoch dominated by matter (and we are still in this epoch). The Universe is composed of a very hot plasma of hydrogen and helium nuclei, electrons and photons (see Section 1.5.2 for a discussion on plasmas). Finally, at the age of 380 000 years occurs the Recombination: the temperature of the Universe is about 3000 K, low enough so that nuclei capture electrons and form neutral atoms. Since then, because of the neutralized electric charge, photons decouple from matter and they start to move freely: the Universe is said to become transparent to light. Before the Recombination, the Universe was opaque (therefore not visible) because photons were continously rescattered by electrons and were trapped in the plasma. After the Recombination, the Universe becomes visible and this residual radiation constitutes the famous cosmic background. As the Universe goes on expanding, the photons energy also decreases accordingly and their wavelength is now in the microwave range, corresponding to a temperature of 2.7 K: this is the so-called Cosmic Microwave Background, which thus provides a fingerprint of the structure of the Universe at the time of the Recombination.

8.1.2 Stellar evolution

The story of matter in the Universe does not end here. Actually, it took a very long time for stars to emerge in the Universe. Large clouds of atoms (essentially 76% hydrogen and 24% helium) appeared about 10^8 years after the Big Bang. It takes about 10^7 years for a gas cloud to condense into a protostar, and 10^8 years to form a star. The very first stars, called 'Population III', consist in H and He atoms only and are said to be 'metal-free'. They were probably extremely massive (between 10^2 and 10^3 solar masses), very hot and very bright, and their lifetime was less than 10^6 years. This is why, given the age of the Universe of 13.7×10^9 years old, no Population III star has ever been observed directly.

The gravitational contraction at the origin of the star causes an increase of its temperature T and its density, especially at the center of the star. When this increase is sufficient, hydrogen atoms are ionized and the core of the star becomes a plasma. Meanwhile, gravitational collapse proceeds further and can overcome the electrostatic repulsion between the positively charged hydrogen nuclei. When T reaches 10^7 K at the core of the star, as in the primordial nucleosynthesis, H nuclei can fuse to produce He nuclei: this is the thermonuclear fusion, which in fact involves complex sequences of nuclear reactions. Classical physics would give a much higher value for the onset temperature of thermonuclear fusion. However, thanks to quantum mechanics and especially the tunnel effect (see Sec 2.5.1), the required temperature is much lower. Let us also mention here, somewhat as a counterexample, the case of brown dwarfs which are protostars with a mass less than 0.08 solar mass ($M_\odot$). In this case, the gravitational contraction is not sufficient to ignite the nuclear fusion of hydrogen.

We shall thus now consider stars with a mass $> 0.08\ M_\odot$: the temperature in the core of such stars is then high enough to produce the thermonuclear fusion. This reaction is highly exothermic and heats the core of the star. Therefore, the created thermal pressure can counterbalance the gravitational contraction and stabilize the star in both thermal and mechanical equilibrium. This mechanism is in fact generic: gravitation provides the necessary confinement to provoke fusion of a given nuclear species, released energy via fusion provides the energy balance to maintain the star in equilibrium. The life of a star is actually an alternation of gravitational contractions and phases of nuclear fusions in its core. But mind that both phases can occur at the same time at different depths inside the star. The first fusion cycle lasts as long as hydrogen nuclei can fuse to produce helium and maintain the star in equilibrium. Meanwhile, a helium-rich core progressively grows. This constitutes the 'main sequence' of a star. Our Sun is approximately half way through its main sequence life (about 10^{10} years). The more massive the star, the shorter its main sequence. For instance, the red hypergiant star VY Canis Majoris which has a radius 1 500 larger than that of the Sun and a mass of more than 17 $M_\odot$, is about 10^6 years old and almost already at the end of its life.

The main sequence ends when all H fuel has been fused into He in the core of the star. With no more thermal pressure produced by thermonuclear reactions, a gravitational collapse of the core begins, bringing H nuclei from outer shells towards the core, where the temperature and the density are high enough to ignite new H fusion. This collapse greatly increases the reaction rates, producing a luminosity (total amount of energy emitted by a star, usually denoted L) $10^2 - 10^3$ the one of the Sun ($L_\odot$) and a huge ejection of matter within a radius $20 - 100$ that of the Sun: the star enters its red-giant phase. Because of the expansion of the star, the temperature at its surface decreases to 3000–4000 K, shifting its color towards the red, whence the name. For the Sun, the red-giant phase will last about 10^9 years. If the star has a mass between 0.4 and 2.25 $M_{\odot}$, gravitation is not sufficient to compress the He core and to reach the 10^8 K needed for helium fusion. Therefore, without a thermal pressure from He fusion, the core of the star finally becomes unstable because of the imbalance between the thermal (outwards) pressure from the H fusion in the shells surrounding the He core and an increasing gravitational (inwards) pressure due to the accretion of newly formed He nuclei. All in all, this causes the implosion of the core of the star. This collapse allows a large part of the He core to fuse and releases a huge amount of energy which is expelled outward in an explosive event on a very short time (a few days). The luminosity of the star is then 10^8 $L_{\odot}$ for a few days, and equals that of the whole Milky Way for a few seconds: this is the core helium flash. A red-giant phase does not necessarily end with such an explosive event. For stars with a mass lower than 0.4 $M_{\odot}$, their red-giant phase ends without any helium flash, leaving at the center an inert He core. On the other side, if the star possesses a mass larger than 2.25 $M_{\odot}$, the accumulated mass of He nuclei in the core of the star is sufficient to reach 10^8 K and to start He burning without the explosive helium flash.

If the mass of the He core is sufficient to ignite fusion into C nuclei, the star enters a second red-giant phase, called the asymptotic-giant-branch: He burns in the core and H in outer shells also fuses into He. These newly He nuclei keep on collapsing towards the core, producing periodically a huge increase of energy which is expelled outwards: these events are thermal pulses or shell helium flashes. Some stars can exhibit several hundreds of such pulses. If the mass of the C core is not large enough for the ignition of C fusion (i.e. more than 1.4 $M_{\odot}$), the star undergoes a final explosive event, forming a planetary nebula with an extremely hot C star. The latter one cools down and becomes a white dwarf (the inert He core at the end of the red-giant phase of very low-mass stars is also a white dwarf). The stability of such a celestial object, which is relying on an infinite gas of electrons, cannot be understood without the help of quantum mechanics, as we will discuss in Section 8.2.

For more massive stars, the collapse of the C core allows the ignition of thermonuclear reactions from C to O, O to Si and finally Si to Fe. The star therefore exhibits an onion-like structure with a Fe core surrounded by successive shells of Si, O, C, He, and H. The thermonuclear fusion ends with Fe because it is the nucleus with the highest binding energy (see Section 7.1.1). In other words, it is energetically unfavorable to fuse two Fe nuclei. When the Fe core reaches about 1.4 $M_{\odot}$, it becomes unstable under gravitation and the phenomenal created density induces electron capture by protons via radioactive processes, then producing neutrons. This final gravitational contraction of the Fe core produces a cataclysmic event called a supernova: the shock wave produced by the collapse of the core expels the material of the star in the interstellar medium in 1 millisecond at 10% the speed of light, leaving in the center a neutron star. As for a white dwarf, quantum mechanics is necessary to understand the stability of a neutron star, which can be viewed as a nucleus of astronomical size, as will be discussed in Section 8.3. A supernova can be as luminous as the Sun during its entire lifetime, but only for a few tens of days. Meanwhile, a huge amount of neutrons are released. And their collisions with the rebounding material allow the production of nuclei heavier than Fe. These elements present on Earth are therefore stardust. The last three observed supernovae have been in 1987, 1604 by J. Kepler and 1572 by Tycho Brahe. The search for an equation of state (EoS) of nuclear matter is thus of great interest not only at low density but also at the high densities that can occur in supernova explosion and in some massive stars.

Let us end this section by coming back to the end of Population III stars. They became supernovae and have enriched the interstellar medium with elements heavier than H and He. In astrophysics, these heavier elements are conventionally referred to as 'metals'. And a star with a significant proportion of C and O is said to be a 'metallic' star, although neither C nor O are true metals. These heavier elements, combined with H and He already present after the Big Bang, form new clouds of dust, precursors of Population II stars. As the latter stars die in turn, even more amount of metals are ejected in the interstellar medium. The further metal-enriched H and He clouds produce Population I star, including our Sun.

8.2 WHITE DWARFS

We focus here on low mass stars which end as white dwarfs. A white dwarf can be considered as an electron fluid that appears in Nature at stellar scale. Indeed, some basic properties of white dwarfs are nicely described in terms of a degenerate electron gas. White dwarfs are the final stage in the evolution of stars whose mass at birth is below about eight Solar masses, namely not high enough to become a neutron star or a black hole. As we shall see below, that does not mean that a white dwarf star is that massive, by far. But one has to keep in mind that during the former evolution of the star, a sizable fraction of its mass can be ejected, as discussed above. The nearest white dwarf is Sirius B, the smaller companion in the Sirius binary star, at 8.6 light years. Our Sun will become a white dwarf after passing through a red-giant stage, which will explode to form a planetary nebula surrounding a white dwarf. In a normal star, the gravitational contraction is balanced by the energy liberated in nuclear fusion reactions at the core of the star. In a white dwarf, with no nuclear reactions taking place, the gravitational collapse is counterbalanced by the quantum pressure due to the degenerate electron gas.

A typical white dwarf has a mass of order the Solar mass $M_\odot$ (with a maximum of $\simeq 1.4M_\odot$, as we shall see later on), and a size of order the Earth's radius. The average mass density is hence very large, of about 10^6 g/cm^3. It is composed by layers of atomic nuclei generated in the successive nuclear fusion stages (see Section 8.1.2) in the original star. Owing to its very high density, one can imagine such a star as a very huge solid, to which one can apply some of the general results discussed in the previous sections. This is how the electron fluid appears at a stellar scale.

Let us make some estimates to characterize the electron fluid. For simplicity, we can assume that there are only helium nuclei involved. The previously indicated average mass density means that the star contains about 10^{29} helium atoms per cm^3. In such an extremely dense medium, there is no room for bound electrons. Using the expressions given in Section 5.2.4, we obtain the values for the Fermi energy $\varepsilon_F \simeq 30$ keV, and the Fermi temperature $T_F \simeq 10^9$ K. Because the Fermi energy is orders of magnitude larger than the ionization potential of He atom, electrons are fully free. And since the associated Fermi temperature is also orders of magnitude larger than the typical temperature in the interior of the white dwarf, about 25 000 K, electrons can thus be considered as a fully degenerate Fermi gas (effective 'zero' temperature). The properties of such a degenerate gas in turn provide severe constraints on the stability and mass of the white dwarf. We will deduce two of these constraints in the simplest case of a spherical white dwarf of radius R and mass M.

8.2.1 The mass-radius relation

If we assume that the star is only made of helium atoms, we have $M = m_{\text{He}} N_{\text{He}}$, namely the product of the mass of the helium atom by the total number of helium atoms. We approximate $m_{\text{He}} \simeq 4m_n$ where m_n is the mass of a nucleon (either a proton or a neutron). As for the total number of electrons N, we have $N = 2N_{\text{He}}$. Finally the mass of the white dwarf reads as $M = 2m_n N = 2m_n V\varrho$, with $V = 4\pi R^3/3$ the volume of the star and ϱ the electron number density.

There is a non-trivial relation between the mass and the radius of a white dwarf which can be deduced by stating that no net pressure exists at equilibrium. We recall that at zero temperature the thermodynamical definition of pressure is given by the derivative $P = -\partial E/\partial V|_N$ of the total energy with respect to volume keeping constant the number of particles. To calculate it, we will use some simple approximations. Consider first the potential energy contribution. We have seen previously that the Coulomb ion-electron attraction is almost completely canceled by the ion-ion and electron-electron repulsions (see Section 5.2.1). For simplicity, we assume that the cancellation is total, and the only interaction we consider is the gravitational attraction. For a homogeneous sphere of radius R and mass M, the potential energy is $E_G = -3/5\, GM^2/R$, where G is the gravitational constant. The associated pressure is

$$P_G = -\left.\frac{\partial E_G}{\partial V}\right|_N = -\frac{3}{20\pi}\frac{GM^2}{R^4} = -\frac{G}{5}\left(\frac{4\pi}{3}\right)^{1/3}(4Mm_n)^{2/3}\,\varrho^{4/3} \quad . \quad (8.1)$$

We now consider the kinetic energy contribution. Let us start with the one of a fully degenerate electron gas (see Section 3.4.1). We remind readers that the Fermi momentum reads $k_F = (3\pi^2\varrho)^{1/3}$ and the Fermi pressure as $P_F = 2\varrho\varepsilon_F/5$. Putting things together, we have:

$$P_F = \frac{\hbar^2}{5m}\left(3\pi^2\right)^{2/3}\varrho^{5/3} = C_F\frac{M^{5/3}}{R^5} \quad \text{, with} \quad C_F = \frac{9}{32}\frac{\hbar^2}{5mm_n^{5/3}}\left(\frac{3}{\pi}\right)^{1/3} \tag{8.2}$$

with m the electron mass.

Let us now estimate the pressure from helium ions. The latters, because of their large mass relative to that of electrons, can be considered as classical points. The pressure they produce is then given by the ideal gas law, $P_{\rm He} = \varrho_{\rm He}k_{\rm B}T$, with $\varrho_{\rm He} = \varrho/2$. Using $T = 25\,000$ K and $\varrho_{\rm He} = 10^{29}$ cm^{-3}, we can estimate the ratio of the pressure of ions to that of electrons as:

$$\frac{P_{\rm He}}{P_F} \simeq 2.3\times 10^{-5} \quad .$$

The classical pressure from ions is thus completely negligible as compared to the quantal pressure of electrons.

We are now ready to compute the equilibrium radius from the condition $P_G + P_F = 0$, which immediately proves a relation between mass and radius, because the two pressure contributions have opposite signs and their dependence on star radius and mass different. This gives:

$$R_{\rm eq} = \frac{2C_F}{GM^{1/3}} \quad . \tag{8.3}$$

We see that the stability of a white dwarf against the gravitational collapse is provided by the quantum electron fluid which produces the quantal pressure. One can check the consistency with the previously mentioned orders of magnitude. Inserting $M = M_\odot$ in Eq. (8.3), we obtain an equilibrium radius $R_{\rm eq} \simeq 8\,000$ km, comparable to the Earth radius.

8.2.2 The maximum mass

The equilibrium condition Eq. (8.3) indicates that the larger the mass, the smaller the radius. This also means an increase of the density and consequently an increase of the electron Fermi momentum. One may therefore face cases where the Fermi energy becomes non negligible anymore with respect to the electron mass energy mc^2. Thus, the Fermi gas description should be extended to the relativistic case. The full relativistic treatment is a bit involved and for the sake of simplicity, we consider the case of the ultra-relativistic regime $\hbar k \gg mc$ which simplifies the expression of electron energies ($\varepsilon \sim \hbar ck$). The kinetic energy can then be calculated as in the non-relativistic case by replacing the term $\hbar^2k^2/2m$ entering the integrand of Eq. (3.26) with $\hbar ck$. The result is $E_{RF}/N = 3\hbar ck_F/4$. To obtain the pressure, we proceed along the same lines as previously in the non-relativistic case, with the result

$$P_{RF} = C_{RF}\frac{M^{4/3}}{R^4} = \frac{1}{4}\left(3\pi^2\right)^{1/3}\hbar c\varrho^{4/3} \quad , \tag{8.4}$$

$$C_{RF} = \frac{3}{64}\left(\frac{3}{\pi m_n^2}\right)^{2/3}\hbar c \quad . \tag{8.5}$$

We observe that, from Eqs. (8.1) and (8.4), the dependence on R is the same for both pressures P_G and P_{RF}, while it differs for M. Therefore, the equilibrium condition $P_G + P_{RF} = 0$ is independent of R and can be only satisfied for a unique critical mass M_{critical} given by

$$M_{\text{critical}} = \left(\frac{20\pi C_{RF}}{3G}\right)^{3/2} \simeq 1.73 M_{\odot} \quad . \tag{8.6}$$

For a slightly larger mass, the gravitational pressure dominates and the white dwarf collapses. For a slightly smaller mass, the electron pressure makes the star to expand, lowering the density and approaching the non-relativistic regime in parts of the interior (see Section 8.2.1), until equilibrium can be obtained at a finite radius.

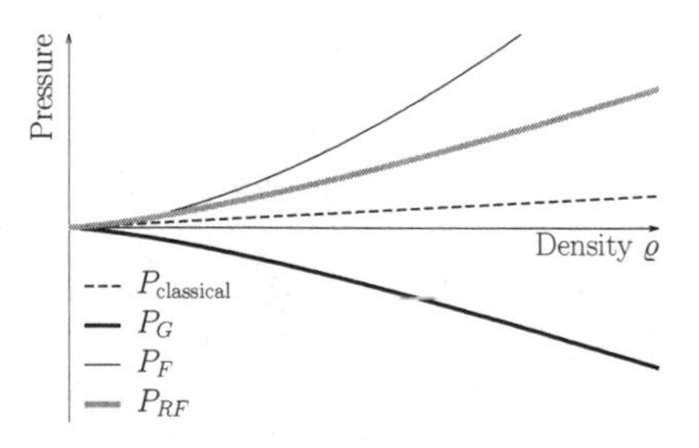

Figure 8.3: Schematic comparison of the electron density dependence of the classical pressure (dashes), the pressure created by gravitation (thick black line), that from a non-relativistic degenerate electron gas (thin black curve), and that from a relativistic degenerate electron gas (gray curve).

In a realistic calculation of the properties of a white dwarf, one must include the full Coulomb interaction, treat correctly relativistic effects, include temperature... The exact critical mass is known as the Chandrasekhar's mass $M_{\text{Ch}} \simeq 1.45 M_{\odot}$. Still, the previous estimate in Eq. (8.6) gives the correct order of magnitude. It is interesting to note that the critical mass does not depend on the electron mass m. Actually the Chandrasekhar mass is the maximum mass that a self-gravitating object may possess. Similar arguments can be immediately applied to a system of degenerate neutrons, leading to the prediction of a critical mass for neutron stars, which is of the same order of magnitude (see Section 8.3).

In spite of the crude approximations used to deduce this relation, the main physical conclusion agrees with more detailed calculations: the basic mechanism responsible for the stability of a white dwarf is related to the pressure originated by the quantum electron liquid. In terms of the density the different contributions to the pressure we have discussed go as follows: $P_G \propto -\varrho^{4/3}$, $P_F \propto \varrho^{5/3}$, and $P_{RF} \propto \varrho^{4/3}$. They are schematically displayed in Figure 8.3, where we have also added the pressure of a classical gas $P_{\text{classical}} \propto \varrho$, which is linear for a given temperature. As a conclusion, the case of white dwarfs constitutes a stellar example of an infinite quantum fermionic liquid in which the results on the degenerate Fermi gas presented in Section 3.4.1 fully make sense.

8.3 NEUTRON STARS

As briefly described in Section 8.1.2, massive stars (typically $> 8M_{\odot}$) develop an iron core at their very center. When the mass of this core reaches the Chandrasekhar's mass, of about $1.4M_{\odot}$, corresponding to a radius of 6,000 km, the electron gas around (mind that matter in a star is in the plasma state and that atoms are highly ionized) does not produce enough degeneracy pressure (as in a white dwarf) to counterweight the gravitational forces in the iron core and there is no new fusion cycle possible to stabilize the star. The core then becomes unstable and starts to shrink. The final collapse of the iron core suddenly stops when the density reaches the saturation density of nuclear matter ϱ_0 and the inner core bounces, creating a shock wave that expels matter and producing the explosion of a supernova. The (hydro)dynamics of this explosion constitutes a field on its own and strongly depends on the nuclear Equation of State (EoS).

8.3.1 Some macroscopic characteristics

The remnant of this fantastic explosion is a proto-neutron star. Indeed, when the density approaches 10^{34} cm^{-3}, electrons are captured by protons via the process $p+e^- \to n+\nu_e$ (inverse β decay). However, a free neutron is unstable and normally disintegrates (15 minutes lifetime in vacuum) by the inverse reaction, $n \to p + e^- + \bar{\nu}_e$. But in the massive star we consider here, the electrons constitute a degenerate gas. Creating a new electron would imply to produce it with a very high energy, and this is not energetically favored. In other words, a free neutron created by an electron capture is stable at very high density. In the collapse of the core of the star, there is thus a massive neutronization of matter. In a proto-neutron star, matter contains almost neutrons only, plus 0.5 % of protons and 0.5 % of electrons (and neutrinos). There are estimations of the minimal mass of a dynamically stable proto-neutron star, at about $0.9 - 1.2\ M_{\odot}$.

Inversely, the maximal mass is one of the parameters that are unknown but strongly depend on the nuclear EoS. Let us give here an estimate of it. General relativity tells us that matter in a star of mass M and radius given by the Schwarzschild radius defined as $R_S = 2GM/c^2$, with G the gravitational constant and c the speed of light, would be so compact that it would collapse into a black hole. Following the saturation property of nuclear matter (see Section 7.1.1) we can write $R = r_1 A^{1/3}$ with A the number of nucleons in the star and $r_1 = 0.5$ fm the distance below which the nuclear force becomes strongly repulsive. The mass of the star being $M = Am_n$ and m_n the mass of the neutron, equating R and R_S allows us to estimate the mass number $A = 3.4 \times 10^{57}$. Injecting this value back into $R = r_1 A^{1/3}$ and $M = Am_n$ gives $R \simeq 7 - 8$ km and $M = 2.5 M_\odot$. It is commonly admitted that the typical radius of a neutron star is about 10 km, and its maximal mass less than $3M_\odot$. The number density is equal to $\varrho = 4/(3\pi r_1^3) = 3.3 \times 10^{39}$ cm$^{-3} \simeq 2\varrho_0$. This value makes this kind of stellar object the most dense in the Universe, after black holes.

In the final collapse of the iron core, temperature can reach tremendous values, up to 10^{11} K ($\sim$ 10 MeV). Then the temperature drops by several orders of magnitude during the following seconds, essentially through the emission of numerous neutrinos produced in the collapse. There are several scenarios of the cooling of a neutron star. Detailing them will bring us too far away from our purpose. Let us just mention that during the first million years of existence of a neutron star, the temperature in the center is about 0.06 MeV (10^9 K) and the cooling is ensured once again by emission of neutrinos, and afterwards, photons take over. Despite this enormous value of the temperature, matter in neutron stars is considered as 'cold' or at least, at very low temperatures. Indeed, regarding the phase diagram of nuclear matter in Figure 8.1, the critical temperature for quark deconfinement is estimated around 170 MeV (see Section 8.4), namely 4 orders of magnitude higher than the temperature at the center of the neutron star. In that respect, studying neutron stars allows us to explore the phase diagram at $T = 0$ and $\varrho > \varrho_0$ (one often talks about 'supranuclear' densities).

Regarding the stability of a neutron star against gravitational collapse, we can do the same reasoning as we did for a dwarf star about the equilibrium between gravitational and quantum pressures, assuming for the latter a non-relativistic Fermi gas of neutrons. Thus, replacing the electron mass with the neutron mass in Eq. (8.3) we obtain the radius-mass relation for a neutron star:

$$R_{\text{eq}} = \left(\frac{9\pi}{4}\right)^{2/3} \frac{\hbar^2}{m_n^{8/3}} \frac{1}{GM^{1/3}} \quad . \tag{8.7}$$

Taking $R_{\text{eq}} = 10$ km in the expression above gives $M = 3M_\odot$.

The calculation above, even enlightning, is however very crude for two reasons. First, we have considered that neutrons in a neutron star constitute a free fermionic gas. But at supranuclear densities, interactions do play a role and make the nuclear matter being a liquid rather than a gas. Henceforth, the simple EoS linking the quantum pressure P and the number density ϱ, written as $P = K\varrho^{4/3}$, is not sufficient here. Instead, microscopic EoS including nuclear interactions, relativistic corrections... are required. Note that experiments on finite nuclei usually explore almost symmetric nuclear matter, with $N \leq 1.5Z$, while here, matter is dominated by neutrons.

8.3.2 The mass-radius relationship

The nuclear EoS influences two important data on neutron stars, namely the maximal mass $M_{\max}$ and the radius of $1.4M_{\odot}$ neutron stars (that is of intermediate mass, also called 'canonical' mass). These two properties are however not known yet. We have already mentioned that $M_{\max} \leq 3M_{\odot}$ as a constraint from general relativity. Moreover, $M_{\max}$ is related to the stiffness of the nuclear EoS: a soft EoS gives $M_{\max} \simeq 1.5M_{\odot}$. As for the radius, it is controlled by the symmetry energy which itself depends on density (see Section 7.1.2).

From an experimental point of view, the observation of a neutron star represents a real challenge: as it is a very compact object with a very low luminosity, this makes its direct observation impossible. For the determination of the mass, the most accurate measurements are obtained in binary systems, from gravitational redshift in spectroscopic measurements. However different spectroscopy analyses can give different masses, illustrating the difficulties of such studies. The situation is even worse for the determination of the radius. There are several possible methods. One of them relies on the fact that a neutron star is, at zeroth order approximation, a black body (see Section 2.1.4). Therefore, by measuring the temperature and the luminous flux of a neutron star, one can deduce the so-called 'radiation radius' R_{∞} which is related to the radius R of the star by $R_{\infty} = R/\sqrt{1 - 2GM/(c^2R)}$. However there are many uncertainties in this method, related to the measured temperature, the distance to the star, etc., and in particular the fact that a neutron star is actually not a perfect black body. There are moreover very few simultaneous precise measurements of the radius *and* the mass for a given neutron star.

Let us nevertheless make a comment on Eq. (8.7), deduced from Newtonian gravity. Because of the very high densities involved here, general relativity cannot be neglected: gravity at the surface of a compact star is 10^{10} that at the surface of our Sun! All in all, the mass-radius relationship in a neutron star is not unique and can exhibit very different behaviors, depending on the considered EoS, but also on the composition of matter in the neutron star. Indeed, we have called the remnant of a supernova explosion a 'proto-neutron' star and not a 'neutron' star. This was done on purpose. One should actually use the term 'compact' star instead of 'neutron' star because the exact composition and structure of such a star is still under debate. One can consider it as composed of neutrons only (neutron stars), of neutrons in an outer shell and a quark matter core (called exotic or hybrid stars), or of exclusively (strange) quark matter (called quark stars). One can even imagine, when going deeper and deeper into the star, that neutrons are in a superfluid state, as ^{3}He atoms at very low temperatures (see Section 9.2.4) and that protons form Cooper pairs, similarly to what happens to electrons in a superconductor (see Section 9.1.3 for more details on the mechanism of Cooper pairing). The existence of these kinds of exotic phases is however still highly speculative.

Let us come back to the possible relationships between the mass and the radius of a compact star. This is illustrated in Figure 8.4 which compares several EoS with two celestial objects, which are admitted to be compact stars, that is the PSR J0348+0432 star of mass $(2.01 \pm 0.04)M_{\odot}$ and the EXO 0748-676 star of mass $(2.10 \pm 0.28)M_{\odot}$ and radius 13.8 ± 1.8 km. The uncertainty

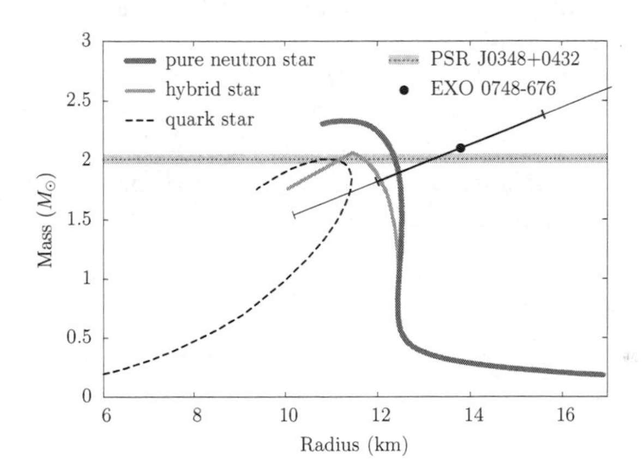

Figure 8.4: Comparison of different equations of state for a compact star: pure neutron matter or hybrid star with a core of u and d quark matter and a mantle of nuclear matter (dark gray and light curves respectively, adapted from T. Klähn, et al., Phys. Lett. B 654 (2007) 170, with permission from Elsevier), and a s quark star (dashes, adapted with permission from P. Mukhopadhyay and J. Schaffner-Bielich, Phys. Rev. D 93 (2016) 083009; copyrighted by the American Physical Society). The black circle stands for the EXO 0748-676 star with 1σ or 2σ error bars as indicated (reprinted by permission from Macmillan Publishers Ltd: F. Özel, Nature vol. 441, p. 1115, copyright (2006)), and the horizontal gray line for the PSR J0348+0432 (from J. Antoniadis, et al., Science 340 (2013) 1233232; reprinted with permission from AAAS).

in the mass and the radius of the observed compact stars cannot discriminate a neutron star from a hybrid star or even a quark star. This emphasizes the interplay between nuclear astrophysics and astrophysical observations: we need observation data of macroscopic quantities to confirm (or infirm) our theoretical modeling of (microscopic) nuclear (quark) matter.

8.4 THE QUARK-GLUON PLASMA

In the previous section, we have briefly discussed objects which lie at high ϱ and at low T in the nuclear phase diagram (see Figure 8.1). We now go to high ϱ and T in this diagram. We know that neutrons and protons are composite particles containing three quarks, which are strongly linked together by the nuclear strong interaction. Quantum Chromodynamics (QCD) provides the theoretical framework to describe strong interaction: quarks interact via the exchange of particles which are called gluons, in much the same way as Quantum Electrodynamics (QED) describes the interaction between charged particles via the exchange of photons. However, there exist only two kinds of electric charges (positive or negative) and one type of photon. In QCD, another type of charge is invoked, the color charge (whence the term 'chromodynamics'). The complication stems from the fact that there exist three different color charges and eight different gluons. Also, at variance with QED, the more the quarks are separated, the higher the number of gluons exchanged between the quarks, and thus the stronger the interaction between them. At our (human) level, quarks cannot be isolated, as they are bound in hadrons. This property is called confinement. Inversely, the strong interaction becomes very weak at very high temperatures, and then exhibits the property of asymptotic freedom. Soon after the discovery of this property, the Quark-Gluon Plasma (QGP) was inferred above a critical temperature T_c^{QGP}, setting the deconfinement phase transition (see the full circle in Figure 8.1). The term 'plasma' draws the parallel with usual plasmas in which electrons are separated from the atomic nucleus at sufficiently high temperatures (see Section 1.5.2). The critical temperature is estimated to be around 160 MeV (or 10^{15} K). The related critical density is estimated at $\varrho_c^{\mathrm{QGP}} \simeq 5\varrho_0$. Above T_c^{QGP}, quarks and gluons, now weakly bound, can move in a volume much larger than that of a nucleon. Note however that the very existence of a critical point between the hadron gas and the QGP is still an open question.

8.4.1 Relativistic heavy ion collisions

As mentioned in Section 8.1.1, it is believed that between 10^{-12} s and 10^{-6} s after the Big Bang, the Universe was filled by a QGP and that a phase transition between the QGP and the hadron gas occured at very low ϱ and 1μs after the Big Bang. In principle, this phase transition should also exist for higher values of ϱ (see thin full curve in Figure 8.1).

The study of the QGP on Earth and the QGP-hadron transition is achieved by colliding nuclei at very high energies. More precisely, one considers nuclei with a large mass number A, called heavy ions, namely atoms from which a large number of electrons have been stripped off to allow easy electromagnetic acceleration. Increasing A allows a huge number of collisions between nucleons of the heavy ions, and therefore a higher probability to create a QGP. Similarly, the larger the collision energy, the higher the probability to exceed T_c^{QGP}, even during a very short time and very locally in space. One finally talks about (ultra)relativistic heavy ion collisions (RHIC). Figure 8.5 schematically depicts the geometry of a RHIC. On the left, the collision is seen at a given instant in the longitudinal xz plane, with z the collision axis. The nuclei appear squeezed because of the Lorentz contraction at these relativistic energies. The impact parameter b is the spatial transverse separation between the centers of the two nuclei. On the right, one sees the collision in the transverse xy plane, with the overlap region called the fireball. This region

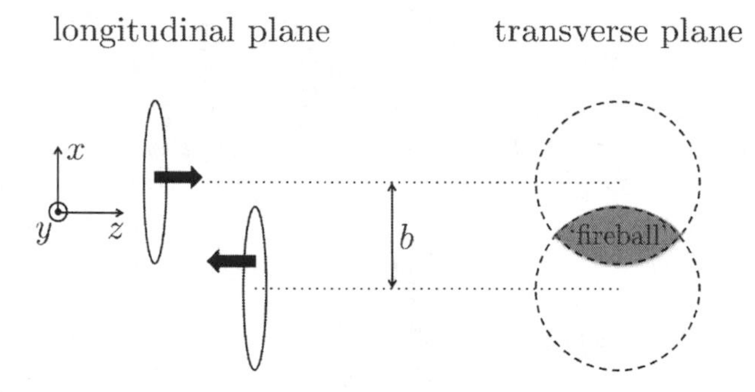

Figure 8.5: Schematic view of two nuclei colliding at relativistic energies at a given impact parameter b, in the longitudinal xz plane (left) and in the transverse xy plane (right) where the overlap region (the 'fireball') appears as a gray shaded area.

has initially an almond shape and is the location for the creation of the QGP. Experimentally, at a given beam energy, one can play with b to control the energy deposited in the collision, and consequently the number of produced particles in the fireball. Indeed, the smaller b, the more central the collision, and the larger the fireball.

The first RHICs ever done appeared in the mid-1980s. The latest experiments are conducted at the Relativistic Heavy Ion Collider in Brookhaven (NY) with Au-Au collisions ($A = 197$), and at the Large Hadron Collider (LHC) at CERN in Geneva with Pb-Pb collisions ($A = 208$). Tremendous physical conditions are reached in these experiments. For instance, at LHC, one expects the following characteristics of the fireball: a temperature as high as 300-400 MeV ($> 10^{12}$ K), a lifetime of 10 fm$/c$ = 3×10^{-23} s, a volume of 4800 fm^3 (to be compared to the volume of a single Pb nucleus $4\pi r_0^3 A/3 \simeq 1290$ fm^3), a density of $100\varrho_0 \simeq 20\varrho_c^{\text{QGP}}$, and an energy density of 15–30 GeV/fm^3. Note that at LHC, the lifetime of the fireball, where the QGP is expected to form (in less than 0.1 fm$/c$), is much shorter than the 1 μs lifetime of the QGP after the Big Bang. Indeed, the matter created in a RHIC possesses much less energy and is therefore much shorter-lived than the one after the Big Bang.

With the energies involved at LHC, there are 30,000 particles created per collision. Such a high number of produced particles allows in principle a statistical treatment (here in the grand-canonical ensemble), if equilibrium, either chemical or thermal, is achieved. The current very precise measurements of hadron yields allow one to verify such an assumption and to extract thermodynamical variables. The thermalized QGP very quickly cools down and when its temperature decreases below the critical temperature of the QGP-hadron phase transition, quarks and gluons reconfine into hadrons. At this moment, the produced particle yields are approximately frozen. This is why this instant is called the freeze-out time ($\simeq 10$ fm$/c$), with a corresponding freeze-out temperature believed to be to the phase transition critical temperature. At LHC, it has been estimated around 164 – 170 MeV.

To increase the significance of the measurements, there are 4,000 Pb-Pb collisions per second. However, the sensitivity of the detectors is of the order of a microsecond, allowing an event-by-event analysis. Still, among the 30,000 created particles per event, how can one be sure that a QGP has been created? We need to identify unambiguous signatures or probes of such a phase.

8.4.2 Production of heavy quarks

The temperatures/energies achieved during a RHIC are sufficiently high to create quarks (and also antiquarks) heavier than those present in normal matter, that is the up and down quarks (think of the famous Einstein relation $E = mc^2$). We are interested in particular in the production of charm (c) and anticharm ($\bar{c}$) quarks (mass $m_c = 1.37$ GeV/c^2). Since in normal nuclear matter, there is no charm quark, in a RHIC, a c quark must be created with a $\bar{c}$. There exist bound states of $c\bar{c}$, called charmonium states, and one of these particles, the so-called J/ψ, is of particular importance. One can estimate the time of formation of $c\bar{c}$ pairs and that of a J/ψ at $\tau_{c\bar{c}} = 0.08$ fm/c and $\tau_{J/\psi} \simeq 1$ fm/c respectively. The produced J/ψ's can collide with the nuclear matter and be absorbed, as any particle travelling through a medium. Such a suppression of the J/ψ is coined 'normal'. An additional suppression has been observed in the first RHICs at the end of the 1990s. This phenomenon was coined 'anomalous', in contrast to the normal suppression. A possible explanation relies on the travel of the J/ψ through a deconfined QGP. Indeed, in such a medium, the potential that binds a c quark to a $\bar{c}$ quark is screened. The free c and $\bar{c}$ can then recombine with other quarks (although the probability that they recombine is very small): the net effect is a melting of the J/ψ. This mechanism for anomalous suppression was predicted in the mid-1980s. However, the first measurements of the anomalous suppression in the 1990s could be also reproduced by theoretical models which did not rely on any presence of a QGP. This thus questioned the reliability of such a signature of the QGP.

Moving to LHC's energies, one can expect to observe a stronger anomalous J/ψ suppression due to the possible creation of a QGP. However, this is by far not the case. Whereas in previous RHIC at lower collision energies the number of produced $c\bar{c}$ pairs per collision was < 1, at LHC this number is estimated at $\simeq 200$, thanks to the high energy density achieved there. In this case, the number of J/ψ is proportional to the total number of possible $c\bar{c}$ pairs and this would result in a huge statistical (re)generation of J/ψ yield. And indeed, the first measurements of the J/ψ yield at LHC reported a strong J/ψ enhancement for the most energetic collisions (that is at small impact parameters). Even if it is now well accepted that this enhancement can constitute a clear signature of the creation of a QGP at LHC, at the side of the theory, the statistical hadronization models and the transport models that had predicted such an enhancement are for the moment not able to reproduce all these data.

New measurements are scheduled during the second run of LHC (2017–2020) and might bring more constraints on the theoretical models.

Note finally that, thanks to the high energy density achieved at LHC, the production of bottomonia, that is bound $b\bar{b}$ states (b is the 'beauty' quark), and the Υ meson in particular, is expected to occur. A suppression pattern of Υ, analogous to that of J/ψ but qualitatively different, has been reported in 2012. The study of this suppression, either experimentally or theoretically, is also another observable to probe the nuclear matter created at LHC.

8.4.3 Signatures of hydrodynamics

The QGP expected to be formed in a RHIC has been for a long time considered as an ideal gas, following the property of asymptotic freedom. A hydrodynamical description is thus well justified here. Indeed, the mean free path λ in the fireball is estimated to be less than 1 fm. This is much smaller than the size of the QGP (radius of $\sim$ 15 fm at LHC, if the QGP would fulfill a sphere), then justifying a description of the QGP as a fluid. Moreover, the density being about $100\varrho_0$ at LHC, the typical distance between constituents is ~ 0.25 fm, of the same order as λ: the QGP is therefore a *liquid* and not a gas. Analyses from recent measurements at LHC even infer that the created QGP behaves as a nearly ideal liquid with a mean free path almost vanishing.

To claim such a statement, we have to discuss an observable which precisely describes a collective behavior of the fireball (see gray almond-shape area in the left panel of Figure 8.6). To that end, we focus on the distribution of momentum in the transverse plane of the beam axis. This momentum is usually denoted by p_{T}. It is believed that, after about 1 fm/c, thermal equilibrium is reached, still keeping a very asymmetrical spatial distribution. Then the fireball evolves, following the laws of relativistic hydrodynamics and the equation of state (which relates volume, temperature and chemical potential). In particular, the gradient of pressure along the small axis of this almond, the x direction in Figure 8.6, is much larger than along the y direction (mind that, outside the fireball, is vacuum). This gradient of pressure pushes the produced particles with a velocity possessing an x component much larger than the y component. This finally produces an asymmetry in the momentum distribution in the transverse plane (schematically indicated by the size of the arrows in the figure).

The elliptic flow, commonly denoted by v_2, is related to the number of particles emitted along the small axis of the almond over the number of particles emitted along the long axis. It quantifies the capability of the QGP to transfer the initial spatial asymmetry into the final momentum asymmetry, and is a trace of collective effects that occur in the fireball. The right panel of Figure 8.6 shows measurements of v_2 as a function of the transverse momentum p_{T} for the lightest particles created after hadronization of the QGP at the LHC, namely pions, protons (p) and antiprotons ($\bar{\mathrm{p}}$). These data are compared to calculations from a numerical code that combines hydrodynam-

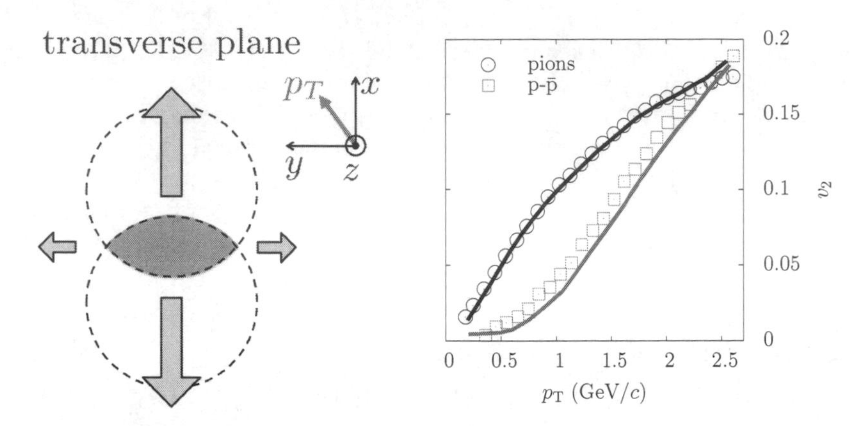

Figure 8.6: Left: relativistic heavy ion collision seen in the transverse plane xy with the gray arrows representing major directions of emission of the particles created in the fireball. Right: elliptic flow v_2 as a function of p_T (momentum in the transverse plane, see Figure 8.5) measured at LHC, for pions (circles) and proton-antiproton pairs (squares), compared to hydrodynamics and transport calculations (curves). Adapted from The ALICE collaboration, J. High Energy Phys. 06 (2015) 190, under the terms of the Creative Commons Attribution License (CC-BY 4.0).

ics at short times (starting at 0.9 fm/c after the collision of the two heavy ions) and transport equations at longer times. The remarkable agreement of the measurements at LHC (but already at the Relativistic Heavy Ion Collider facility at Brookhaven, NY, with lower collision energies) with hydrodynamical calculations constitutes a clear signature of the creation of the QGP in RHICs.

We now address the notion of 'ideal liquid'. We have seen in Chapter 1 that a fluid flow can be quantified by the Reynolds number Re (see Eq. (1.11)), which compares a typical characteristic velocity v and a typical length scale L to the kinematic viscosity $\eta/(m\varrho)$, m the mass of the constituents and η the shear velocity. The larger Re, the 'better' the fluid in the sense that it justifies the use of hydrodynamics equations which correspond to the first order of an expansion in terms of 1/Re. In a relativistic fluid, one replaces the mass density $m\varrho$ by the enthalpy density, that is sT with s the entropy density. As for L, we can take the size of the fireball created in a RHIC. All in all, we have Re $\propto s/\eta$. Therefore, the smaller η/s, the more 'ideal' the fluid.

One can estimate a lower bound for η/s. Indeed, we have $\eta \sim m\varrho v\lambda = \varrho p\lambda$. In a RHIC, we can consider that the velocity of a produced particle is almost the light speed c. If $\tau_{\rm coll}$ is the mean free path time, we have $\lambda \simeq \tau_{\rm coll}c$. Since the energy of a (ultra)relativistic particle is $E = pc$, we can write $\eta \sim \varepsilon\tau_{\rm coll}$ with ε the energy density. On the other hand, the entropy density is $s \sim k_{\rm B}\varrho$, giving $\eta/s \sim \varepsilon\tau_{\rm coll}/(k_{\rm B}\varrho)$. Finally, ε/ϱ being the energy per particle, the Heisenberg inequality tells us that $\varepsilon/\varrho \times \tau_{\rm coll} \geq \hbar$. In other words, $\eta/s \geq \hbar/k_{\rm B}$.

A more precise calculation gives:

$$\eta/s \geq \alpha \frac{1}{4\pi} \frac{\hbar}{k_B} \simeq 0.0796\, \alpha \frac{\hbar}{k_B} \quad , \tag{8.8}$$

α being a constant of order 1. There are various theoretical estimations of α: one can find $\alpha = 1$ or $16/25 = 0.64$. There is no consensus on that lower bound for the time being.

What about the QGP? In the hydrodynamic calculations used to fit the measurements of the elliptic flow in the right panel of Figure 8.6, the value $\eta/s = 0.16\ \hbar/k_B$ has been used. As it should be, this value is greater than the minimal value of η/s given in Eq. (8.8), even with $\alpha = 1$. But it turns out that it is the smallest value ever measured. Therefore, it seems that the QGP constitutes the best quantum fluid ever produced on Earth. We will come back to that point in the conclusion of the book by comparing this value of η/s with those in the other liquids considered in this book.

8.5 CONCLUDING REMARKS

The natural place to switch from finite to infinite quantum liquids is, maybe surprisingly, in hugely macroscopic systems, namely stars. Indeed some specific stars offer thermodynamical conditions such that one can observe true 'infinite' phases of quasi degenerate quantum liquid phases. White dwarfs are stable thanks to the pressure of its electrons which, because of the density, behave as a close to perfectly degenerate Fermi gas. It is in that case the Fermi pressure, namely the pressure due to Fermi motion (or Pauli exclusion principle) which allows a stabilization of the star which would otherwise collapse. Neutrons, in turn, constitute a dominant species in neutron stars which correspond to the latest stage of evolution of very massive stars. They involve huge densities, even in terms of nuclear matter density. The detailed understanding of the structure and stability of neutron stars, though, remains debated.

We have concluded this chapter by exploring another aspect of very dense (and hot) matter, as attained in the very early instants of the Universe and/or as (re)created in the course of very energetic heavy-ion collisions. We reach here the limits of the quantum liquids we have focused on in this book as we have attained situations in which quarks may be deconfined. This implies a change in the nature of the particles under study (from nucleons to quarks) and this also means entering a domain in which interactions become extremely large. The physics of the thus formed quark gluon plasma is correlatively extremely complex and rich. It furthermore involves specific difficulties linked to the highly dynamical context in which such states of matter can be observed only indirectly. But such explorations have, in turn, brought us back to hydrodynamical concepts, typical of the motion of the liquids we are familiar with.

CHAPTER 9

Superphases

We have seen in Chapter 3, that as temperature is increased quantum statistics, both Bose-Einstein or Fermi-Dirac, tend to classical. High temperatures blur out quantum effects in a system, which can be safely described in classical terms. In this chapter, we will consider some systems which manifest quantum properties at macroscopic level, but at very low temperatures. We recall that helium is the only substance which remains liquid at low temperature, down to absolute zero (see Chapter 6), a phenomenon related to the quantum behavior of its constituent atoms. Liquid helium was instrumental to the discovery that some metals are able to sustain an electrical current for long periods of time without any driving force. This phenomenon is named superconductivity, and is also a macroscopic manifestation of the quantum behavior of the electrons in metals. Later on it was discovered that liquid helium itself shows a similar property, named superfluidity, consisting in its ability to flow through a capillary without viscosity. Hence, a persistent flow of liquid can be maintained without any driving force. While the quantum agents of superconductivity are electrons, which are spin-1/2 fermions, those of superfluidity are ^{4}He atoms, which are spin-0 bosons. There are other systems able to display similar behavior as liquid ^{3}He, atomic nuclei and nuclear matter. In fact, superfluidity and superconductivity are new phases of matter related to the bosonic character of ^{4}He atoms or of pairs of fermions (either electrons, ^{3}He atoms or nucleons), the same as Bose-Einstein condensation in ultracold atomic gases. The temperatures where these new phases manifest range from 165 K for superconductivity down to a few nano-Kelvin for Bose-Einstein condensation.

In this chapter we will start from the discovery of superconductivity and several manifestations of this phenomenon. BCS is the name of the microscopic theory describing superconductivity, in which an essential ingredient is the pairing of electrons to form a bosonic entity. We then discuss in some detail superfluidity manifestations in liquid ^{4}He, and relate superfluidity in liquid ^{3}He with BCS. We will also see that superfluidity is not exclusively manifested in bulk systems, but also appears in quantum droplets. In this respect, we have seen that pairing is a basic property of atomic nuclei. Is this pairing related to some superfluidity/superconductivity of nucleon systems? Finally, we will address some questions regarding both types of phenomena, BEC and BCS.

9.1 SUPERCONDUCTIVITY

The electrical conductivity describes how well a material allows valence electrons to travel through it. The inverse quantity is the electrical resistivity, which for most materials at room temperature is of order 10^{-8} Ω m, and shows a rough linear dependence with T. By lowering down temperature, one should expect the electric resistance to decrease. What happens at the absolute zero of temperature? At the beginning of the 20th century, several possibilities were discussed. One of them predicted an infinite resistance at some low temperature, due to the absence of any motion at absolute zero. Obviously, quantum mechanics was not developed at that time and the zero-point motion was ignored. The alternative was a steady drop to zero at $T = 0$, or to a small constant value reached at some low temperature. The former was expected for a perfect lattice, and the second for a real metal with defects and impurities.

Experimental research on this topic led eventually to the discovery of superconductivity, one manifestation of which is the absence of electrical resistivity. We do not aim at being exhaustive here and rather focus on some specific features of this very specific electron fluid.

9.1.1 Physics at low temperatures

Kammerlingh Onnes succeeded in liquefying helium (see Section 6.1) at a few K, thus reaching the lowest temperatures ever obtained. In 1911, he decided to give an experimental answer to the issue regarding the electrical resistivity of metals at very low temperatures. Measurements with gold showed a constant value of its electric resistance below $\sim$ 10 K. Then a sample of mercury was employed to get rid of impurities. This metal is liquid at room temperatures and can be repeatedly distilled to make it as pure as possible. The resistance of mercury dropped continuously with T. But a sudden drop was observed from about 0.1 Ω at 4.3 K down to 3×10^{-6} Ω at 4.2 K. The results are displayed in Figure 9.1. Mercury below 4.2 K was said to be a *superconductor* since electric current seemed to flow without any resistance. Subsequently, it was observed that it was not necessary to purify the metal: superconductivity is an effect intrinsic to mercury. The sudden change from finite to zero electrical resistance at the superconducting critical temperature T_c represents a thermodynamic phase transition, from a 'normal' state to a 'superconducting' one. It is a second order phase transition, since there is no latent heat during the transition (see Section 1.1.3). However, the transition is of first order in the presence of a magnetic field.

The superconducting phase transition is not a rare phenomenon at all and occurs for many chemical elements. Note however that some excellent conduc-

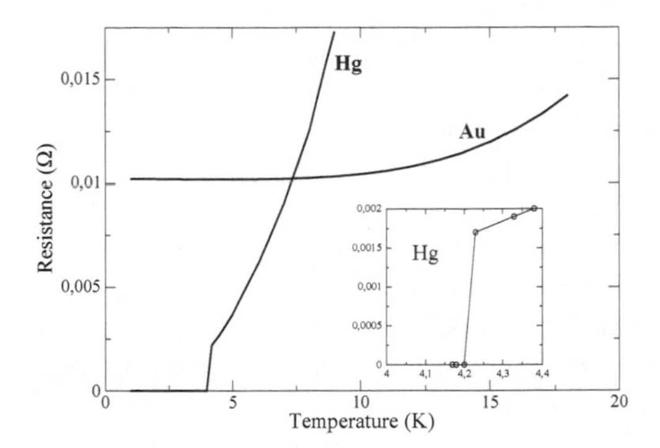

Figure 9.1: Electrical resistance at low temperatures of Au and Hg. The insert displays the Hg around the critical temperature.

tors of electricity, as Cu, Ag and Au, are not superconductors. Some elements become superconductors at atmospheric pressure, while others require high pressure to exhibit such a behavior. There are a few exceptions for which superconductivity does not manifest itself in bulk but in very specific geometries: C in nanotubes, Cr in thin films, Pd thin films with a special type of defects created by irradiation of α particles, and Pt in compact powders. In general, critical temperatures are very low, from 3×10^{-4} K for Rh up to 9.3 K for Nb. The presence of paramagnetic impurities can also lower the critical temperature. A paramagnetic material is made of atoms or molecules with spin unpaired electrons and these electrons create a permanent magnetic moment, allowing the material to be attracted by an external magnetic field if applied.

Alloys can also become superconductors, even if the constituent metals are not. That is the case of gold with bismuth in 4% proportion. There are compounds, oxide based materials, ceramics, organic materials... which can become superconductors. All in all, there are nowadays several hundreds of known superconducting materials. Before 1986, the highest known critical temperature was 23 K, in the case of the Nb_3Ge compound. That year, Bednorz and Müller discovered that ceramics made of barium, lanthanum, copper and oxygen become superconductor at $T_c = 30$ K. The same year, it was found that replacing lanthanum by yttrium, the critical temperature raises to $T_c = 90$ K. This constituted a key result because it is higher than the liquefaction temperature of nitrogen (77 K). The present record is held by a thallium-doped mercury-cuprate, with $T_c = 138$ K at atmospheric pressure, and 166 K at 23 GPa. It is conventionally written as $Hg_{0.8}Tl_{0.2}Ba_2Ca_2Cu_3O_{8.33}$, where fractional subindices indicate the average occupation of atoms in some given site of the lattice, that is, Hg and Tl share one of these sites, and the site of O is sometimes vacant. To date, no superconductor with critical temperatures above room temperature has been found. We should also mention that the mechanical properties of ceramics are very different from those of metals, rendering the fabrication of wires made of ceramics much more complex than with copper, which is an excellent electrical conductor... at room temperature.

In the superconducting state, the resistivity is exactly zero. In practice, because of experimental limits, it is not an easy task to measure a quantity to be equal to zero. As already stated above, the resistance in a circuit has an impact on the current which circulates in the circuit: the higher the resistance, the faster an electrical current is damped. Experimentally, to measure a tiny resistivity of a material, one considers a close loop made of a wire of the studied material. For simplicity, let us take a circular ring. We explore the fact that applying a magnetic field which changes in time creates an induced current in the ring. Inversely, a current which circulates inside the ring creates a magnetic field parallel to the axis of the ring. We come back to the measurement of the resistivity of a loop. In the normal state of the material ($T > T_c$), we apply an external constant magnetic field perpendicular to the plane of the ring. This creates a certain amount of magnetic flux, which roughly quantifies how much magnetic field flows through the ring. In a second step, we cool down the ring below T_c. Finally, we suddenly switch off the external magnetic field, thus creating, during a very short period of time, a time-dependent magnetic field which, in turn, induces an electrical current inside the ring. The amount of induced current is determined by the conservation of the initial magnetic flux. This induced current creates its own magnetic field which can be measured. Now, even a tiny resistance in the ring implies a gradual dropping in the electrical current (therefore in the induced magnetic field), and one can measure it merely by waiting long enough. In the case of a normal metal, the current dies out in a small fraction of second. In a superconductor instead, there were experiments in which no change of induced magnetic field has been observed during... two years! From these experiments, an upper value of the resistivity of a superconducting loop has been estimated to 10^{-26} Ω.m, that is 18 orders of magnitude smaller than that of copper at room temperature. Such currents created in a superconductor loop are termed *persistent* currents. They are one of the characteristics of a superconductor. In Section 9.2, we shall see that persistent currents also exist in connection with the phenomenon of superfluidity in liquid helium.

We end this short review with the specific heat of a superconductor. We have seen in Section 5.2.3 that at very low temperatures the heat capacity of a (normal) metal is linear in T. That of a superconductor, denoted by $c_{\rm sc}$, behaves very differently. First, at $T = T_c$, one can show analytically that there is a finite difference between the normal $c_{\rm n}$ and the superconducting heat capacities: $(c_{\rm sc} - c_{\rm n})|_{T=T_c} \propto T_c$. Second, for $T < T_c$, the heat capacity exhibits an exponential behavior $c_{\rm sc} \propto e^{-E_g/(2k_{\rm B}T)}$. Such a temperature dependence is rather characteristic of an insulator or a semi-conductor, and not a metal, with an energy gap above the Fermi surface equal to E_g. Measurements of the heat capacity at very low temperatures on various elements (e.g.

aluminium, gallium and vanadium) indeed show a discontinuity at $T = T_c$ and the existence of an energy gap in the superconducting phase, and thus, a very specific behavior of electrons in this phase. Some elements of explanation for the existence of such a gap will be given in Section 9.1.3.

9.1.2 Magnetic properties of superconductors

A perfect conductor should have a zero electrical resistivity. Is a superconductor identical to a perfect conductor? The answer is no: superconductivity is indeed associated to a phase transition. Actually, many specific features of a superconductor appear when an external magnetic field is applied. We will here describe briefly some of these properties.

The Meissner effect

To better identify the phase transition mentioned above, let us consider the field lines of a magnetic field $\mathbf{H}_a$ applied on a sample. This is illustrated in Figure 9.2. In the absence of $\mathbf{H}_a$, one cannot actually differentiate a material

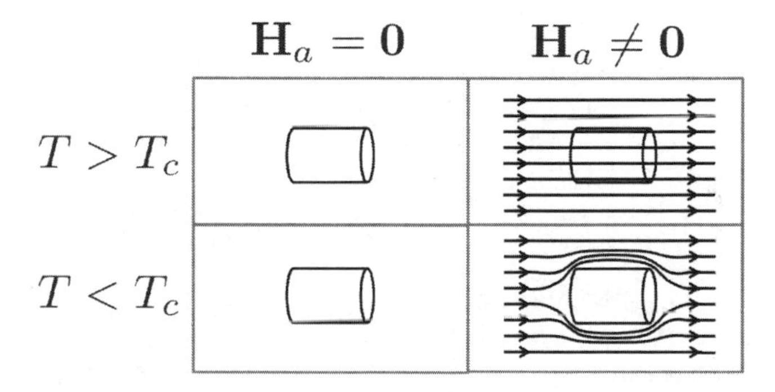

Figure 9.2: Cylindrical sample of a material in the normal phase (top row, $T > T_c$) or in the superconducting phase (bottow row, $T < T_c$), without (left column) or with (right column) the presence of a magnetic field $\mathbf{H}_a$.

in its normal phase (above T_c) from its superconducting phase (left column). On the contrary, when a superconductor is placed in a region with a magnetic field (right column) and then cooled below its T_c, the magnetic field lines are abruptly expelled from the sample. This is currently referred to as the Meissner effect. This phase transition at $T = T_c$ cannot be explained by the mere assumption of a null resistivity and thus makes a superconductor fundamentally different from a perfect conductor.

A material which repels an external magnetic field $\mathbf{H}_a$ is said to be *diamagnetic*. Microscopically, the orbital motion of electrons in atoms creates very small current loops, and then very small magnetic fields. An external magnetic field tends to align all these small magnetic fields, hence creating a magnetization $\mathbf{M} = \chi_{\mathrm{m}}\mathbf{H}_a$, where the dimensionless constant χ_{m} is the magnetic susceptibility of the material. A macroscopic induced magnetic field is created inside the sample, opposite to $\mathbf{H}_a$ ($-1 \leq \chi_{\mathrm{m}} < 0$). All materials are inherently diamagnetic. But it is in practice a very small effect ($|\chi_{\mathrm{m}}| \sim 10^{-5}$). It is also often hidden by paramagnetism (for which materials possessing unpaired electrons are attracted by an external magnetic field) or ferromagnetism (for which materials possess a permanent magnetization, regardless the presence of an external magnetic field). Coming back to a superconducting sample, the fact that it completely expels an applied magnetic field $\mathbf{H}_a$ from its interior means that the induced magnetic field is exactly $-\mathbf{H}_a$. In other words, $\chi_{\mathrm{m}} = -1$: a superconductor is therefore a perfect diamagnet. Note however that neither a perfect conductor nor a perfect diamagnet are necessarily superconductors. However, a superconductor is both a perfect conductor and a perfect diamagnet. An amazing consequence of this enforced expulsion of magnetic fields from a superconductor is that it can levitate above a magnet. Indeed, the repulsive force from the strong induced magnetization can compensate the weight of the superconducting sample, originating its levitation.

We also mention that, in addition to a critical temperature, there exists a critical magnetic field H_c (and therefore a critical supercurrent I_c) above which the superconducting state is destroyed, even if $T < T_c$. It has been experimentally observed that $H_c(T) \propto 1 - (T/T_c)^2$. In other words, the lower T, the higher H_c and I_c. At a given temperature, the passage from the superconducting to the normal state by increasing H above H_c is reversible by decreasing again H below H_c. We will come back to the microscopic origin of H_c in Section 9.1.3. The value of H_c is not an intrinsic characteristic of an element but depends on the geometry of the sample.

The penetration and coherence lengths

In the Meissner effect, the applied magnetic field H_a is not abruptly expelled from the superconductor, but is exponentially damped as one penetrates inside the system from an external surface. The distance over which the magnetic field decay is called the penetration length λ. Since the magnetic field is strictly 0 inside the superconductor, there is no current as well. The fact that H_a does penetrate the sample means that a current in the superconductor, if any, can only flow in a layer at the surface with a width $\sim \lambda$. One can estimate λ by assuming a two-fluid model, which we shall encounter later on in connection with liquid helium (see Section 9.2.2). In this model, the fluid of valence electrons is viewed as two intertwined fluids: the 'normal' one, which cannot carry electric current without dissipation, and the 'superconducting' fluid at the origin of the superflow. The number density is then written as $\varrho = \varrho_{\mathrm{n}}(T) + \varrho_{\mathrm{sc}}(T)$, with the obvious requirements that $\varrho_{\mathrm{sc}}(T = 0) = \varrho$ and $\varrho_{\mathrm{sc}}(T = T_c) = 0$. From some thermodynamical arguments, one can show that $\varrho_{\mathrm{sc}}(T)/\varrho = 1 - (T/T_c)^4$. In the two-fluid model and considering the equations satisfied by the magnetic field and the surface current, the penetration depth λ reads as:

$$\lambda = 4.19 \left(\frac{r_s}{a_0}\right)^{3/2} \left(\frac{\varrho}{\varrho_{\mathrm{sc}}}\right)^{1/2} \mathrm{nm} \quad , \tag{9.1}$$

where r_s is the Wigner-Seitz radius and a_0 the Bohr radius (see Section 5.2.4). To estimate λ, one can assume for instance that all free electrons in a metal are superconducting ($\varrho_{\rm sc} = \varrho$). For tin, we have typically $\varrho = 1.5 \times 10^{23}$ cm^{-3}, thus giving $\lambda = 14$ nm. Compared to sizes of typical macroscopic samples, this value is virtually zero. One had to wait until the 1950s to find experimental evidence of λ in thin films and in powders, for which the ratio surface-volume is small enough so that the penetration effect is observable. For tin, the experiments indicated a penetration length of $\lambda = 52$ nm. Notice that, by assuming a smaller proportion of superconducting electrons as $\varrho_{\rm sc} \simeq 0.1\varrho$, Eq. (9.1) gives a result in agreement with the experimental observation.

There is another relevant length scale ξ to be considered. Indeed, only electrons around the Fermi level can contribute to the superconducting state. More precisely, we have seen that in a degenerate Fermi gas at a finite temperature T, the electrons which participate to the heat capacity possess an energy $\varepsilon_F \pm k_{\rm B}T$ (see Section 3.4.2). Analogously, we consider an electron of kinetic energy $p^2/(2m)$. This electron participates to the superconducting transition at $T = T_c$ if $p = p_F \pm \Delta p$ with the uncertainty Δp such that $p\Delta p/m \leq k_{\rm B}T_c$. Replacing p by $p_F = mv_F$, we get $\Delta p \leq k_{\rm B}T_c/v_F$. Due to the Heisenberg uncertainty principle $\Delta p\, \Delta x \geq \hbar$, this leads to a distance $\Delta x \geq \xi$, where

$$\xi = \frac{\hbar v_F}{k_B T_c} \tag{9.2}$$

is called the coherence length. This is the length scale over which the supercurrent does not change much in a spatially varying magnetic field.

Superconductors of type I and II

By comparing the penetration length λ and the coherence length ξ, it can be shown that there exist two types of superconductor, by regarding the variation of the magnetization M as a function of the applied magnetic field H_a. To measure the magnetization of a superconductor, one can place a long cylinder of the material in a magnetic field aligned with the cylinder axis. As the magnetic field is varied, samples of pure elements exhibit the behavior displayed in the left panel of Figure 9.3. We have seen above that even below T_c, there is a critical field H_c above which the superconducting state disappears. At low H_a, we have a perfect diamagnet. It survives up to H_c, the superconductor is of type I. Most of pure metals are type-I superconductors. On the contrary, alloys and other materials have very different magnetization curves, with two critical values for the magnetic field, as shown in the right panel of Figure 9.3. These materials are called type II superconductors. Below the lower critical field H_{c1}, the magnetization curve is like that of a type I superconductor. For $H_{c1} < H_a < H_{c2}$, the flux density in the superconductor is not zero, the Meissner effect is said to be incomplete. Finally, above H_{c2}, we recover the normal state.

We can qualitatively explain these different behaviors in terms of the coherence length and the penetration depth. At $T < T_c$, the superconducting phase is the most stable. Let us make an energy balance. On the one hand,

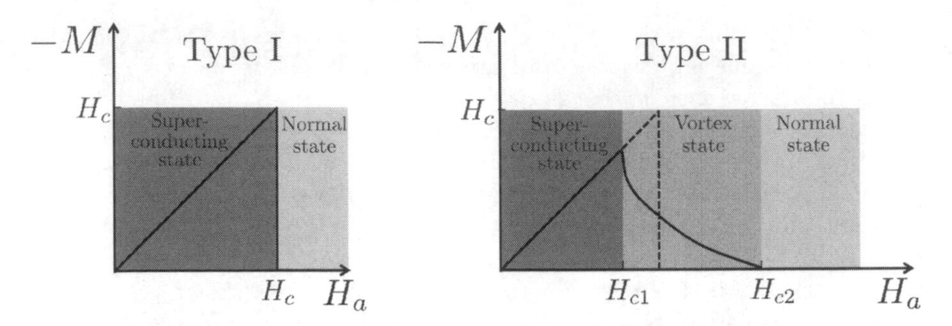

Figure 9.3: Magnetization M versus applied magnetic field H_a ($M = \chi_m \mathbf{H}_a$) for a bulk superconductor of type I with one critical magnetic field H_c (left panel) and type II with two critical magnetic fields H_{c1} and H_{c2} (right panel).

there is an energy cost to expel a magnetic field from a superconductor, as one approaches an interface between the superconducting and the normal states. The coherence length ξ characterizes this small region. On the other hand, the penetration of the magnetic field represents a gain in energy. This happens in a small superficial region over a distance of order the penetration length λ. Therefore, if $\xi > \lambda$, the energy cost of destroying superconductivity near the interface overweighs the energy gain when the magnetic field penetrates a bit. This is the situation for type I superconductors, which exhibit a complete Meissner effect. Typical values in type I superconductors are $\xi \sim 1$ μm and $\lambda \sim 50$ nm. On the contrary, in type II superconductors, we have $\xi < \lambda$ and the energy saving due to the penetration of the magnetic field overweighs the cost of destroying superconductivity. In the region between H_{c1} and H_{c2}, there is the creation of a certain number of interfaces between the superconducting and normal phases to save energy. The material is threaded by magnetic flux lines and is said to be in the vortex state (also called intermediate or mixed state). Type II superconductors are much more common than type I superconductors. For instance, Nb_3Sn is of type II and possesses $\xi \sim 3.5$ nm and $\lambda = 80$ nm. Certain alloys have a critical magnetic field H_{c2} much larger than the H_c values found in pure elements. It is very interesting because if a superconductor can withstand a larger magnetic field, it can also carry more electrical current before superconductivity is lost.

Quantized magnetic flux

Another feature of superconductors appears when there are holes in the sample. Let us consider a superconducting material with a cylindrical hole and apply a magnetic field parallel to the symmetry axis of this hole. It is observed that the magnetic flux Φ, is quantized. We recall that Φ roughly represents the amount of magnetic field flowing through a section of the cylinder.

To understand the origin of this quantization, predicted by F. London in 1950, we start with a unique wave function $\Psi(\mathbf{r}) = \sqrt{\varrho_{\mathrm{sc}}}\exp[\mathrm{i}\theta(\mathbf{r})]$ describing *all* superconducting electrons. In other words, we assume they all move in phase throughout the sample and Ψ thus constitutes a *macroscopic* wave function. We now consider a closed loop inside the superconductor and encircling the hole. Without loss of generality, we can take a circle with the same axis as the cylinder. A wave function is necessarily single-valued, otherwise it would not correspond to a single quantum state. This means that the phase θ is a continuous function. The consequence is that, when one moves along the circle, after a complete turn, θ has necessarily changed by a multiple of 2π. On the other hand, from vectorial analysis, one can show that the magnetic flux through a surface $\mathcal{S}$ bordered by the circle is proportional to this change of θ (hence $\propto 2\pi$), with a proportionality constant equal to $\hbar c/q$, where q denotes the charge of the supercurrent carriers. By the Meissner effect, the magnetic field is expelled from the interior of the superconductor but can still go through the hole. The magnetic flux above thus reduces to the flux Φ through a section of the cylinder. By putting all things together, one finally ends with the result that

$$\Phi = n\Phi_0 \text{ with } \Phi_0 = \frac{hc}{|q|} \tag{9.3}$$

where n is an integer, and the quantum of magnetic flux Φ_0 is called a 'fluxon'.

If we put $|q| = e$, the value of the fluxon is 4.14×10^{-15} T.m^2. This very small value corresponds to the flux of the magnetic field of the Earth through a human red blood cell (with a diameter of about 7 μm). This is why it took more than a decade to experimentally verify the theoretical prediction Eq. (9.3). And it actually constituted a surprise at that time: if one indeed observes that the magnetic flux through holes of a superconducting sample can take only discrete values, all multiples of a flux unit, this flux quantum was half the value calculated above, that is:

$$\Phi_0 = \frac{hc}{2e} = 2.07 \times 10^{-15} \text{ T.m}^2 \quad . \tag{9.4}$$

Note here the effective charge $q = 2e$ of superconducting carriers which are thus *pairs* of electrons. We will come back to this in Section 9.1.3.

In the vortex state of type II superconductors, since the Meissner effect is imperfect for an applied field H_a between H_{c1} and H_{c2}, the magnetic field can penetrate the superconducting sample through tubes made of normal phase. These tubes arrange themselves into a triangular lattice in most cases. Each tube contains a fluxon. The electric current flows around each tube to shield the superconducting region from the penetrating magnetic field. Because of this electric current circulation, these tubes are called vortices, drawing the parallel with the turbulent regime in a classical fluid (see Section 1.4.2). We can give estimates of the two critical fields H_{c1} and H_{c2} by using the fluxon Φ_0, the penetration depth λ and the coherence length ξ. Indeed, the magnetic field inside a vortex will spread into the superconducting region over a distance given by λ. Therefore, the formation of a single vortex occurs for $H_{c1} \simeq$

$\Phi_0/(\pi\lambda^2)$. On the other hand, when H_a approaches H_{c2}, the vortices are more and more packed, the limiting distance between two vortices being ξ. Each vortex still carrying a flux quantum Φ_0, we thus have $H_{c2} \simeq \Phi_0/(\pi\xi^2)$. We therefore observe that the larger the ratio λ/ξ, the larger the difference between H_{c2} and H_{c1}. Note also that in practice the penetration length λ and the coherence length ξ depend on the purity of the normal phase. Indeed, if the metal possesses some impurities, the scattering length is smaller than in a very pure metal and ξ thus decreases. By introducing impurities in a controlled manner, one can play on the parameter $\kappa = \lambda/\xi$ and progressively go from a type-I superconductor to a type-II one.

Let us finally mention that superconductivity having some similarities with superfluidity, lattices of vortices can also appear in superfluid helium (see Section 9.2.3).

9.1.3 The Bardeen-Cooper-Schrieffer theory

For years numerous theoretical and experimental contributions to the understanding of superconductivity remained at a phenomenological level and relied on a free gas of electrons, either in the normal or superconducting state, and completely neglected the role of the lattice ions, although essential to guarantee the charge neutrality of the material. The phenomenological approach brought by Ginzburg and Landau has considerably helped to understand many properties of superconductors with the introduction of only two phenomenological parameters. The first microscopic theory of superconductivity was proposed in 1957 by Bardeen, Cooper and Schrieffer (BCS). Its great success was to identify the relevant interaction and mechanism for superconductivity, especially the role of the lattice. The full derivation of BCS theory requires advanced methods of the many-body problem; we only sketch here the most important ingredients and discuss their relation with the properties of a superconductor described previously.

The isotope effect

The starting clue for BCS was related to the role of the lattice ions. A crystalline modification below the critical temperature was first considered. But very detailed crystallographic studies done with X rays have never demonstrated such a modification. This corroborated that superconductivity was essentially driven by electrons. On the other hand, it has been observed that for a given superconductor element, the value of T_c varies with its isotopic mass A, roughly following the law $T_c A^\alpha \simeq$ constant with $\alpha \simeq 1/2$ for most superconductors. This law is known as the isotope effect. It is an indication that lattice vibrations are involved in superconductivity. We recall that the scattering of electrons by phonons partly explains the resistivity in normal metals (see Sections 5.2.4 and 5.2.5). They are actually also implied in superconductivity, as we will see below.

Cooper pairs

We recall that the measurement of the flux quantum already suggested that the supercurrent carriers were pairs of electrons, and not individual electrons (Section 9.1.2). The BCS theory explains the microscopic formation of such pairs, the famous Cooper pairs. To understand the existence of such objects, consider a particular point of the lattice and a conduction electron moving within the lattice. This motion induces a lattice deformation nearby the electron's trajectory via Coulomb attraction. This displacement leaves a local positive charge. The time at which the ionic displacement is maximal is given by $1/\nu_D$ where ν_D is the Debye frequency. We recall that ν_D is the maximal allowed frequency of phonons introduced by Debye to explain the heat capacity of metals at low temperatures (see Section 5.2.3). On the other hand, electrons have typically the Fermi velocity v_F. Now, if a second electron with opposite spin and opposite momentum passes by this slight excess of positive charge at time $1/\nu_D$, it feels an attraction by the lattice greater than if it were not deformed. The net effect is a transfer of momentum and energy from one electron to the other, mediated by the lattice deformation or phonon. In other words, both electrons are now correlated, producing a Cooper pair.

The separation distance actually corresponds to the coherence length ξ. From the values of $\nu_D \sim 10^{13}$ Hz and $v_F \sim 10^8$ cm/s, we can estimate $\xi = v_F/\nu_D \sim 100 - 1000$ nm. This value is much larger than lattice parameters, typically a fraction of nm. From an energetic point of view, the energy of phonons is of the order of a few tens of meV. For electrons separated by 1 μm, the (screened) Coulomb interaction is precisely of the same order. Therefore, the short-range electron-phonon interaction may lead to a long-range electron-electron attractive interaction. Instead of the dissipative electron-phonon coupling which explains electrical resistivity, we are now in presence of a virtuous electron-phonon-electron coupling which allows for the formation of electron pairs, eventually leading to a superconducting state. This also explains why the best conductors are not superconducting. Indeed, in such metals, the electron-phonon interaction is too small to compensate the Coulomb repulsion between electrons, even if this repulsion is screened.

The superconducting energy gap

We have already mentioned in Section 9.1.1 the existence of an energy gap in the electronic structure of a superconductor evidenced by the measurement of its heat capacity. This gap is actually related to the very existence of Cooper pairs. Indeed, let us consider a fully degenerate Fermi gas and isolate two electrons of opposite momenta and close to the Fermi sea level, that is with an energy such that $\varepsilon_F - h\nu_D \leq \varepsilon \leq \varepsilon_F + h\nu_D$. It can be shown that in the presence of an attractive potential, whatever the value of this potential, there exists a solution of the Schrödinger equation for these two electrons with an energy smaller than the energy of two noninteracting electrons (that is $2\varepsilon_F$). This means that this electron pair is bound. Such a bound state is called a Cooper pair.

In the case of a zero range and small attractive interaction $-V_0$ (with

$V_0 > 0$), we have the energy of the pair given by:

$$E_{\text{Cooper}} = 2\varepsilon_{\text{F}} - \Delta \quad , \quad \Delta = \Delta(0) = 2h\nu_D \exp\left[-\frac{2}{V_0\mathcal{D}(\varepsilon_F)}\right] \quad , \tag{9.5}$$

with $\mathcal{D}(\varepsilon_F)$ the density of states at the Fermi level (see Section 4.2.3). The quantity Δ is the superconducting energy gap and it separates the ground state and the excited states in the electronic structure of a superconductor. Note also that the expression of Δ given above is that at 0 K. We will see below that it depends on temperature. The quantity $-V_0$, which corresponds to the electron-electron attraction, stems from the electron-phonon interaction needed to form Cooper pairs. It can thus be evaluated from the resistivity of the sample at room temperature. Typical values of Δ are in the range of 0.1–1 meV.

The expression (9.5) of the superconducting energy gap constitutes one of the major achievements of the BCS theory. It can be related to all phenomenological quantities discussed in Secs. 9.1.1 and 9.1.2, such as the penetration length λ, the coherence length ξ, the critical temperature and magnetic field T_c and H_c.

For instance, let us consider a plane wave representing a Cooper pair, with a wavevector k which is scattered by the lattice, resulting in a plane wave with a wavevector $k+q$ (see discussion of an electron traveling in a periodical lattice in Section 5.2.5). The kinetic energy of both waves reads:

$$\frac{\hbar^2 k^2}{2m} + \frac{\hbar^2(k+q)^2}{2m} \simeq \frac{\hbar^2 k^2}{m} + \frac{\hbar^2}{m}kq \quad , \tag{9.6}$$

where we have neglected a correction term in q^2. This kinetic energy exhibits a correction term linear in q relative to the kinetic energy of the plane wave. It brings two pieces of information. First, when the applied magnetic field is increased, its expulsion from the interior of the superconductor by the Meissner effect correlatively increases the induced supercurrent at the surface of the sample. This means that the velocity of the Cooper pairs, and thus the wavevector q above, becomes larger and larger. For too high a value of q, the kinetic energy of the Cooper pair can exceed the energy gap $E_g = 2\Delta$, leading to the destruction of the Cooper pair. This therefore explains the existence of a critical field H_c (see Section 9.1.2), even if $T < T_c$. Second, the critical value q_c can be defined as the one that provides a correction term in Eq. (9.6) equal to E_g with an initial wavevector $k = k_F$. If we estimate the typical size of a Cooper pair through the coherence length $\xi = 1/q_c$, we obtain $\xi = \hbar v_F/(2\Delta)$. A more precise calculation as provided by the BCS theory gives the following relation:

$$\xi = \frac{\hbar v_{\text{F}}}{\pi\Delta} \quad . \tag{9.7}$$

This has to be related to the previous estimate (9.2), $\xi = \hbar v_F/(k_{\text{B}}T_c)$, leading to $k_{\text{B}}T_c = \pi\Delta$. A more rigorous calculation in the framework of the BCS theory gives:

$$\Delta(0) = 1.76\, k_{\mathrm{B}} T_c \quad , \tag{9.8}$$
$$\Delta(T)/\Delta(0) = 1.74\sqrt{1 - T/T_c} \quad . \tag{9.9}$$

In particular, the superconducting energy gap, hence the coherence length ξ, depend on temperature. Since $\Delta \to 0$ when $T \to T_c$, ξ increases accordingly and diverges at the superconductivity phase transition, thus providing a smooth transition to the normal state.

Let us come back to the experimental evidence for Δ. On the one hand, from Eqs. (9.8) and (9.9), if T_c is known, one can calculate Δ. On the other hand, in Section 9.1.1, we have mentioned that the specific heat capacity of a superconductor $c_{\mathrm{sc}} \propto e^{-E_g/(2k_{\mathrm{B}}T)}$ would reflect an energy gap E_g in the band structure *if the carriers were single electrons.* We now know that the carrier of the supercurrent are Cooper pairs. Let us assume that electrons in excited states in the superconducting phase are 'normal' (single) electrons. This means that to promote one electron from the ground state of a superconductor to an excited state, one cannot actually promote a single electron but a pair of electrons instead since an unpaired electron cannot be left in the ground state. In other words, it costs $E_g = 2\Delta$ to access the first excited state. This thus gives $c_{\mathrm{sc}} \propto e^{-\Delta/(k_{\mathrm{B}}T)}$. Using this expression, one can extract the value of Δ from measurements of c_{sc} and compare it to that obtained from Eqs. (9.8) and (9.9). Note that the obtained value is consistent with other measurements, as microwave absorption spectra or tunneling of electrons at the interface of a normal metal and a superconductor. This demonstrates at the same time the existence of a gap and of pairs of electrons in the ground state of a superconductor.

Another success of Eq. (9.5) is the fact that Δ is proportional to the Debye frequency ν_{D}. This thus naturally explains the isotope effect discussed above. However, one should also assume that V_0 is constant, whatever the considered isotope. This is probably not true since V_0 takes its origin in the electron-phonon interaction and should then depend on the isotopic mass. However, in the BCS Hamiltonian, the derivation of the electron-electron interaction from that between an electron and a phonon is not straightforward and relies on a certain number of assumptions that we do not aim at describing here.

As a final result, Eq. (9.8) taking into account the typical values of the superconducting gap of about a fraction of meV, the critical temperature cannot exceed $\sim$ 30 K, still far below the liquefaction temperature of N_2 at 77.36 K, this gas being less costly than helium (liquefaction temperature at 4.2 K) frequently used in low temperature physics. This is why materials that are well described by the BCS theory are usually coined 'conventional'. This theory is unable to account for the so-called high-T_c superconductors. Actually, the search for a unique theory to explain high-T_c superconductors is still underway.

A macroscopic quantum state

One has to realize that electrons in a Cooper pair are coherently bound and that all pairs are described by a single wave function. Indeed, let us estimate the average distance between Cooper pairs. To this end, we consider a Fermi gas at 0 K. We have seen that for two electrons at the Fermi level, the formed Cooper pair has an energy equal to $2\varepsilon_F - \Delta$. Consider now more and more electrons at the Fermi sea level that pair up. The Fermi sea gets depleted and the Fermi sea level decreases until it falls at $\varepsilon_F - \Delta/2$. Indeed, a pair of electrons at that level already has an energy of $2\varepsilon_F - \Delta$. There is thus no energy gain to form a Cooper pair and electron pairing stops. The number of formed Cooper pairs is then given by half the number of electrons (with spin up and spin down) in a band $[\varepsilon_F - \Delta/2; \varepsilon_F]$, that is $\frac{1}{2} \times \mathcal{D}(\varepsilon_F)\Delta/2$ where $\mathcal{D}(\varepsilon_F)$ is the density of states at the Fermi level. We have seen for a fully degenerate Fermi gas that $\mathcal{D}(\varepsilon_F) = 3N/(2\varepsilon_F)$; see Section 3.4.1 and Eq. (3.28). Therefore, the number of Cooper pairs is $N_{\text{Cooper}} = 3N\Delta/(8\varepsilon_F)$. Finally, the distance d between two Cooper pairs is $d \simeq (V/N_{\text{Cooper}})^{1/3} \simeq 30a$ with a the lattice parameter. If we take $d = 10$ nm for simplicity, with the typical length of a Cooper pair of $100 - 1000$ nm, we end with a thousand to a million of Cooper pairs in the same region! One can easily imagine then that their wave functions strongly overlap. They actually add up constructively and result in a unique wave function of macroscopic size.

9.2 SUPERFLUIDITY OF HELIUM

After a short review on the superphase consisting of a superconducting state, we now turn to another superphase, namely superfluidity. We have already encountered it in Section 6.1. We recall that historically, superfluidity was first observed in ^{4}He. Helium possesses a very characteristic phase diagram with the so-called lambda transition at $T_\lambda = 2.17$ K, separating the normal liquid helium (termed He-I) and the superfluid helium (termed He-II), where it can flow with no viscosity. First explanations relied on the analogy between a BEC and He-II. But we have already pointed out in Sections 6.1.3 and 6.1.4 that, even if they share some common features, they are not strictly identical. We now enter into more details on this superphase. Some words about the case of ^{3}He, which is also superfluid at very low temperatures, and the possibility to observe superfluid droplets will be given at the end of this section.

9.2.1 Superfluidity and elementary excitations

One manifestation of the absence of viscosity is the fact that a superfluid liquid flows through a capillary with no pressure dropping along the tube. This macroscopic property can be qualitatively understood at a microscopic level, by linking the energy exchanges (between helium and capillary) with the possible excitations of the liquid. Landau guessed the form of the relation between the energy ε and the wavevector k of an elementary excitation of the superfluid, which is named the dispersion relation. He then derived a simple criterion of superfluidity, based on general arguments of energy and momentum conservation and Galilean invariance.

Critical velocity for dissipation

Let us consider a superfluid flowing in a capillary at velocity $\mathbf{v}$. In the liquid reference system, and at zero temperature, the liquid is at rest in its ground state of energy E_0. Consider the laboratory frame, with the capillary at rest and the fluid flowing at velocity $\mathbf{v}$. In absence of excitations, the momentum of the liquid is $M\mathbf{v}$ and its energy $E_0 + Mv^2/2$, where M is the liquid total mass. The existence of viscosity produces a change of motion of the liquid, which starts by a gradual occurrence of the internal excitations in the liquid. If an elementary excitation with momentum $\hbar\mathbf{k}$ and energy $\varepsilon(\mathbf{k})$ appears in the liquid, the momentum of the liquid is shifted to $\hbar\mathbf{k} + M\mathbf{v}$, and its energy to $E_0 + \varepsilon(\mathbf{k}) + \hbar\mathbf{k}\cdot\mathbf{v} + Mv^2/2$. Therefore in the laboratory frame, the change in energy of the liquid due to the appearance of an elementary excitation is $\varepsilon(\mathbf{k}) + \hbar\mathbf{k}\cdot\mathbf{v}$.

Viscosity means dissipation. In turn, a dissipation corresponds to a loss of energy from the fluid to the capillary. An elementary excitation allows dissipation if the energy change in the laboratory frame is negative. If θ is the relative angle between $\mathbf{k}$ and $\mathbf{v}$, the scalar product $\hbar\mathbf{k}\cdot\mathbf{v} = \hbar kv\cos\theta$ has a minimum value attained when $\mathbf{k}$ and $\mathbf{v}$ are collinear and in opposite directions ($\theta = \pi$). We further assume rotational invariance, that is ε independent of the spatial direction of $\mathbf{k}$. We then simply write $\varepsilon(\mathbf{k}) = \varepsilon(k)$ in the following. A necessary condition for dissipation is then provided by $\varepsilon(k) - \hbar kv < 0$. This defines a critical velocity as the minimum with respect to k of the excitation energy/momentum ratio

$$v > v_c = \mathrm{Min}_k \, \frac{\varepsilon(k)}{\hbar k} \quad . \tag{9.10}$$

If the liquid moves with velocity $v > v_c$ relative to the capillary, it is energetically favorable to produce an excitation, then the flow of the liquid is unstable and the kinetic energy is transformed into heat. On the contrary, if the relative velocity is $v < v_c$, no excitation is possible and the liquid flows without dissipation. Consider now an ideal gas, with no interaction. The elementary excitations are those of single particles, with an energy $\varepsilon(k) = \hbar^2k^2/2m$. The minimum of $\varepsilon(k)/\hbar k$ is obviously zero, reached at $k = 0$. The critical velocity v_c from Eq. (9.10) is thus 0 in this case. Consequently, at any value of the relative velocity, the gas always exchanges energy and momentum with the capillary, hence dissipates energy.

The existence of v_c constitutes another illustration of the limitations of identifying He-II and an ideal BEC. The existence of a condensate alone does not imply superfluidity. Whereas the assumption that superfluidity is related to the fact that liquid helium consists of identical bosons is essentially correct, the interaction between bosons induces dramatic deviations with respect to an ideal BEC. In particular, the interaction changes the excitation spectrum from that of an ideal gas. An elementary excitation is in fact a collective excitation of atoms in the liquid. For low excitation energies, the elementary excitations behave as a set of individual particles. The latters are called 'quasiparticles' because they should not be identified with single atoms.

Dispersion relation

The form of the elementary excitation energy $\varepsilon(k)$ as a function of momentum $\hbar k$ is known as the dispersion relation. Landau suggested that there are two kinds of elementary excitations in ^{4}He. In the region of low k, they correspond to the ordinary acoustic waves in a liquid and the dispersion relation is linear $\varepsilon(k) = \hbar k c_1$, with the slope $c_1 = 238$ m/s being the speed of ordinary sound (see Table 6.1). Landau called 'phonons' these excitations, as they are analogous to phonons in a solid (see Section 5.2.3). At higher values of k, the pattern of $\varepsilon(k)$ deviates from a linear one. Its precise form depends on the details of the interaction and it is not an easy task to obtain its general form. Based on general thermodynamical properties, in particular the measure of the specific heat, Landau figured out the existence of another type of excitation, which he called 'rotons', assuming for them a parabolic dispersion relation $\varepsilon(k) = \Delta + \hbar^2(k - k_0)^2/2\mu$ around a certain momentum k_0. The parameters (Δ, k_0, μ) are fitted to experimental results. This delivers a photon-roton dispersion relation which qualitatively agrees with the experimental determination of excitations in He-II, currently measured by neutron scattering, and displayed in Figure 9.4. We note that the phonon-roton dispersion relation is

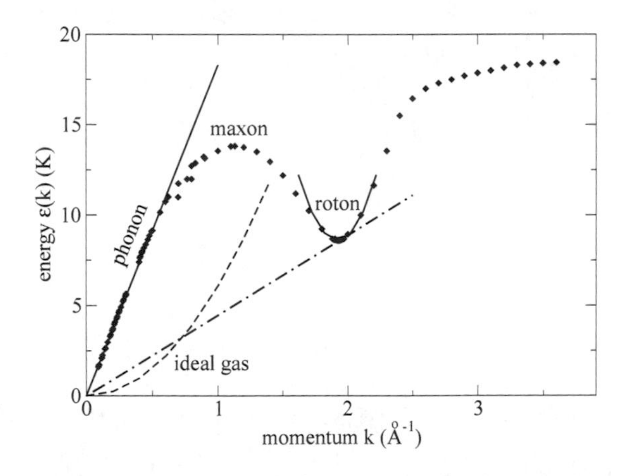

Figure 9.4: Dispersion relation of liquid ^{4}He that is $\varepsilon(k)$, given in temperature units, as a function of momentum k, given in angstrom. Diamonds: experimental points from R. J. Donnelly and C. F. Barenghi, J. Phys. Chem. Ref. Data 27, 1217 (1998). Solid and dashed-dotted lines are tangents from the origin to the experimental curve. The dashed curve is the excitation energy of an ideal gas.

very different from the quadratic k^2-behavior of a noninteracting gas (dashed curve in the figure). Indeed, it starts with a linear part at low wave vectors,

$k \leq 0.8$ Å^{-1}, which is the phonon region. It is followed by a maximum at $\simeq 1$ Å^{-1}, called the maxon region, and then by a pronounced minimum (the roton region) at $\simeq 2$ Å^{-1}. There is a final steep rise at large k. Measurements of thermal properties below $\simeq 0.6$ K (as the low-T behavior of the specific heat) suggest that phonons are the only excitation modes. Rotons need a minimum energy (denoted by Δ in the previous reasoning) to be excited, and therefore they require higher values of T and/or k.

Let us come back to the critical velocity given in Eq. (9.10). To determine this velocity, we consider a straight line passing by the origin and tangent to the curve $\varepsilon(k)$. We identify v_c with the minimal slope of this line. From Figure 9.4, in the phonon region, v_c is naturally equal to $c_1 = 238$ m/s (see full straight line) and the liquid can flow without viscosity for undersonic velocities $v \leq c_1$. In the roton region, the minimum slope (see dotted-dashed straight line) fixes a smaller value of the critical velocity $\simeq 58$ m/s. However, the actual critical velocity is much smaller than this value because other modes of excitation, such as vortices, can also be produced. Anyway, Landau's criterion means that, below a certain gap in the excitation energy, the system is superfluid.

The dispersion relation was first microscopically explained by Feynman in terms of a wave function completely symmetric with respect to the exchange of any pair of bosons, and the atom-atom interaction. The dispersion relation takes the form $\varepsilon(k) = \hbar^2 k^2 / 2m\, S(k)$ where $S(k)$ is the structure factor of the system. We recall that $S(k)$ is related to the Fourier transform of the two-body correlation function $g(r)$ (see Section 1.3.3). The structure factor is experimentally determined by either X-rays diffraction or neutron scattering experiments.

Figure 9.5 displays the structure factor $S(k)$ as measured by neutron scattering, together with the correlation function $g(r)$ deduced by inverse Fourier transformation. $S(k)$ starts by increasing linearly with k, rises to a maximum

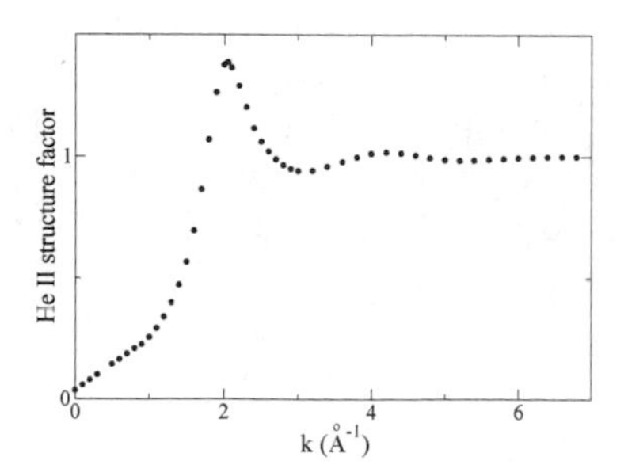

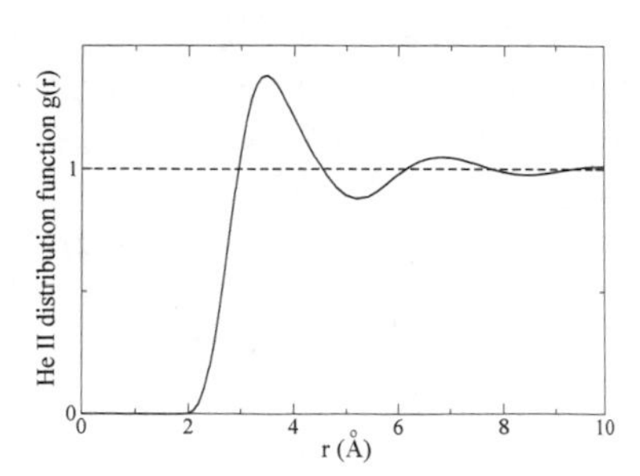

Figure 9.5: Left: Structure factor $S(k)$ of liquid ^{4}He determined by neutron scattering at $T = 1$ K. Data extracted from R. J. Donnelly and C. F. Barenghi, J. Phys. Chem. Ref. Data 27, 1217 (1998). Right: the deduced correlation function $g(r)$.

at $k_{\max} \simeq 2$ Å^{-1}, and falls afterwards to approach the limit unit, with small amplitude oscillations. Consequently, $\varepsilon(k) = \hbar^2 k^2/2m\, S(k)$ also starts linearly with k, exhibits a maximum followed by a dip with a minimum at k_0, and rises again as $\hbar^2 k^2/2m$. Therefore, the phenomenological guess by Landau is qualitatively explained in microscopic terms, by solving the Schrödinger equation.

9.2.2 The two-fluid model

A two-fluid model has been discussed in the context of superconductivity (see Section 9.1.1), in particular to introduce the penetration length. An analogous model provides a useful and effective description of the hydrodynamics of a superfluid. The basic idea of this model is that He-II may be thought as a mixture of two independent fluids that interpenetrate without apparent interaction. Actually there is one and the same fluid: it makes no sense to consider some atoms that belong to the one or to the other fluid, since all ^{4}He atoms are identical and indistinguishable. But the model provides an intuitive guide to understand a certain number of properties of helium. The total particle density is divided into two components $\varrho = \varrho_s + \varrho_n$, normal and superfluid, in a proportion depending on temperature. The superfluid component ϱ_s flows with no viscosity and no entropy, while the normal component ϱ_n behaves as a viscous fluid and carries all thermal energy and entropy of the liquid. At zero temperature, the whole He-II is superfluid, while the normal fluid appears in an increasing proportion as temperature increases, and eventually at T_λ, there is no superfluid component and the whole fluid becomes He-I.

The model is consistent with experiments showing that He-II flows through a capillary with no viscosity. Indeed, if the capillary tube is too narrow, the normal fluid would feel too much friction and will remain at rest. Only the superfluid component flows with no friction. There are however some other experiments which seem to imply the presence of viscosity in He-II, thus providing a direct way to measure the normal component. Such effects can be observed by measuring the moment of inertia of a torsion pendulum inside the fluid. This is at the root of Andronikashvilii's experiment in 1956. The torsion pendulum used in this experiment was made of a stiff wire on which was attached a dense stack of aluminum discs very close to each other. The oscillation frequency of the pendulum depends on the torsional stiffness of the wire and on the moment of inertia of the disc stack. If a solid body rotates about a given spatial direction, its angular momentum L along this direction is the product of the angular velocity Ω and the moment of inertia $\mathcal{I}$ of the body. $\mathcal{I}$ depends on the mass distribution with respect to the rotation axis. For a given value of L, the less spread the mass in a plane normal to the rotation axis, the smaller $\mathcal{I}$, the larger Ω and the faster the body spins.

The moment of inertia of the torsion pendulum used in the experiment depends on the mass of the liquid in between the discs. The idea was to measure the oscillation frequency, and to deduce this moment of inertia, as a function of temperature. In a normal liquid ($> T_\lambda$), oscillations are very quickly damped, due to the viscosity of the liquid between the discs. On the contrary, below T_λ, the pendulum swings more freely, its moment of inertia approaching the value of the free pendulum. Knowing $\mathcal{I}$ allows one to deduce the normal and superfluid fractions which are schematically plotted in Figure 9.6. We also

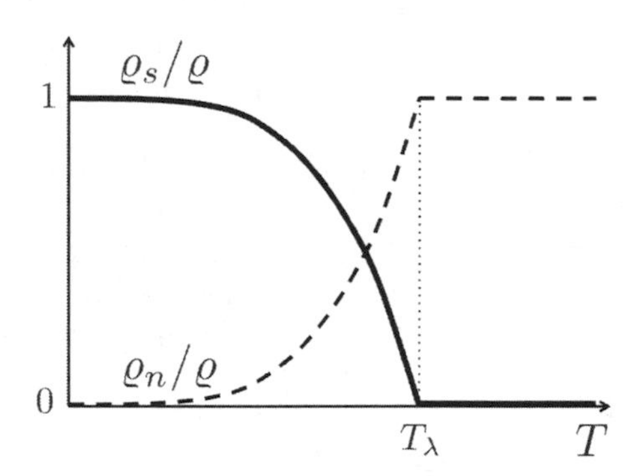

Figure 9.6: Superfluid fraction (full curve) and normal fraction (dashes) of He-II as a function of temperature T.

mention that the temperature dependence of the superfluid fraction goes as T^4, in contrast with the $T^{3/2}$ behavior of the condensed fraction of a BEC; see Eq. (3.20a).

The two-fluid model explains a certain number of observations, including the thermomechanical effect. Consider two vessels joined by a fine capillary, filled with He-II, initially at the same temperature and standing at the same level. If the temperature of a vessel is raised by a small amount ΔT, one observes a flow from the other vessel. This results in the raise of the level in the hottest vessel. The phenomenon is also known as the fountain effect, because in one of the earliest experiments, the value of ΔT was so high that the liquid was thrown up in a jet. The capillary inhibits any flow of the normal fluid, and only the superfluid may pass freely to establish equilibrium.

The model has also led to predict new phenomena, which have been experimentally confirmed later on. The existence of two fluids permits the occurrence of two different modes of longitudinal waves. The two fluids can oscillate in phase, resulting in a density wave which produces the ordinary sound, also called 'first sound'. As in any other fluid, it propagates at a velocity c_1 related to the variation of pressure with respect to density at constant entropy by $mc_1^2 = \partial P/\partial\varrho$. In a two-fluid system, there exists a 'second sound', related to the oscillation of the two fluids with opposite phases. This produces a temperature wave, analogous to the pressure waves of the ordinary sound, so that variations of T do not propagate according to the usual Fourier equation. Therefore, one can produce temperature waves with a heater in a resonance tube, and detect stationary waves using a thermometer as a detector. In summary, a pressure gradient will tend to drive both fluids in the same direction, while a temperature gradient tends to drive the superfluid component towards the high temperature, and the normal component in the opposite direction. Interestingly, these thermal waves propagate at a velocity which depends on the ratio ϱ_s/ϱ_n, and henceforth, from the observation of the second sound, one can deduce the proportion between superfluid and normal fluids. In case of restricted environments, such as surfaces or narrow capillaries, the standing waves are referred to as third, fourth, ... sound modes.

9.2.3 Superfluid in rotational motion

Since He-II can flow through a capillary with no apparent viscosity, it can be imagined that the flow could be maintained indefinitely along an infinite capillary. More realistically, one can consider helium in a toroidal container (a doughnut shape), packed with a fine powder so that only the superfluid flows freely through the pores. Starting at a temperature above T_λ, the container is set into rotation about its central axis with a constant angular velocity $\mathbf{\Omega}$. The container and the normal He-I liquid go into a solid-body rotation. While rotating, the temperature is reduced below T_λ. If the container's rotation is now stopped, one expects that the superfluid helium keeps rotating and stores the rotational angular momentum, so that the system should behave as a kind of gyroscope. If the axis of this 'superfluid gyroscope' is tilted, it would precess about its original direction. The experiment shows that at low rotating velocities and at temperatures well below T_λ, the superfluid flow current persists for an indefinite amount of time. Note that the superfluid gyroscope is reminiscent of experiments with persistent electric currents in superconductor rings in a magnetic field (see Section 9.1.1). This provides another analogy between superfluidity and superconductivity.

Rotating liquid

Another interesting phenomenon appears in a cylindrical vessel filled with helium and set into rotation at a constant angular velocity $\mathbf{\Omega}$ with respect to its symmetry axis, denoted by z (see left panel of Figure 9.7). Starting at a temperature above T_λ, one sets He-I in rotation. Cooling it down below T_λ, one obtains a rotating He-II. There are however some interesting and non-standard phenomena occurring in the superfluid helium.

Before describing and explaining them, let us first describe the free surface of a normal rotating liquid. When a classical liquid is forced to rotate, after a transient regime, equilibrium establishes and the liquid rotates as a rigid body. This is accompanied with the formation of a concave shape at the free surface as shown in the second panel from left of Figure 9.7. This shape is called a paraboloid of revolution whose equation reads $2gz = r^2\Omega^2$, where (z, r, ϕ) are the cylindrical coordinates and g the acceleration of gravity. Note that, owing to rotational invariance, there is no dependence on the polar angle ϕ (measured in a horizontal plane normal to the z direction). We have seen in Section 1.4.3 that such a rigid body rotation is due to a non-vanishing vorticity. We recall

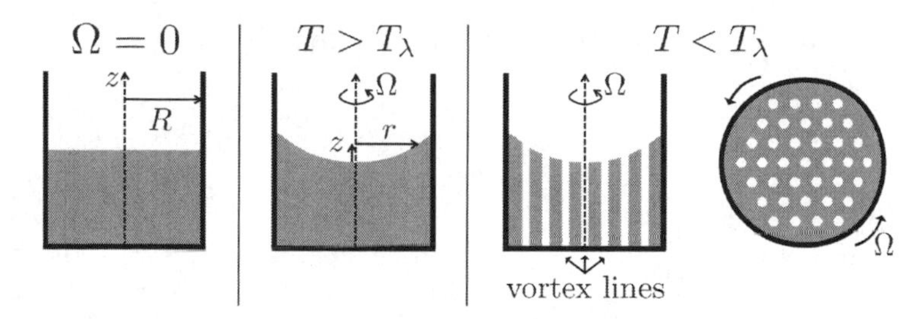

Figure 9.7: Cylindrical vessel of radius R and symmetry axis z containing liquid helium. Left panel: at rest. Second panel from left: vessel set to rotate about the vertical axis at an angular velocity Ω, above the lambda transition temperature T_λ. Third panel from left: vessel at $T < T_\lambda$ and rotating with Ω greater than $\Omega_0 = h/(2\pi R^2 m)$ (see text for details) with creation of vortex lines. Right panel: top view of the rotating vessel at $T < T_\lambda$.

that vorticity is related to the way elements of fluid locally flow with respect to each other in a rotational motion. A non-vanishing vorticity means that the fluid shears and cannot exist without any viscosity. In brief, a classical liquid cannot exhibit a free surface of paraboloid shape without a vorticity, namely without viscosity.

Rotating He-II

In the case of He-II, one would expect the normal component to follow the same pattern as the classical liquid, while the absence of viscosity in the superfluid component would hinder a rigid body rotation. The free surface of the rotating superfluid would thus remain flat. This would result in a reduction of the height of the paraboloid by a factor ϱ_n/ϱ (ratio of superfluid density over total density). Since this ratio is temperature-dependent (see Section 9.2.2) the height of the paraboloid should also be temperature-dependent. However, this is not observed experimentally: the free surface of rotating He-II keeps the same shape as that of a normal liquid, and it is independent of T. The answer to this apparent paradox lies in the appearance of an array of vortex lines (see white tubes in the third panel from left of Figure 9.7).

To explain this phenomenon, we have to understand how a superfluid can keep a non-vanishing vorticity, while possessing no viscosity. To this end, we exploit again the similarity between superfluidity and superconductivity. Indeed, in Section 9.1.2, we have briefly explained why the flux of magnetic field through a hole inside a superconducting material was quantized. This flux roughly represents the amount of magnetic field that flows through a section of the hole. The quantum of magnetic flux is $\Phi_0 = hc/(2e)$. We here draw a parallel between a vortex line which forms in He-II and a cylindrical hole in a superconductor. What is quantized here is the flux of vorticity (mind that vorticity is a vectorial quantity) through a section of the vortex line. More precisely, the quantum of vorticity flux is equal to

$$\Phi_0 = \frac{h}{m} \quad , \tag{9.11}$$

where m is the mass of a ^{4}He atom (we will not demonstrate this result but rather admit it in the following).

The appearance of vortices thus brings a non-vanishing vorticity into a superfluid (hence non-viscous) liquid. This is actually the only way a superfluid can acquire an angular momentum. The number n of vortices depends on the overall vorticity imposed by the rotation of the vessel. Without entering into vectorial analysis that would bring us too far away from our topic, let us admit that for a cylindrical vessel of radius R and rotating at an angular velocity Ω about its symmetry axis, the total flux of vorticity of the fluid flowing in the vessel is $\Phi = 2\pi R^2\Omega$. To conserve this total flux, the rotating superfluid must contain the right number n of vortices as $\Phi = n\Phi_0$. Using Eq. (9.11), we deduce the number of vortices per unit area in the rotating superfluid as:

$$\frac{n}{\pi R^2} = 2m\frac{\Omega}{h} \simeq 2000\ \Omega\ \mathrm{cm}^{-2} \quad . \tag{9.12}$$

The appearance of vortices thus hinders liquid He to exhibit a flat free surface at the lamdba transition and allow He-II to keep a concave shape. These quantized vortices are therefore another signature of superfluidity.

We can also proceed the other way around: we start with He below T_λ, hence superfluid helium, and set the rotation by increasing very slowly the angular velocity. At the beginning, as expected, the superfluid first remains at rest with a flat free surface. By increasing the angular velocity, vortices appear at precise values of $\Omega = n\Omega_0$ with $\Omega_0 = h/(2\pi R^2 m)$ according to Eq. (9.12), and the free surface acquires a paraboloid shape. At small rotation rates, vortices in He-II tend to form an ordered triangular array, as sketched in the right panel of Figure 9.7. This array is analogous to the array of flux lines in a type-II superconductor discussed at the end of Section 9.1.2.

Quantized vortices in a BEC

To end this section, we come back to the comparison of He-II and a BEC. As already mentioned at several places before, an ideal BEC cannot be superfluid, because the presence of interactions is a necessary condition for the appearance of superfluidity. Atoms in a real BEC do interact between themselves, and this weak interaction produces visible effects. One can expect the appearance of quantized vortices in a real BEC, when the confining trap (the equivalent of the superfluid container) rotates. This is indeed the case, and increasing the angular velocity similarly produces an increasing number of vortices. More than one hundred vortices have been observed in some experiments. They form a regular hexagonal lattice, a structure which is very similar to that produced by the magnetic lines in type-II superconductors (see Section 9.1.2). In other words, while an ideal BEC cannot be superfluid, a real BEC can be, and the observation of quantum vortices in a real BEC is a direct proof of its superfluid character.

9.2.4 Superfluidity of ^{3}He

For years, superfluidity has been associated to the bosonic isotope ^{4}He. But when in 1957 appeared the BCS theory of superconductivity (see Section 9.1.3), things were seen differently. Similarly to the electrons, the fermionic ^{3}He atoms could indeed form pairs, analogous to Cooper pairs, at sufficiently low temperatures. Instead of superconductivity, related to charged particles, the possible phase transition for neutral particles is superfluidity. The main objective of earlier experiments with ^{3}He was then to detect a superfluid behavior in the liquid. But no superfluidity was observed at temperatures as low as a few hundredths of kelvin, and it was thought that such a property was apparently specific to boson systems, until the first experimental evidence of superfluid ^{3}He was observed in 1972. In Figure 9.8, we display again the phase diagram of ^{3}He, completing the more schematic one shown in Figure 6.3.

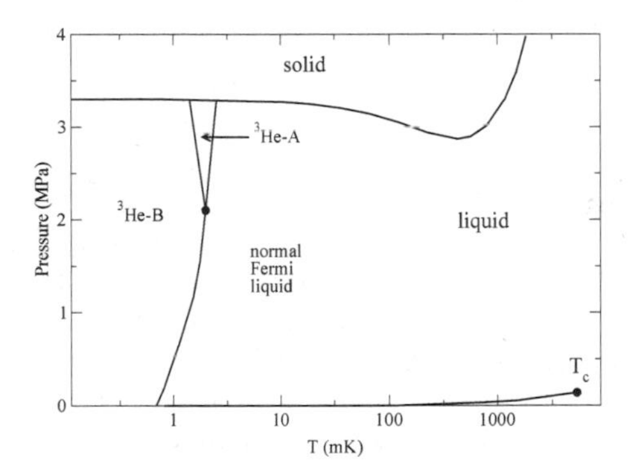

Figure 9.8: Phase diagram in the pressure-temperature map of ^{3}He.

To reach temperatures below a few mK, the experiments were done near the melting curve, slowly increasing and decreasing pressure at constant rate. Along the melting curve, solid and liquid are in equilibrium, and thus the value of pressure determines the value of temperature. Plotting pressure as a function of time, two discontinuities were observed at about 2.7 and 1.8 mK, which were denoted as points A and B, and taken at first as evidence for the searched new phases in solid. New measurements showed the existence of magnetic changes in the liquid, which were correlated with these transitions. Therefore, there are actually two superfluid phases, labelled A and B in the phase diagram. Moreover, the complete phase diagram should be represented in 3D, adding a new axis for an external magnetic field. A third phase manifests at varying the magnetic field, it is located between normal and A phases and is known as phase A_1.

Measurements of the specific heat show a pronounced peak at a critical temperature T_c, whose shape seems at first view to be similar to the λ-curve of liquid ^{4}He (see Figure 6.2). There are however some important differences between both curves. In the ^{3}He case, the specific heat decreases linearly above T_c, the jump at T_c is finite, instead of an almost logarithmic one in the ^{4}He case. This behavior is characteristic of a second order transition and, interestingly, is very similar to that of some superconductor metals around the transition temperature (see Section 9.1.1). The discussed curve corresponds to the transition between normal and A phases. If the measurements are done at values of pressure in the interval $2.1 - 3.2$ MPa, one can also see the transition between A and B phases, which appears as a change of slope.

Microscopically, the superfluidity in ^{3}He cannot arise from a BEC mechanism because we are dealing with fermions. Instead, as mentioned above, it was immediately suggested that it could be related to the BCS theory of superconductivity. Some modifications were however needed to pair electrically neutral particles instead of (negatively charged) electrons. Note that in ^{3}He, there is no analog to the electron-phonon interaction to create Cooper pairs (see Section 9.1.3), so that the attraction to form pairs of atoms must arise directly from the He-He interaction. The basic theory was established by Leggett in the next years following the discovery of superfluidity. We shall give here some basic ideas.

The problem is that, as we have seen in Section 6.1.2, the van der Waals interaction is strongly repulsive at short distances and the weak attractive part cannot bind the analog of the Cooper pairs. Instead, one can consider an effective interaction between quasiparticles which includes effects arising from nuclear spin fluctuations. A quasiparticle with spin 1/2 causes a local polarization of the neighboring quasiparticles via the Pauli exchange interaction. The result is a tendency to favor parallel alignment of the spins, very similar to the mechanism producing the ferromagnetism in some metals. As a consequence, Cooper-like pairs can be produced, but the actual mechanism is more involved and we shall not detail it here. A full microscopic description of superfluidity in liquid ^{3}He and a completely satisfying explanation of some experimental facts are still missing, despite the numerous investigations and the enormous progress made in the last two decades.

9.2.5 Superfluid droplets

Helium droplets can be doped with nearly any kind of atomic or molecular species, forming new complexes (see Section 6.3). This property has been used to perform experiments on superfluidity in ^{4}He droplets doped with weakly interacting probes. The excitation spectrum of the guest molecule can indeed bring some insight into the properties of the host. This is the equivalent of the Andronikashvili's experiment in a finite system (see Section 9.2.2). The torsion pendulum made of a small metal disk stark is replaced here by a linear molecule, and superfluidity manifests in the appearance of well resolved rotational lines, indicating that the molecules rotate freely inside the droplets. The first experiments were done with sulfur hexafluoride (SF_6) and carbonyl sulfide (OCS) molecules, and have been since then confirmed for several dozens of other small molecules.

A series of experiments with the linear OCS molecules provided evidence that superfluidity appears with only a few tens of ^{4}He atoms. More precisely, the infrared spectrum of single OCS molecules, associated to rotational modes of the molecule, was first measured inside large pure ^{4}He droplets and pure ^{3}He droplets, both consisting of about 10^4 atoms. In the ^{4}He droplets, sharp rotational lines were observed, whereas in ^{3}He only a broad, unresolved peak was found. This difference between both spectra is interpreted as the fact that narrow rotational lines, which imply free rotations, are a microscopic manifestation of superfluidity. Afterwards, ^{4}He atoms were added to large ^{3}He droplets doped with OCS molecule, and the spectrum measured for increasing number of ^{4}He atoms. Sharp rotational lines were found upon addition of 60 ^{4}He atoms, which was considered as an upper limit for the minimum number needed for superfluidity.

Another important observation made for heavy molecules like OCS embedded in superfluid ^{4}He droplets is the increase of the effective moment of inertia by about a factor 2.8 as compared to the free OCS molecule. Upon addition, ^{4}He atoms surround the OCS molecule, intending to solvate it. The increased moment of inertia can be explained by the formation of a symmetric donut ring of ^{4}He atoms around the waist of the OCS molecule, which participate in the end-over-end rotations of the molecule. Theory has identified the donut atoms as a non-superfluid fraction as opposed to the other surrounding superfluid atoms which do not rotate with the molecule.

As described in Sections 9.1.2 and 9.2.3, quantized vortices are one of the clearest hallmarks of superfluidity or superconductivity. The observation of vortex arrays in BECs has shown that superfluidity can manifest in finite systems confined into traps. Theoretical studies have predicted that vortices may also exist in ^{4}He nanodroplets, and that these vortices can be stabilized by pinning impurities to the vortex lines. The first direct evidence of quantum vortices was observed in 2014 in isolated ^{4}He droplets with radius about $10^2 - 10^3$ nm, containing about $10^8 - 10^{11}$ atoms. The inhomogeneous flow of the liquid through the nozzle is believed to originate rotating droplets, at speeds that could be very high. Evaporative cooling decreases the temperature down to 0.37 K, and eventually one has a beam of superfluid droplets. Rotational motion in a superfluid implies the formation of vortices, that the experimentalists expected to observe. The idea was to use a coherent X-ray beam of very high brightness produced by a free-electron laser, to visualize individual droplets and study their dynamical properties. The analysis of their diffraction images provided information about the shape of the droplets. The droplets are either spherical or spheroidal (in proportions of about 60-40 %, respectively). The observed shapes are therefore very different from the peanut-like structure that a normal liquid drop would adopt, as they rotate faster. A difficulty to observing vortices is raised by their low-density core, which does not provide sufficient contrast to image them. It was overcome by doping the droplets with Xe atoms. Indeed it is known that impurities pin vortex lines, so that they might guarantee stability for a time long enough to permit detection, acting at the same time as a contrast agent.

Approximately 50% of doped droplets contain vortex lattices. The identification of quantum vortices thus provided direct evidence of the superfluidity of helium droplets. Up to about 200 vortices were observed in a single droplet. The vortex density can be as high as 10^{10} cm^{-2}, associated to a rotation speed of about 10^6 s^{-1}. We recall that in bulk ^{4}He, the vortex density is 2000Ω cm^{-2} with Ω the angular velocity in s^{-1} (see Section 9.2.3). These figures are several orders of magnitude larger than those observed in experiments with bulk superfluid helium, which indicates a different regime of rotational excitations. The appearance of triangular vortex arrangements agrees with previous observation of triangular arrays of quantum vortices in rarified BEC.

Finally, let us come back to the question raised in Section 6.1.4 about the possibility of observing superfluidity in substances other than helium. Molecular parahydrogen pH_2 (see Section 6.1.4) seems a good candidate in spite of being a solid below $\simeq$ 13 K, but no evidence of superfluidity has been detected in the experiments performed insofar. However, theoretical calculations indicated that small pH_2 clusters can be superfluid: below about 2 K, the superfluid fractions in clusters containing $N = 13$ and 18 pH_2 molecules become large. These results inspired an elaborate experimental research based on the same idea which had provided evidence of superfluidity in ^{4}He clusters, that is to use a linear OCS molecule as a probe for measuring the effective moment of inertia. In the experiments, a large ^{4}He droplet was doped with a OCS molecule, and then with pH_2 molecules. All in all, an OCS molecule was surrounded by $N =$ 14-16 pH_2 molecules within a large ^{4}He droplet, the latter providing a temperature $\simeq$ 0.38 K. By coating the whole with a thick outer layer of ^{3}He atoms, the temperature is lowered at 0.15 K. The changes observed in the infrared spectra of the OCS-$(pH_2)_N$ cluster upon lowering the temperature indicated a large decrease of the effective moment of inertia. The results are consistent with the onset of superfluidity, thereby providing the first experimental evidence for superfluidity in a liquid other than helium, although this exception to the rule requires very special conditions.

9.3 NUCLEAR SUPERFLUIDITY

Since nucleons are fermions of spin 1/2, nuclear systems could be expected to share some generic properties of other fermion fluids composed of either electrons or ^{3}He atoms. In particular, a nuclear fluid could exhibit both superfluidity (involving neutrons) and superconductivity (involving charged protons). When we have discussed the liquid-drop mass formula in Chapter 7, we have added an additional term to take into account that most of stable nuclei are doubly even, that is to say an even number of both neutrons and protons.

The name pairing given to that term reflects the general feature that identical nucleons (either neutrons or protons) have a tendency to form coupled pairs in a state of zero angular momentum (spin up and spin down). These pairs of nucleons might be viewed as similar to the Cooper pairs encountered in superconductivity of metals or superfluidity of liquid ^{3}He. We will however see in Section 9.4.3 that some words of caution are necessary in this analogy. In the following, we will start with the case of finite nuclei and then move to infinite nuclear matter in neutron stars.

9.3.1 Superfluid nuclei

We here discuss two manifestations of superfluidity in nuclei, first in terms of separation energies, and second in relation with nuclei in rotation.

Separation energy of neutron

Experimental evidence supporting the existence of pairs of nucleons is provided by looking at the energies required to extract one or two nucleons out from a given nucleus. These energies are defined in terms of the binding energy $B(N, Z)$. For instance, consider the isotopes of a given element with atomic number Z. To extract a single neutron one requires an energy $S_n(N, Z) = B(N, Z) - B(N-1, Z)$, while to extract two neutrons the energy is $S_{2n}(N, Z) = B(N, Z) - B(N-2, Z)$. One defines analogously the separation energies of one and two protons for a series of isotones (nucleus with the same value of N).

In the left panel of Figure 9.9 are plotted these two separation energies as a function of the mass number A for isotopes of calcium ($Z = 20$). One can see that the energy $S_n(N, Z)$ has a clear saw-tooth shape. This behavior is sometimes called 'staggering'. It indicates that it requires more energy to remove a neutron from a nucleus when the number of neutrons is even than when it is odd. Indeed, to remove a neutron from an even N isotope, one has first to break a coupled pair, and that implies an additional energy cost. Notice also that this energy value varies smoothly with N, except for mass between $A = 40$ and 41, and $A = 48$ and 49 (respectively $N = 20$ and $N = 28$), that is when the number of neutrons coincides with a magic number. This explains the larger steps observed in separation energies, typical of a shell closure mechanism. We have already encountered such shell closure effects in the ionization potential of atoms (see Figure 2.6) and potassium clusters (see Figure 5.8). When looking at the separation energies $S_{2n}(N, Z)$ of two neutrons, the odd-even staggering is smoothed out, but the discontinuities at magic numbers remain.

The formation of pairs of nucleons is thus a well-established phenomenon, confirmed by a number of experimental observations. For instance, with no exception, all doubly-even nuclei have ground states with zero angular momentum, and their excitation energies spectra are different from those of odd-A nuclei. In the right panel of Figure 9.9 are displayed the energy spectrum for isotopes of tin ($Z = 50$) as a typical example. For doubly-even nuclei, the

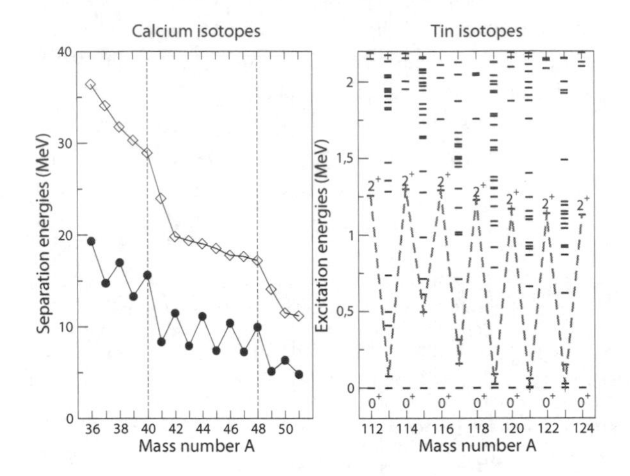

Figure 9.9: Left: separation energy of one (closed circles) or two neutrons (open diamonds) as a function of the mass number A for isotopes of calcium ($Z = 20$). Right: energy spectrum (short horizontal lines) for tin isotopes ($Z = 50$) with mass number A between 112 and 124. In doubly-even isotopes, the ground state is labeled 0^+ and the first excited state 2^+. The dashed line joints the first excited states.

first excited state has always an angular momentum 2^+, and lies at about 1.2 MeV above the ground state 0^+. We note that the second excited state lies far above the 2^+ state as well. The situation is very different for odd-A nuclei, which have more low-lying states and a denser spectrum. The dashed line in the figure joints the first excited states of the considered isotopes, and displays the characteristic odd-even staggering, similar to the separation energy of one neutron in the left panel of the figure.

Backbending

We have seen in Section 9.2 that a typical superfluid behavior can be identified in some experiments involving rotations. The rotation of nuclei, which we have discussed in Section 7.2.2, is another way to evidence nuclear superfluidity. We recall that, for a rigid rotor set to rotation about a given axis, say z, the moment of inertia $\mathcal{I}_z$ quantifies the mass distribution of the system in a plane normal to the axis z. If we now compare the rotation of a droplet made of normal liquid to that of a droplet of same shape but with a superfluid core (see right part of Figure 9.10), the moment of inertia of the former is greater than that of the latter. This is because the rotation of a solid body in a liquid is conditioned by viscosity, and superfluidity is precisely the absence of viscosity.

Experimentally and theoretically, the moments of inertia of deformed nuclei can be deduced from the excitation energies of the rotational spectra.

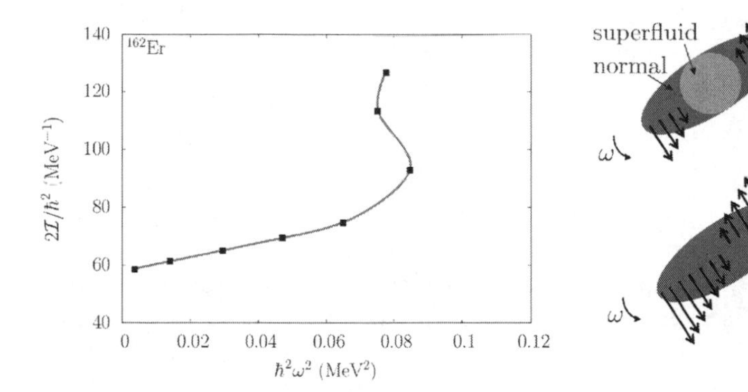

Figure 9.10: Left: nuclear moment of inertia $\mathcal{I}$, normalized to $\hbar^2/2$, as a function of $\hbar^2\omega^2$ with ω the angular velocity, for ^{162}Er ($Z = 68$). The dots are experimental points and the line is here to guide the eye. Adapted from A. Johnson, H. Ryde, and S. A. Hjorth, Nucl. Phys. A 179, p. 754, Copyright (1972), with permission from Elsevier. Right: schematic deformed nucleus filled of normal Fermi liquid (bottom) or with a superfluid core (top), set to rotation about an axis normal to the figure plane at an angular velocity ω. The arrows represent the local velocities which scale with the moment of inertia.

Some nuclei exhibit moments of inertia which depend on the total angular momentum J. In turn, J can be expressed as a function of the angular velocity ω. It has been observed that $\mathcal{I}$ varies approximately as ω^2. Therefore, if one plots $\mathcal{I}$ as a function of ω^2, one expects a linear behavior. The example of ^{162}Er is shown in the left part of Figure 9.10. Note that, since the total angular momentum J is quantized, $\mathcal{I}$ is not a continuous function of ω^2 and is measured at discreted values of J (see closed circles). Erbium ($Z = 68$) is the first nucleus in which superfluidity has been evidenced in the 1970s. The linear behavior of $\mathcal{I}$ with ω^2 was verified, up to $\hbar^2\omega^2 \simeq 0.06$ MeV^2. However, above this value, $\mathcal{I}$ suddenly jumps at a higher value. It has been even observed a bending backwards of $\mathcal{I}$, hence the term 'backbending' (see gray curve).

Various physical origins can explain such a behavior. This can be a sudden change of shape. But it can also be due to a phase transition from a superfluid state at low ω, energetically preferred, to a normal state at higher ω. Schematically, one can imagine a pair of nucleons which is more and more stretched as long as ω increases, until the pair breaks. $\mathcal{I}$ at the backbending angular velocity typically increases by a factor 2. We can draw the parallel with molecules in rotation in a ^{4}He droplet. In Section 9.2.5, we describe the case of an OCS molecule and report an increase of its moment of inertia by a factor 2.8 with respect to the moment of inertia if the OCS molecule is isolated. This has been explained by the formation of a 'donut' of non-superfluid ^{4}He around the OCS molecule. In some sense, the increase of the moment of inertia in some nuclei can be related to the destruction of paired nucleons, that is the appearance of more 'normal' nucleons (opposed to 'superfluid' nucleons when they are paired), inducing a higher moment of inertia.

9.3.2 Superfluidity in neutron stars

We now turn to bulk nuclear fluid such as in neutron stars. In 1959 following a detailed analysis of rotations of finite nuclei, it was suggested that nuclear matter could manifest superfluidity. Therefore, the possibility that nuclear matter could be superfluid and superconducting was raised a long time ago before the discovery in 1967 of pulsars, which are neutron stars spinning extremely rapidly.

We recall that neutron stars are extremely dense objects that form when massive stars run out of nuclear fuel and collapse on themselves (see Section 8.1.2). The enormous pressure within the star forces almost all of the protons and electrons together to form neutrons. One should actually use the term 'compact' star instead of 'neutron' star because the exact composition and structure of such a star is rather complex and still under debate. It contains neutrons, protons, electrons and also finite nuclei formed during the nuclear fission process of the original star, and during the collapsing process.

We give here a few more details on the internal structure of a compact star. There is a tiny atmosphere (a few cm width) of a hot plasma composed of ionized hydrogen, helium or carbon atoms. There is then a crust of $\simeq 1$ km depth, containing around 1% of the total mass. It consists of a lattice of neutron-rich nuclei immersed in a sea of unbound neutrons and a few percentage of protons and relativistic electrons. Neutrons and protons can form Cooper pairs, and thus superfluid and superconducting phases could coexist. More details on this pairing, in connection with the BCS-BEC crossover, will be addressed in Section 9.4.3. Going further inside the star, before the nuclei dissolve completely into the liquid of the core, the nuclear matter can develop other exotic configurations as well, i.e., rods, plates, tubes, and bubbles (sometimes called 'nuclear pasta'). Finally, the inner core composition is purely speculative. It could contain particles possessing strange quarks and a deconfined quark-gluon plasma. One can even imagine that identical quarks, being spin-1/2 fermions, could form Cooper pairs, leading to a superconducting phase. We recall that the 'color' charge of quarks is the equivalent of the electric charge (see Section 8.4). There could be different color-superconducting phases, depending on the quarks involved in this matter. Since there are 6 different quarks, each of them carrying 1 of the 3 different color charges, one easily realizes that there are a lot of different types of superconducting phases. For instance, if there are only u and d quarks, one talks about 2SC (2-flavor superconducting) phase. In spite of its small size and mass, the properties of inner crust matter, especially its superfluid properties, have important consequences for the dynamics and the thermodynamics of neutron stars.

We end this brief review on nuclear superfluidity by asking how superfluidity in neutron stars can be detected. Let us mention some famous pulsars as Vela, located at 100 light years from Earth, which is the remnant of a supernova that exploded 11400 years ago. The youngest pulsar ever observed is Cassiopeia A, born 330 years ago at a distance of about 11000 light years. The period of rotation of pulsars ranges from milliseconds to seconds. There are delays of a few milliseconds per year at most, hence providing the most accurate clocks in the Universe. However, irregularities have been detected in long-term pulsar timing observations. These pulsars exhibit sudden increases in their rotational frequency and spin-down rate, which are followed by a relaxation over days to years. This behavior can be associated to superfluidity and superconductivity. We have seen in Section 9.2.3 that a rotating superfluid is threaded by a regular array of quantized vortex lines, each carrying a quantum of angular momentum. Similarly, neutron stars are expected to contain a huge number of vortices, of order 10^{17} for a pulsar like Vela. The superfluid is supposed to be weakly coupled to the crust by mutual friction forces and to thus follow its spin-down via the motion of vortices away from the rotation axis unless vortices are pinned to the crust. In this case, the superfluid would rotate more rapidly than the crust. The lag between the superfluid and the crust would induce a rotational effect on the vortices thereby producing a crustal stress. At some point, the vortices would suddenly unpin, the superfluid would spin down and, by the conservation of angular momentum, the crust would spin up leading to a glitch.

9.4 BCS-BEC CROSSOVER

In this chapter, we have discussed two superphases, namely superfluidity of helium and superconductivity in metals. The origin of superfluidity in the bosonic ^{4}He is different from the one in the fermionic ^{3}He. BCS theory has been suggested to describe the superfluidity of ^{3}He by the pairing of two ^{3}He atoms of opposite spins, in the same way as superconductivity is related to the pairing of two electrons of opposite spins. This is why superconductivity is sometimes coined 'charged' superfluidity (in the case of electrons) or 'Fermi' superfluidity (in the case of ^{3}He). Also, at various places in this chapter, we have mentioned the similarities between a superconductor and a superfluid, as for instance the appearance of quantized vortices under certain conditions. Soon after the advent of the BCS theory, there were a lot of theoretical developments to figure out if BCS and BEC were not the two faces of the same metal, that is the two limits of a more unified description. This is the aim of this section to briefly address this fascinating question.

9.4.1 The need for a BCS-BEC crossover theory

The BCS approach is essentially a theory of arbitrarily weak interaction. Cooper pairs are not rigid objects: they rather correspond to transient structures which form and disappear constantly. Their size ξ is also much larger than the average distance between electrons. The overlap between Cooper pairs is therefore very high.

At the same time that BCS theory was successfully applied, some theoretical works considered the possibility of forming 'Bose molecules', that is *local* pairs of electrons of size ξ much smaller than the distance between molecules, which can scarcely overlap. These works demonstrated that a gas of interacting (charged) Bose molecules made of charged fermions can condense and become 'charged' superfluid. Of course, BCS theory was not able to describe such a system. In this context emerged the idea of a crossover between the BCS regime of large-size, heavily overlapping and weakly bound Cooper pairs, and the BEC regime of small-size, very little overlapping and strongly bound Bose molecules. These two systems are schematically depicted in the extreme sides of Figure 9.11, which will be explained in more detail later on. Then the

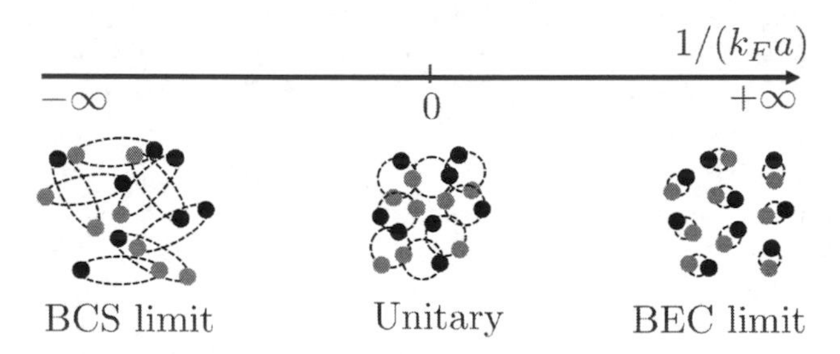

Figure 9.11: Schematic picture of an atomic Fermi superfluid (BCS limit) and a molecular BEC (BEC limit), separated by a unitary Fermi gas. A pair of fermions of opposite spins (black and gray dots), is schematically bound by dashes.

question arises, whether it is possible for a fermion system to evolve from a BCS superfluid state to a molecular BEC superfluid state. This evolution is called the BCS-BEC crossover.

To produce the BEC-BCS crossover, one should be able to change at will the interaction between fermions. This cannot be done in a versatile way in superconductors, liquid ^{3}He, nuclear matter... but in an ultracold Fermi gas trapped in a magnetic field it can be. We recall that the trapping magnetic field provides a tunable length, the s-wave scattering length a. From this length, one can build a dimensionless parameter by multiplying it by a momentum. In a Fermi system, the natural momentum scale is the Fermi momentum k_F. One then defines the so-called 'scattering parameter' as $1/(k_F a)$. An experiment on ultracold Fermi gases is amazingly versatile in the sense that, by playing with the magnetic field, one can change at the same time the value of a but also its *sign*. The scattering parameter $1/(k_F a)$ is represented in the upper horizontal axis of Figure 9.11. The BCS limit corresponds to $1/(k_F a) \to -\infty$ while the BEC limit is attained for $1/(k_F a) \to +\infty$. In between, there is a wide range of values of the scattering parameter for which one explores the BCS-BEC crossover region. In this region, the case of vanishing $1/(k_F a)$ is of particular interest: it is called the unitary limit. We will describe it in more detail in Section 9.4.4. For the time being, we summarize the characteristics of these three cases in Table 9.1.

Table 9.1: Comparison of physical quantities of a BCS, a BEC and a unitary system.

BCS	Unitary	BEC
Weak coupling $\frac{1}{k_F a} \ll -1$	$1/(k_F a) = 0$	Strong coupling $\frac{1}{k_F a} \gg +1$
No bound state	Quasibound state	Bound state
Large pair size $\xi \gg \frac{1}{k_F}$	Pair size $\xi \sim \frac{1}{k_F}$	Small pair size $\xi \ll \frac{1}{k_F}$
Pairing in momentum space		Pairing in real space
Strongly overlapping Cooper pairs	Strongly interacting gas	Tightly bound Bose molecules

The evolution from BEC to BCS or vice versa is not a phase transition, hence the name of crossover. As a matter of fact, BCS and BEC do represent the two endpoints of a continuum of quantum mechanical behavior. By varying the scattering parameter, one smoothly goes from one limit to the other one, via the unitary regime. We finally mention that the physics of the BCS-BEC crossover is in principle universal. It should be observed in any system of interacting fermions, regardless of the nature of the constituents: liquid ^{3}He, nuclear matter, neutron stars, electrons in a crystal lattice, quark-gluon plasma, ... This is why experiments in ultracold Fermi gases are of high interest since their results can be exportable to other systems not experimentally accessible.

9.4.2 BCS-BEC crossover in atomic Fermi gases

In Figure 9.11, the upper horizontal scale concerns the variation of the scattering parameter $1/(k_F a)$, allowing to go from the BCS limit to the BEC one. Because $k_F \propto \varrho^{1/3}$, this parameter can be varied by either changing a or ϱ. The latter solution will be addressed in Section 9.4.3. We now discuss the first option by the variation of a as usually done in ultracold Fermi gases.

Experiments are mainly performed with ^{40}K and ^{6}Li isotopes, which contain odd numbers of electrons, protons and neutrons, that is an odd number of fermions. The experimental achievement of reaching very low temperatures using evaporative cooling presents an additional difficulty for fermions, because the s-scattering length a is zero for atoms in the same hyperfine state, due to Pauli principle. We recall that hyperfine states arise from the interaction of the spin of the electron with the one of the atomic nucleus. To overcome the difficulty related to the hyperfine structure, one can mix atoms in different hyperfine states. The first experiments with the spin-9/2 ^{40}K atoms used mixture of atoms, some in the 7/2 hyperfine state and the other ones in the 9/2 hyperfine state. In other experiments, instead of mixing fermions from different hyperfine states, one mixes Fermi atoms with Bose atoms. This procedure is referred to as sympathetic cooling: bosons are cooled evaporatively and fermions through collisions with the bosons. Specifically, (bosonic) ^{87}Rb atoms were added to (fermionic) ^{40}K atoms, and the spin-3/2 fermion ^{6}Li atoms have been mixed either with (bosonic) ^{7}Li atoms or ^{23}Na atoms.

An important step forward was the discovery that by simply applying an external variable magnetic field, it is possible to arbitrarily modify the interaction between atoms, both in strength and sign. The mechanism for the variation of the interaction between atoms is based on the so-called Fano-Feshbach resonances which require that a pair of atoms can be formed in different spin states with different magnetic moments.

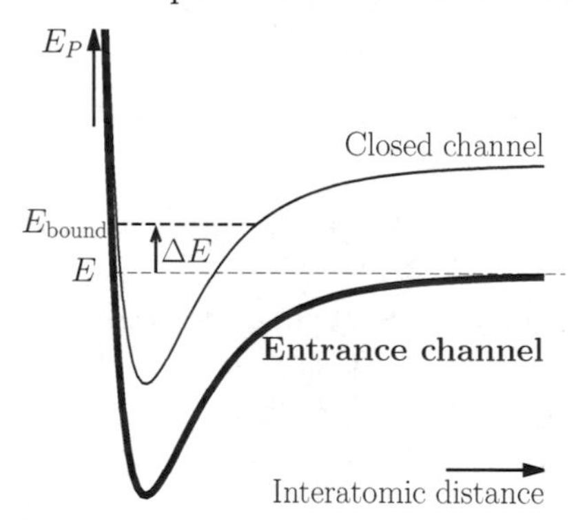

Figure 9.12: Potential energy of a pair of colliding atoms, in the entrance channel (thick curve) with an initial kinetic energy E, and in the closed channel (thin curve) with a bound state at E_{bound}, as a function of the distance between the two atoms.

The mechanism of a Fano-Feshbach resonance can be understood with the help of Figure 9.12. In the absence of an external magnetic field, we prepare two atoms far apart in a given spin state and with a given energy E, and we allow them to collide. For the sake of simplicity, only elastic collisions are considered (as most of the collisions in a Fermi gas experimentally are). When they approach each other, they feel an interatomic interaction whose potential energy is shown as the black curve in Figure 9.12 and coined 'entrance channel' (also often called 'open channel'). We suppose that there exists another channel, corresponding to an excited state of the pair of atoms, but in another spin state. This channel possesses a bound state at the energy E_{bound}. If $E_{\text{bound}} > E$, as in the figure, this channel is coined 'closed channel' because it is not accessible by the colliding atoms possessing an initial energy E. The mismatch in energy is denoted by $\Delta E = E_{\text{bound}} - E$. By applying an external magnetic field B, one can modify the value and the sign of ΔE. In particular, for a finite value of B_0 (see below), one can make the energy of the bound state in the closed channel match the energy in the entrance channel, hence the word 'resonance'. Coming back to Table 9.1, the BCS limit corresponds to $\Delta E > 0$ (no accessible bound state), the BEC limit to $\Delta E < 0$ (an accessible bound state), and the unitary limit to the resonance $\Delta E = 0$ (a quasibound state). The tuning of the Fano-Feshbach resonance by varying B produces an effective s-wave scattering length a_{eff} such that $a_{\text{eff}}/a \simeq 1 - \Delta/(B - B_0)$, where Δ and B_0 are two parameters characterizing the resonance. In short, varying the applied magnetic field B allows one to change the scattering length, henceforth the scattering parameter, and to go from the BCS limit to the BEC limit via the BCS-BEC crossover region, in the same physical ultracold Fermi gas.

The theoretical investigations of the BCS-BEC crossover region, partly motivated by the microscopic understanding of high T_c superconductors, have led to the numerical estimate of the critical condensation temperature of a Fermi gas. A kind of phase diagram of the crossover is shown in Figure 9.13. A reduced temperature, that is T normalized by the Fermi temperature T_F,

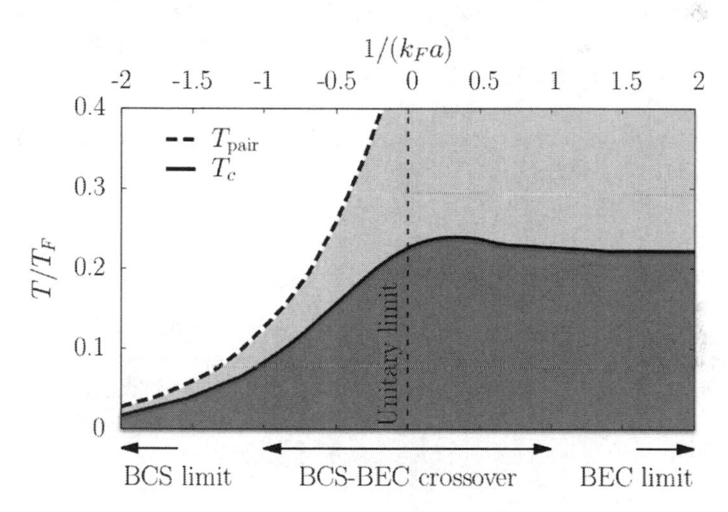

Figure 9.13: Reduced temperature T/T_F where T_F is the Fermi temperature of a noninteracting Fermi gas, as a function of $1/(k_F a)$ with k_F the Fermi momentum and a the s-scattering length. See text for details. Adapted from C. A. R. Sá de Melo, M. Randeria, and J. R. Engelbrecht, Phys. Rev. Lett. 71, 302. Copyrighted (1993) by the American Physical Society.

is plotted as a function of the scattering parameter $1/(k_F a)$. There are two curves in this figure. The dashed line corresponds to the critical temperature T_{pair} below which pairs of fermions can form (light gray area). Above T_{pair}, the system is a normal gas of unbound fermions. The solid curve shows the critical temperature T_c for the Bose-Einstein condensation, below which the system is a BEC (dark gray area).

We observe that T_c is typically a fraction of T_F, the highest T_c being equal to $\simeq 0.24 T_F$. This maximum is not attained at the unitary limit but slightly beyond (various numerical estimates give T_c/T_F varying between 0.15

and 0.17). Interestingly, $T_{\rm pair} \simeq T_c$ in the BCS limit, meaning that Cooper pairs form and condense at the same temperature. $T_{\rm pair}$ starts to significantly differ from T_c only when one approaches the BCS-BEC crossover region and beyond. For $1/(k_F a) \gtrsim -1$, there is a range of temperature between $T_{\rm pair}$ and T_c in which the system is a mixture of individual fermions and preformed pairs. The Bose-Einstein condensation is then obtained only at much lower temperatures ($< T_c$). As already mentioned in Section 6.4.1, ultracold atomic gas experiments are performed at a fraction of μK. This temperature is indeed very low at a human scale, hence the term 'ultracold'. However, in terms of T_F, such a system is quite 'hot' in the unitary and BEC limits since $T_c/T_F \sim 0.2$, significantly larger than the value in superconductors ($T_c/T_F \sim 10^{-3}$ for conventional ones, $\sim 1/30$ to $\sim 1/20$ for high-T_c ones).

As an illustration of the experimental exploration of the BCS-BEC crossover, we mention here that lattices of quantized vortices have been observed in the same ultracold ^{6}Li gas in the three regimes, BCS, unitary and BEC, by varying the scattering parameter. We have seen that such quantized vortices are a clear signature of either superconductivity in the BCS limit (see Section 9.1.2) or superfluidity in the BEC limit (see Section 9.2.3). The fact that quantized vortices are also observed in the unitary limit demonstrates its superfluid character.

9.4.3 BCS-BEC crossover in nuclei and nuclear matter

We have seen that ultracold Fermi gases provide a unique table-top experimental setup to explore the BCS-BEC crossover with the possibility of varying the s-wave scattering length a. We have nevertheless mentioned that another option for the variation of the scattering parameter $1/(k_F a)$ consists in changing the density of the Fermi gas since $k_F \propto \varrho^{1/3}$. This has been indeed used in ultracold Fermi gases with density going from 10^{13} cm^{-3} down to 10^{10} cm^{-3}. The variation of the density is also possible in ^{3}He liquid under pressure. Note that in high-T_c superconductors, the variation of the scattering parameter is performed by doping (that is inclusion of specific atoms in the material) which directly modifies the Fermi energy, hence the Fermi momentum. The range of variation however remains limited, as compared to the high controllability of ultracold Fermi gases. We here discuss the case of nuclei and nuclear matter in which one can also vary the density and then explore the BCS-BEC crossover to some extent.

The first question to ask is the validity of the BCS theory in nuclear matter. In Section 9.3, we have discussed pairing effects in nuclei and in neutron stars but we have not related them to the existence of weakly bound Cooper pairs. The fact that BCS can be applicable in nuclear matter might sound surprising, though, because among the four fundamental interactions, the strong nuclear one is the most intense force in the Universe. However, some estimates of the scattering length in nuclear matter give $a = -18$ fm for a neutron-neutron pair and $a = 5$ fm for a proton-neutron pair. We recall that the typical distance between nucleons is about $d \simeq 2$ fm. We are therefore in the regime $|a| > d$: the size of a fermion pair (especially a neutron-neutron pair) is significantly larger than the internucleon distance. There is thus a strong overlap between the various pairs. And this is in this context that BCS is applied in low-density nuclear matter. Such a low density might be found in the inner crust of a neutron star. Indeed, we have mentioned in Section 9.3.2 that in this crust, there is probably a neutron liquid (and small proportions of protons and electrons) immersed in a lattice of neutron-rich nuclei. The density of this neutron liquid might be low enough to allow the formation of Cooper pairs. And this is of course in relation with the fact that a neutron star might embed a very large number of quantized vortices, the latters being again a signature of superfluidity.

As also mentioned, matter in a neutron star mostly contains neutrons, but also some protons. One can then expect to have three types of Cooper pairs: neutron-neutron, neutron-proton (forming a deuteron) and proton-proton. Henceforth, a superfluid phase in nuclear matter would be a mixture of condensates, analogous to mixtures of ^{4}He and ^{3}He. Note however that in the case of helium, only two kinds of condensate can exist, namely the one from ^{4}He (condensate of atomic Bose gas) and the one from ^{3}He (condensate of molecular Bose gas). In nuclear matter, we have in the three cases a condensate of 'molecular' Bose gas.

The case of finite nuclei can also be considered in the study of the BCS-BEC crossover. Indeed, in 'exotic' nuclei (with a great imbalance between neutrons and protons) which can be produced experimentally, neutron-rich nuclei—very far away from the stability line—exhibit the presence of neutron pairs (referred to as 'halo' or 'skin') at large distance from the nuclear core. In particular ^{6}He, ^{8}He, ^{11}Li, ^{14}Be are known to be halo nuclei. In the neutron halo, the density can be as low as 10% or even 1% the saturation density ϱ_0. This allows one to explore the crossover region to some extent. Note however that even with $\varrho = 0.01\varrho_0$, one should not expect to reach the unitary limit in nuclei's halos.

We finally mention the case of α clustering in nuclei. We recall that the α particle is the nucleus of a ^{4}He atom, consisting of 2 neutrons and 2 protons. The fact that the α particle presents a very high binding energy (see Section 7.1.2) is a clear manifestation of nuclear pairing. The idea that α particles are the building blocks of nuclei makes sense if one thinks about α radioactive decay of heavy nuclei. Henceforth, instead of considering a homogeneous distribution of neutrons and protons inside a nucleus, one can imagine a cluster of α particles and consider interactions between α particles instead of interactions between nucleons. Such a nucleus with a cluster's structure would be, in some sense, a very small condensate of 'molecular' Bose gas. There are however many difficulties to describe, at the theoretical level, α clusters, for instance in accordance with nuclear collective modes. There is also a lack of model-independent observables of such clustering to conclude on the existence of α clustering in nuclei containing a number of nucleons multiple of 4.

9.4.4 The unitary regime for atoms in a trap

We end this section on the BCS-BEC crossover by giving more details on the unitary limit, where the scattering length a becomes divergent. It is attained by exploiting resonant scattering (and thus very large cross sections corresponding to very large scattering lengths). Such a situation may even occur in a regime where the system is very dilute, leading to the apparent paradox of a very dilute but strongly interacting system. Because the system is dilute, the range of the interatomic potential becomes small as compared to the interatomic distance. But at the same time, the effective interaction length becomes much larger than the interatomic distance. This has the consequence that specific length scales associated to interactions do not enter the problem anymore, so that the system is expected to exhibit a truly universal behavior.

A unitary Fermi gas can be rather simply described by integrating correlations into an energy density functional which allows one to obtain a surprisingly simple description. The key to this simplicity lies in the fact that, in the unitary limit, one can establish a simple scaling behavior because one length scale, the s wave scattering length a, effectively disappears from the picture. At zero temperature, the energy of the system can then be written as:

$$\frac{E}{N} = \frac{3}{5}\frac{\hbar^2 k_{\mathrm{F}}^2}{2m}(1+\beta) \quad . \tag{9.13}$$

The above form thus appears as a simple renormalization of the Fermi gas expression Eq. (3.27): interaction effects are included effectively through the β parameter. This parameter can be estimated by more elaborate calculations, predicting a value of $\beta = -0.58$.

For the sake of simplicity and because it is a widely used potential, one can consider, a harmonic external potential. It is then relatively simple to obtain the ground state of the system by minimizing its total energy composed of the 'enhanced' kinetic energy Eq. (9.13) and the potential energy due to the harmonic trap. In this context, an analytical expression for the density is relatively easy to establish. In the case of a unitary ^{6}Li gas, the agreement between the simple picture briefly drawn above and the experiment is excellent. We also mention that the comparison with the free Fermi gas demonstrates a strong deviation with respect to the experimental data. This demonstrates that a unitary gas is a strongly interacting system.

To further illustrate the role of interactions in the unitary limit, we consider a Fermi gas confined in a 2D trap and suddenly release the trap. One can look at the time evolution of the spatial distribution. We have seen in Section 8.4.3 that in a relativistic heavy ion collision, there is initially a strong spatial anisotropy which is converted into a momentum anisotropy via pressure effects. The elliptic flow v_2 is a measure of the efficiency of this conversion. Such a quantity has also been studied in ultracold Fermi gases, as is illustrated in Figure 9.14 for a ^{6}Li gas. Experimental data are compared with two theoretical calculations, namely ideal and viscous hydrodynamics.

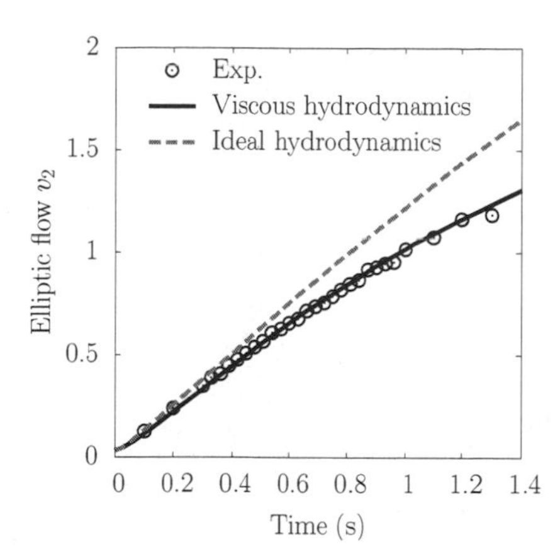

Figure 9.14: Elliptic flow of a unitary ultracold dilute gas of ^{6}Li, as a function of time after release of the trap. Adapted with permission from K. Dusling and T. Schäfer, Phys. Rev. A 84, p. 013622, copyrighted (2011) by the American Physical Society.

There is obviously the need to include dissipative effects through a shear viscosity η in the theory to be able to fit the experimental measurements. This is thus a clear indication that a unitary Fermi gas can be at the same time very dilute and strongly interacting. Comparisons between experimental data and viscous hydrodynamics allow one to extract values of the ratio η/s (with s the entropy density), as in the quark-gluon plasma produced in a relativistic heavy ion collision. We recall that η/s is related to the inverse of the Reynolds number (see Section 8.4.3), and that the smaller η/s, the more 'ideal' the liquid. For ^{6}Li gases, we have $\eta/s \lesssim 0.5\ \hbar/k_{\mathrm{B}}$. This upper value is only one order of magnitude above the lower theoretical bound, which is either $1/(4\pi) \simeq 0.0796$ or $16/25/(4\pi) \simeq 0.0509$, making an ultracold Fermi gas in the unitary regime not very far from the ultimate ideal quantum liquid.

As a final word, we would like to note that the issue of BCS-BEC crossover theory allows a fertile cross-talk among physicists from various fields, from nuclear physics, astrophysics, atomic physics, to condensed matter physics, and even quantum gravity.

9.5 CONCLUDING REMARKS

The term superphase we have entitled this chapter with is a construction trying to point out the specific properties of quantum liquids in some peculiar conditions. By this, we have meant the capability of some quantum fluids to flow better than a fluid in its 'normal' phase. The two paragons of such behaviors are superconductivity and superfluidity. In the former, the electron fluid flows with no resistance, a dream for technical application involving current transport. In superfluidity, helium has lost all viscosity and flows perfectly, even against gravity. Both superconductivity and superfluidity are attained at very low temperatures and dominated by subtle quantum effects. Both have attracted numerous investigations and raised numerous challenges to understand them. There remain intriguing questions attached in particular to the new generation of superconductors existing at relatively high temperatures. The field thus remains largely open.

We have also explored superfluidity effects in finite systems. Numerous investigations have been led in droplets of helium and para-hydrogen. But the most dominant share concerns nuclei which exhibit well established pairing effects. The latters are responsible for numerous specific nuclear properties, at the side of separation energies in particular, and more generally in collective motion, such as in rotational motion. This aspect is in fact central in nuclear physics.

We have concluded this chapter by addressing a more formal and highly studied question concerning the nature and the relationship of the superphases (superconductivity and superfluidity) addressed in this chapter. This issue is again addressed in nuclei and high-T_c superconductors. But this is in atomic Fermi gases that this question is fully explored since, thanks to their flexibility, they constitute a remarkable playground for fundamental investigations.

Conclusion

The diversity of quantum liquids

Quantum liquids are everywhere. This is a basic message we have tried to deliver throughout this book. But realizing the existence of quantum liquids is by no means the end of the story. An even more important issue is the fact that the various systems we have discussed in this book share common properties, precisely linked to their liquid nature. And realizing the existence of such generic trends in such different systems is a key issue by itself, as it provides a synthetic overview of whole fields of physics. In this concluding chapter, we would like to recall a few basic aspects somewhat scattered all over this book and put them again in perspective, taking for illustration a few key examples.

A first aspect to be mentioned again is the fact that quantum liquids, as analyzed here, cover a huge range of systems and associated spatial, energy and time scales. We have mostly focused our discussions on three kinds of systems: electrons as observed in metals or clusters, atoms (especially helium) as explored at very low temperatures and nucleons as constituents of nuclei and neutron stars. Recent years have also seen the emergence of studies on 'artificial' systems, such as in particular quantum dots and systems made of trapped cold atoms. Without entering details again, widely documented in dedicated chapters, let us briefly recall that these various systems, themselves constituted of different constituents, cover an impressive range of scales. Leaving aside white dwarfs and neutron stars, we reach 7 orders of magnitude for length scales, from nucleons (down to 10^{-15} m) to quantum dots (up to the 10^{-8} m). In terms of energies, the range may be even larger from MeV/GeV in tightly bound nuclear systems, through eV (6-9 orders of magnitude difference) in softer bound metals, down to K$\sim 10^{-4}$ eV (10-13 orders of magnitude difference) in hardly bound helium and even less in quantum dots (mK) and atomic traps (nK). We have less discussed time scales which, by definition, depend on the actual dynamical scenario. But they also cover an impressive range from fm/c ($\sim 3 \times 10^{-24}$ s) in nuclear collisions to astronomical times in the billion of years range.

The immediate question stemming up here is how such apparently different systems may be associated to a common concept. The answer of this question was precisely the aim of this book. A simple clue to resolve this apparent paradox is provided by the characteristic length r_s existing in most of these systems, and known as the Wigner-Seitz radius. This quantity is the key to the scaling of the radius R of a finite system of N fermions as $R \sim r_s N^{1/3}$. This saturation property provides the same average density of such systems for all sizes. The effect is well known in nuclei but also observed in metal clusters and helium droplets, these three systems already covering most of the energy/size range under consideration here. The Wigner-Seitz radius furthermore provides a natural energy scale via the Fermi energy ε_F of the least bound fermion, with associated Fermi momentum $\hbar k_F$. This momentum in turn provides another interesting scale, namely de Broglie wavelength in the ground state, $\lambda_B \sim 2\pi/k_F \sim \pi r_s$, which gives an estimate of the quantal character of the system. In all considered fermion systems, the typical distance between constituents is typically of order 1–2 r_s, smaller than λ_B. Their quantal character therefore fully expresses itself.

Before proceeding to more detailed comparisons, there remains one point which deserves some discussion, namely the nature of the dominating interaction. Let us omit the extreme cases such as neutron stars, which heavily involve gravitation, to focus on the smaller range between nuclei and helium droplets. The dominant interactions are then the nuclear one (in nuclei) and the Coulomb one (in nuclei and in any electronic system), the latter being furthermore well known to dominate in most physical systems, from the atom to the human body. It should be noted that both nuclear and Coulomb interactions are primarily singular at short distance. This should make their handling delicate. In fact, it turns out that for example in nuclei the net interaction between two nucleons is strongly suppressed at short distance by virtue of the Pauli exclusion principle. In a somewhat similar way, valence electrons ensuring binding in a metal cluster 'see' no truly bare ionic charges but a rather smoothed ionic density. In helium droplets, binding stems from van der Waals interactions, and thus corresponds to a strongly reduced Coulomb interaction. There remains of course the net repulsion between electrons in a material or between protons in a nucleus. All in all, these various systems can therefore be viewed as moderately interacting systems without involving singular interaction processes. This furthermore allows one to approximately describe them as quasi-independent particles inside a common average potential (whence the success of mean-field approaches used in many systems).

Reasoning on the strength of the interactions alone is however not sufficient to infer that the system is liquid. Indeed, classically, the stronger the interaction between constituents of a system, the more likely the system is in a solid state, even if this interaction is described at a quantal level. We have shown that the competition between interactions and quantal effects, as the zero point motion and the quantum statistics, can create disorder such as the system is indeed in a liquid state (and sometimes, in a 'superliquid' state in the case of a superconductor or a superfluid).

To summarize the above considerations, we could indeed conclude that in spite of their apparent diversity, quantum liquids exhibit 'similar' properties which allow rather direct comparisons between them. This is the aim of the following discussion to address such comparisons in some more detail.

Potentials and densities

Following the above discussion, we are now in a position to compare various different systems, using the common spatial and energy scales stemming from the Wigner-Seitz radius r_s. A first comparison consists in considering ground state potentials and associated densities, properly scaled (length scale in r_s, energy scale in ε_F).

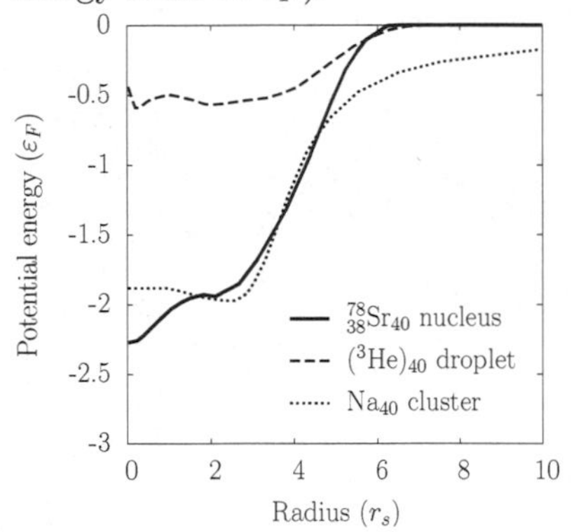

Figure 1: Mean-field potentials for a nucleus, a helium droplet, and a sodium cluster, all with 40 constituting particles. Note that only the neutron part is shown in the nuclear case. Natural units are used on the basis of Wigner-Seitz radius: lengths in units of r_s and potentials in units of ε_F.

We compare in Figure 1 three emblematic systems, namely an atomic nucleus, the nucleus ^{78}Sr (38 protons and 40 neutrons), the Na_{40} cluster and a helium droplet with 40 ^3He atoms. The first important point to note is that the three systems do fit into one single figure. This means that they have comparable scales when expressed in natural units. They exhibit the same spatial extension, which reflects the underlying saturation property introduced by r_s. Differences can also be spotted, for example in the depth of the potential wells and in the asymptotic behaviors. The cluster and the nucleus exhibit a comparable potential depth, not surprisingly larger than the helium one. This mostly reflects the (relative) faintness of the interaction between two He atoms.

Conversely, asymptotic behaviors of the nuclear and helium cases are both exponential, reflecting the dominance of a short range interaction, while the cluster case exhibits a typical long-range Coulomb behavior. Still, in spite of these small differences, the striking feature remains the remarkable qualitative and, to a large extent quantitative, similarity between the three considered cases.

Shell closures

The occurrence of shells in the energy spectrum of finite fermion systems is a generic feature. The quantization of energy levels directly stems from the boundary conditions set by a potential to the wave functions, so that only selected (and separated) energies become possible. In general, these quantized energy levels tend to gather in bunches of more or less degenerate values forming energy shells. The effect tends to disappear for very large energies or fermion numbers because of the occurrence of too dense spectra. This is exactly the mechanism for the appearance of valence and conduction bands in bulk metal for instance. Apart from this extreme case, typical finite fermion systems exhibit a well marked shell structure associated to a sequence of magic numbers at shell closure. The precise values of these numbers depend on the characteristics of the confining potential.

The existence of shell structure is essential to explain many physical and chemical properties. Indeed, the existence of an energy gap between two shells implies an enhanced stability of species at shell closure. This shows up in many observables such as ionization potentials and radii in atoms and clusters, or in nuclear abundances, to cite a few. The point is illustrated in Figure 2 which compares similar physical quantities (ionization potentials, separation energies) as a function of the system size. The sequence of nuclear shell closures

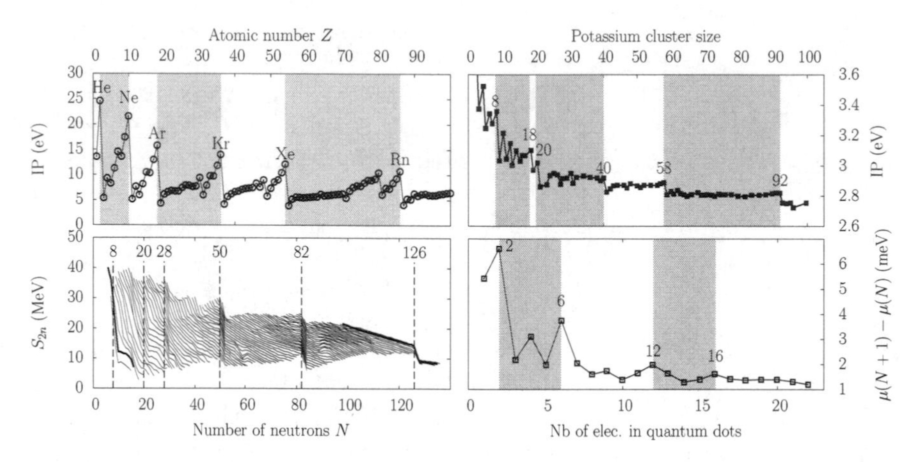

Figure 2: Illustration of magic numbers associated to shell closures in finite fermion systems as a physical quantity displayed as a function of the system size. Top left: ionization potentials in atoms. Top right: ionization potential in potassium clusters. Bottom left: separation energies of two neutrons (analog to ionization potential) in nuclei. Bottom right: differences in the chemical potential (corresponding to ionization potentials) of disk-shaped quantum dots (figure adapted with permission from S. Tarucha et al., Phys. Rev. Lett. 77 (1996) 3613; copyrighted by the American Physical Society).

at 2, 8, 20, 28,... leads to marked steps in the separation energies (here, of two neutrons). This case is structurally different from the atomic case because of the self-bound nature of nuclei, in which the actual potential stems from nucleon-nucleon interactions with no external factor. The atomic case is very similar to that of simple metal clusters (here potassium) which is illustrated in the upper right panel of Figure 2. In that case, the confining potential is a balance between ionic background attraction and purely electronic effects, in particular exchange. This balance provides shell closures at 2, 8, 20, 40, 58, and 92. Finally the case of quantum dots is illustrated in the lower right panel of the figure. In quantum dots, the confining potential, a harmonic oscillator, is exclusively external and the associated shell closure sequence is then at 2, 6, 12, ...

Transport

We have discussed above properties of the ground state and emphasized the common aspects in the quantum liquids discussed in this book. We end here with a more dynamical quantity, archetypal of a fluid behavior. We have seen in Chapter 1 that one characteristic of a fluid is the shear velocity η. From one fluid to another one, η can vary by several orders of magnitude. For instance, $\eta = 1.2 \times 10^{-6}$ Pa.s for ^{4}He at 2 K and under 1 bar, while for water, $\eta = 2.9 \times 10^{-4}$ Pa.s at standard pressure and temperatures but decreases to 6.0×10^{-5} Pa.s under 226 bars and at 650 K. However, in terms of dynamical behavior, one should rather consider the Reynolds number Re (see Eq. (1.11)), which depends on the ratio of η over the entropy density s. The smaller η/s, the larger the Reynolds number and the 'better' the fluid flow.

In Section 8.4.3, we have given an estimate of the lower bound for η/s, that is $\alpha/(4\pi)\,\hbar/k_\mathrm{B}$ with $\alpha = 1$ or 16/25. At LHC, the analysis of the elliptic flow allows to deduce $\eta/s = 0.16\hbar/k_\mathrm{B}$ for the quark-gluon plasma. For ^{4}He at 2 K and 1 bar and for water at 226 bars and 650 K, we have $\eta/s =$ 1.9 $\hbar/k_\mathrm{B}$ and 2.0 $\hbar/k_\mathrm{B}$ respectively, therefore significantly higher than this lower bound. On the contrary, as we have seen in Section 9.4.4, in ultracold dilute gases of ^{6}Li ($P = 12 \times 10^{-9}$ Pa and $T = 23 \times 10^{-6}$ K), it has been measured $\eta/s \lesssim 0.5\ \hbar/k_\mathrm{B}$. Figure 3 compiles the measurements of η/s in various liquids, and especially in the quantum liquids explored in this book. The ratio η/s is plotted as a function of temperature, itself measured relative to the critical temperature of the considered system. The first noticeable point from this figure is the fact that the various considered cases enter the same figure, as does the potential depicted in Figure 1. This is again an important message emphasizing similarities. Still, a word of caution is necessary here as the vertical scale is logarithmic. This scale somewhat separates the various cases, although not really much more than by an order of magnitude. Looking in more detail at the figure nevertheless shows that the evolution is rather gradual, starting with the largest values in the two classical fluids H_2O and N_2, down to the very particular case of the bosonic ^{4}He superfluid and a unitary ultracold ^{6}Li gas. Finally the quark-gluon plasma measurement delivers the smallest value of the ratio, typically 2–3 times smaller than in the ^{6}Li case and rather close to theoretical values. One thus sees that the quark-gluon plasma provides the smallest value of η/s which makes it the best fluid ever experimentally realized, in spite of the high energies and densities involved to create it.

To conclude this chapter, and consequently this book, we would like to say a few words on an aspect which has been addressed at a few places but which has also been pending in many other discussions. It concerns the actual sizes of the systems we have considered. By this, we do not mean actual scales but rather the number of constituents of the systems. And a correlated question is under what form(s) a given system can exist. This addresses the possibility, for given finite systems, of the existence of a corresponding 'bulk' material, i.e. the arbitrary scalability of a system. Clusters, and in particular metal clusters,

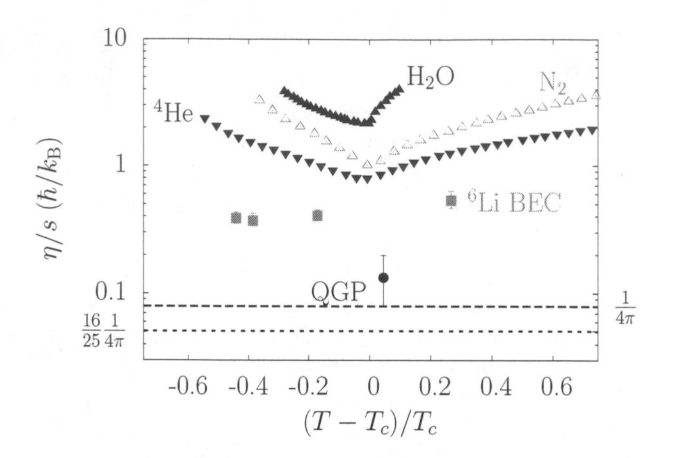

Figure 3: Ratio of shear viscosity η over density of entropy s, in units of $\hbar/k_B$, for H_2O, N_2 and ^4He (data from NIST Chemistry WebBook, http://webbook.nist.gov), for a Bose-Einstein condensate of ^6Li and for the quark-gluon plasma formed at LHC (adapted from A. Adams, et al., New J. Phys. 14 (2012) 115009 under the terms of the Creative Commons Attribution-NonCommercial-ShareAlike 3.0 licence). The theoretical values at $1/(4\pi)$ and $16/25/(4\pi)$ are reprinted (abstract) with permission respectively from P. K. Kovtun, D. T. Son, and A. O. Starinets, Phys. Rev. Lett. 94, 111601 (2005) and from M. Brigante, H. Liu, R. C. Myers, S. Shenker, and S. Yaida, Phys. Rev. Lett. 100, 191601 (2008); copyright (2005 and 2008) by the American Physical Society.

are the ideal scalable systems which can be produced at any size from the small molecule up to bulk metal. Helium droplets, being also clusters, can be extended from few units of atoms up to the bulk liquid helium. The case of nuclear matter is also quite interesting in this respect. It is often considered as the bulk limit of nuclei, while nuclei are strongly limited in size. This is conceivable once avoiding the Coulomb interaction which sets an upper limit of nuclear stability. This is achieved in neutron stars where a virtually infinite phase of nuclear matter exists. Fermion traps, in turn, are artificial systems which can be tuned, in principle, to any size.

To a large extent, the quantum systems explored in this book may appear as finite droplets and/or infinite bulk phases... of liquid of course. An interesting aspect in these systems is to possibly interpolate between fully microscopic systems and macroscopic materials. And this is probably not the least surprise to realize how important quantum mechanics appears in both cases, in microscopic systems, as usually expected, and in macroscopic systems, as certainly less expected.

Appendices

A. CONSTANTS, UNITS AND PREFIXES

Gas constant: $R = 8.314\,460\ \mathrm{J\,mol^{-1}\,K^{-1}}$
Avogadro's number or constant: $\mathcal{N}_A = 6.022\,141 \times 10^{23}\ \mathrm{mol^{-1}}$
Electron charge magnitude: $e = 1.602\,176 \times 10^{-19}$ C
Electron-Volt: 1 eV $= 1.602\,176 \times 10^{-19}$ J
Angstrom: 1 Å $= 10^{-10}$ m
Planck's constant: $h = 6.626\,070 \times 10^{-34}$ J s
Reduced Planck's constant: $\hbar = h/2\pi = 1.054\,571 \times 10^{-34}$ J s
Speed of light in vacuum: $c = 2.99\,792 \times 10^{8}\ \mathrm{m\ s^{-1}}$
Boltzmann's constant: $k_{\mathrm{B}} = 1.380\,649 \times 10^{-23}\ \mathrm{J\,K^{-1}} = 8.617\,332 \times 10^{-5}\ \mathrm{cV\,K^{-1}}$
Electron mass: $m_e = 9.109\,383 \times 10^{-31}$ kg $= 0.510\,999$ MeV/c^2
Proton mass: $m_p = 1.672\,622 \times 10^{-27}$ kg $= 938.272$ MeV/c^2
Neutron mass: $m_p = 1.675\,303 \times 10^{-27}$ kg $= 939.565$ MeV/c^2
Bohr radius: $a_0 = \hbar^2/m_e e^2 = 0.29\,177 \times 10^{-10}$ m
Rydberg: 1 Ry $= 13.605\,804$ eV
Absolute zero of temperature: 0 K $= -273.15$°C

For electromagnetic properties, the Gaussian system of units is used.

Standard conditions of temperature and pressure: 25°C and 1 atm $= 1.013 \times 10^5$ Pa. Normal conditions of temperature and pressure: 20°C and 1 atm.

Table of multiples and submultiples, prefixes and symbols.

prefix	value	symbol	prefix	value	symbol
femto	10^{-15}	f	deca	10	da
pico	10^{-12}	p	hecto	10^{2}	h
nano	10^{-9}	n	kilo	10^{3}	k
micro	10^{-6}	μ	mega	10^{6}	M
milli	10^{-3}	m	giga	10^{9}	G
centi	10^{-2}	c	tera	10^{12}	T
deci	10^{-1}	d	peta	10^{15}	P

B. THE PERIODIC TABLE OF CHEMICAL ELEMENTS

The periodic table of chemical elements, also known as the Mendeleev's table, is shown below. The figures in the table indicate the atomic number Z.

1 H																	2 He
3 Li	4 Be											5 B	6 C	7 N	8 O	9 F	10 Ne
11 Na	12 Mg											13 Al	14 Si	15 P	16 S	17 Cl	18 Ar
19 K	20 Ca	21 Sc	22 Ti	23 V	24 Cr	25 Mn	26 Fe	27 Co	28 Ni	29 Cu	30 Zn	31 Ga	32 Ge	33 As	34 Se	35 Br	36 Kr
37 Rb	38 Sr	39 Y	40 Zr	41 Nb	42 Mo	43 Tc	44 Ru	45 Rh	46 Pd	47 Ag	48 Cd	49 In	50 Sn	51 Sb	52 Te	53 I	54 Xe
55 Cs	56 Ba	★	72 Hf	73 Ta	74 W	75 Re	76 Os	77 Ir	78 Pt	79 Au	80 Hg	81 Tl	82 Pb	83 Bi	84 Po	85 At	86 Rn
87 Fr	88 Ra	#	104 Rf	105 Db	106 Sg	107 Bh	108 Hs	109 Mt	110 Ds	111 Rg	112 Cn	113 Nh	114 Fl	115 Mc	116 Lv	117 Ts	118 Og

Lanthanide series ★	57 La	58 Ce	59 Pr	60 Nd	61 Pm	62 Sm	63 Eu	64 Gd	65 Tb	66 Dy	67 Ho	68 Er	69 Tm	70 Yb	71 Lu
Actinide series #	89 Ac	90 Th	91 Pa	92 U	93 Np	94 Pu	95 Am	96 Cm	97 Bk	98 Cf	99 Es	100 Fm	101 Md	102 No	103 Lr

C. ELEMENTARY PARTICLES AND INTERACTIONS

The present knowledge about elementary particles and fundamental interactions is provided by the so-called 'standard model'. It excludes gravitation, for which a quantum theory is still lacking. The basic ingredients of the standard model are summarized in the following two tables.

	I	II	III
Quarks	up (u) $q = 2/3$ $\simeq 2.3$ MeV/c^2	charm (c) $q = 2/3$ $\simeq 1.3$ GeV/c^2	top (t) $q = 2/3$ $\simeq 173$ GeV/c^2
	down (d) $q = -1/3$ $\simeq 4.8$ MeV/c^2	strange (s) $q = -1/3$ $\simeq 95$ MeV/c^2	bottom (b) $q = -1/3$ $\simeq 4.2$ GeV/c^2
Leptons	electron (e^-) $q = -1$ 0.511 MeV/c^2	muon (μ^-) $q = -1$ 105.7 MeV/c^2	tau (τ^-) $q = -1$ 1.777 GeV/c^2
	electron neutrino (ν_e) $q = 0$ < 2.2 eV/c^2	muon neutrino (ν_μ) $q = 0$ < 0.17 MeV/c^2	tau neutrino (ν_τ) $q = 0$ < 15.5 MeV/c^2

The elementary particles are grouped in three generations (or flavors), as shown in the table above. Electric charges q (in unit of the elementary charge

e) and masses (in MeV/c^2 or GeV/c^2) are given for each particle. Any of them has its antimatter counterpart, or antiparticle, which has the same mass but the opposite electric charge, and adds the prefix 'anti' to the original particle (e.g. anti-up quark, anti-muon, etc.). For historical reasons, the anti-electron is also named as positron.

Modern physics describes an interaction between two particles of matter by the exchange of a force carrier (also called gauge boson). The most known one is the photon which is the mediator of the electromagnetic interaction. The quantum theory of the electromagnetic interaction is Quantum Electro-Dynamics (QED). Gluons (8 in total) are the mediators of the strong nuclear force and bind the quarks together. The photon and gluons have no mass and no electric charge. Gluons possess however another type of charge, the color charge, which allows physicists to model their interactions with quarks. The quantum theory of quarks and gluons is Quantum ChromoDynamics (QCD). The Z^0, W^- and W^+ are the carriers of the weak nuclear force. They possess a very large mass and can be exchanged between any particle of matter. At very high energies ($>$ 100 GeV), the weak force and the electromagnetic interaction can be unified into the electroweak interaction. The electroweak theory and QCD constitute the so-called Standard Model. In this model, the Higgs boson interacts with fermions and the carriers of the weak interaction and endows them a mass. The unification of the electroweak and the strong interaction, coined Grand Unified Theory, goes beyond the Standard Model. Among the various attempts lies the Supersymmetry Theory. As for gravitation, it is described classically, in the sense that it is not quantized, by general relativity which is approximated either by special relativity for weak gravitational fields or Newtonian gravity for velocities much smaller than the speed of light. Theoreticians try to achieve quantum gravity (also called supergravity) and seek its mediator, the graviton. Finally, merging a Grand Unified Theory and a quantum gravity would lead to a Theory of Everything. String theory is a candidate for such a theory.

The table below summarizes some physical characteristics of the four fundamental interactions. Their relative strength is estimated with respect to the gravitational force.

Interaction	Strong	Electro-magnetic	Weak	Gravitation
Current theory	QCD	QED	Electroweak	General relativity
Force carriers and mass	Gluons 0	Photon 0	Z^0, 91.2 GeV/c^2 $W^\pm$, 80.4 GeV/c^2	Graviton (?) 0 (?)
Range (m)	10^{-15}	∞	10^{-18}	∞
Relative strength	10^{38}	10^{36}	10^{28}	1

Bibliography

Annett, J.F. (2004) *Superconductivity, Superfluids and Condensates.* Oxford University Press.

Ashcroft, N.W. and Mermin, N.D. (1976) *Solid State Physics.* Thomson Learning.

Close, F. (2009) *The New Cosmic Onion: Quarks and the Nature of the Universe.* CRC.

Cottingham, W.N., Greenwood, D.A. (2001) *An Introduction to Nuclear Physics.* Cambridge U. Press.

Dinh, P.M., Reinhard, P.G. and Suraud, E. (2013) *An Introduction to Cluster Science.* Wiley.

Eisberg, R. and Resnick, R. (1985) *Quantum Physics of Atoms, Molecules, Solids, Nuclei and Particles*, J. Wiley.

Gasiorowicz, S. (2003) *Quantum Physics*, Wiley-WCH.

Heyde, K. (2004) *Basic Ideas and Concepts in Nuclear Physics. An Introductory Approach.* CRC.

Huang, K. (1963) *Statistical Mechanics.* J. Wiley.

Kittel, Ch. (2005) *Introduction to Solid State Physics.* Wiley.

Kittel, Ch., Kroemer, H. (1980) *Thermal Physics.* Freeman.

Krotscheck, E. and Navarro, J. Eds. (2002) *Microscopic Approaches to Quantum Liquids in Confined Geometries.* World Scientific.

Lipparini, E. (2003) *Modern Many-Particle Physics.* World Scientific.

March, N.H., Tossi, M.P. (2002) *Introduction to Liquid State Physics.* World Scientific.

Oxtoby, D.W., Gillis, H.P., Campion, A. (2008) *Principles of Modern Chemistry.* Thomson Brooks/Cole.

Pitaevskii, L. and Stringari, S. (2016) *Bose-Einstein Condensation and Superfluidity.* Oxford University Press.

Tipler, P.A. and Llewelyn, R.A. (2008) *Modern Physics.* Freeman.

Trevena, A. (1993) *Statistical Mechanics. An Introduction.* Ellis Horwood.

Wilks, J. and Betts, D.S. (1987) *An Introduction to Liquid Helium.* Oxford University Press.

Index